Applying Computational Intelligence

Applying Computational Intelligence

Arthur K. Kordon

Applying Computational Intelligence

How to Create Value

 Springer

Dr. Arthur K. Kordon
Data Mining and Modeling Capability
Corporate R&D, The Dow Chemical Company
2301 N. Brazosport Blvd.
B-1226, Freeport
TX 77541
USA
akkordon@dow.com

ACM Computing Classification (1998): I.2, I.6, G.3, G.1, F.1, I.5, J.1, J.2, H.1.

ISBN: 978-3-642-42426-7   ISBN: 978-3-540-69913-2 (eBook)
DOI 10.1007/978-3-540-69913-2
Springer Heidelberg Dordrecht London New York

Cover design: KuenkelLopka GmbH, Heidelberg, Germany

Printed on acid-free paper

Springer is part of Springer Science+Business Media (www.springer.com)

To my friends

Preface

In theory, there is no difference between theory and practice. But, in practice, there is.

Jan L.A. van de Snepscheut

The flow of academic ideas in the area of computational intelligence has penetrated industry with tremendous speed and persistence. Thousands of applications have proved the practical potential of fuzzy logic, neural networks, evolutionary computation, swarm intelligence, and intelligent agents even before their theoretical foundation is completely understood. And the popularity is rising. Some software vendors have pronounced the new machine learning gold rush to "Transfer Data into Gold". New buzzwords like "data mining", "genetic algorithms", and "swarm optimization" have enriched the top executives' vocabulary to make them look more "visionary" for the 21st century. The phrase "fuzzy math" became political jargon after being used by US President George W. Bush in one of the election debates in the campaign in 2000. Even process operators are discussing the performance of neural networks with the same passion as the performance of the Dallas Cowboys.

However, for most of the engineers and scientists introducing computational intelligence technologies into practice, looking at the growing number of new approaches, and understanding their theoretical principles and potential for value creation becomes a more and more difficult task. In order to keep track of the new techniques (like genetic programming or support vector machines) one has to Google or use academic journals and books as the main (and very often the only) sources of information. For many practitioners, the highly abstract level and the emphasis on pure theory of academic references is a big challenge. They need a book that defines the sources of value creation of computational intelligence, explains clearly the main principles of the different approaches with less of a focus on theoretical details and proofs, offers a methodology of their integration into successful practical implementations, and gives realistic guidelines on how to handle the numerous technical and nontechnical issues, typical for real-world applications.

Motivation

Applying Computational Intelligence is one of the first books on the market that fills this need. There are several factors that contributed to the decision to write such a book.

The first is to emphasize the different forces driving academic and industrial research (see Fig. 0.1).

Value is the basis of the different modes of operation between academic and industrial research. The key objective of academic research is to create new knowledge at any level of Nature (from quantum to cosmos) and the definition of success is the quality and quantity of publications. In contrast, the key objective of industrial research is to create value by exploring and implementing knowledge from almost any level of Nature. The definition of success is increased profit and improved competitive position in the market.

Since the value factor in industry is a question of survival, it has a dominant role and dictates different ways of doing research than in the academic world. University professors can satisfy their curiosity at any level and depth of the knowledge ladder even without funding. Industrial researchers don't have this luxury and must concentrate on those levels of the knowledge ladder where the profit is maximal. As a result, assessment of value creation in almost any phase of industrial research is a must. Unfortunately, this important fact is practically ignored in the literature and one of the goals of this book is to emphasize the decisive role of value creation in applying emerging technologies, like computational intelligence, in practice.

The second factor that contributed to the decision to write this book, is to clarify the drastic changes in doing applied science in the last 10 years. Once upon a time there were industrial laboratories, such as Bell Labs, with a broad scientific focus and almost academic mode of operation. Not anymore. Globalization and the push of shareholders for maximal profit has significantly transformed applied research

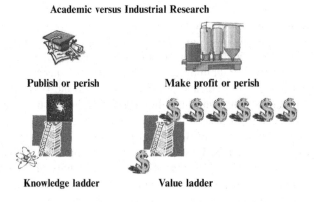

Fig. 0.1 Different driving forces between academic and industrial research

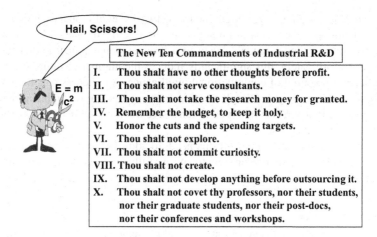

The New Ten Commandments of Industrial R&D

I.   Thou shalt have no other thoughts before profit.
II.   Thou shalt not serve consultants.
III.   Thou shalt not take the research money for granted.
IV.   Remember the budget, to keep it holy.
V.   Honor the cuts and the spending targets.
VI.   Thou shalt not explore.
VII.   Thou shalt not commit curiosity.
VIII. Thou shalt not create.
IX.   Thou shalt not develop anything before outsourcing it.
X.   Thou shalt not covet thy professors, nor their students,
       nor their graduate students, nor their post-docs,
       nor their conferences and workshops.

Fig. 0.2 The new Ten Commandments of Industrial R&D

into the direction of perpetual cuts and short-term focus. A satirical view of the current *modus operandi* is presented in the new Ten Commandments of Industrial R&D, shown in Fig. 0.2.

The cost reduction crusade imposed a new strategy for applying emerging technologies, which assumes minimal exploratory efforts. As a result, the new methodology for applied research must significantly reduce the risk and the time for introducing new technologies. Unfortunately, the known solutions are *ad hoc*, and there's little experience of developing adequate application strategies. One of the goals of the book is to offer a systematic approach for practical implementation of emerging technologies with low total-cost-of-ownership, appropriate in the current environment of lean industrial R&D.

The third factor that contributed to the decision to write this book is the broad experience of the author in applying computational intelligence in a global corporation, such as his employer, The Dow Chemical Company. With the collective efforts of talented researchers, visionary managers, and enthusiastic users, Dow Chemical became one of the industrial leaders in opening the door to the advantages of computational intelligence. In this book, the key learning from this accumulated knowledge is shared with the academic and industrial community at large – although I will skip details that may endanger Dow Chemical's intellectual property. In addition, all given examples are based on publicly available sources.

Purpose of the Book

Computational intelligence is relatively new to industry. It is still a fast- growing research area in the category of emerging technologies. On top of that, computational intelligence is based on a smorgasbord of approaches with very different theoretical bases, such as fuzzy logic, neural networks, evolutionary computation, statistical learning theory, swarm intelligence, and intelligent agents. Promoting it in a consistent and understandable way to a broader nontechnical audience is a challenge. Unfortunately, this significantly reduces the number of future users and the application potential of computational intelligence. Another issue is the unsystematic and *ad hoc* ways of implementing most of the known applications. In summary, it is not a surprise that a comprehensive book on applied computational intelligence is not available.

The purpose of the book is to fill this need, to address these issues and to give guidelines to a broad audience on how to successfully apply computational intelligence. The key topics of the book are shown in the mind-map in Fig. 0.3 and are discussed next:

1. *How to broaden the audience of computational intelligence?* The first main topic of the book focuses on the ambitious task of broadening the audience of potential users of computational intelligence beyond the specialized communities, as is now the case. The main computational intelligence methods will be explained with minimal mathematical and technical details, and with an emphasis on their unique application capabilities.
2. *How to create value with computational intelligence?* The second key topic of the book clarifies the most important question in practical applications – the issue of value creation. The answer covers: identifying the sources that make computational intelligence profitable; defining the competitive advantages of the technology relative to the most widespread modeling methods in industry; and analyzing the limitations of the approach.
3. *How to apply computational intelligence in practice?* The third key topic of the book covers the central point of interest – the application strategy for

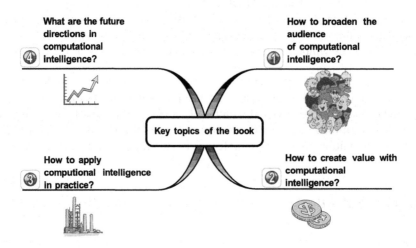

Fig. 0.3 Main topics of the book

computational intelligence. It includes methodologies for integration of different technologies, marketing computational intelligence, specific steps for implementing the full application cycle, and examples of specific applications.

4. *How to evaluate the future directions in computational intelligence?* The fourth main topic of the book addresses the issue of the sustainability of applied computational intelligence in the future. The potential for growth is based on the promising application capabilities of new approaches, still in the research domain, and the expected increased demand from industry.

Who Is This Book for?

The targeted audience is much broader than the existing scientific communities in computational intelligence. The readers who can benefit from this book are presented in the mind-map in Fig. 0.4 and are described below:

- *Industrial Researchers* – This group includes scientists in industrial labs who create new products and processes. They will benefit from the book by understanding the impact of computational intelligence on industrial research, and by using the proposed application strategy to broaden and improve their performance. In addition, they will know how to leverage the technology through proper marketing.
- *Practitioners in Different Businesses* – The group consists of the key potential final users of the technology, such as process engineers, supply-chain organizers, economic analyzers, medical doctors, etc. This book will introduce the main

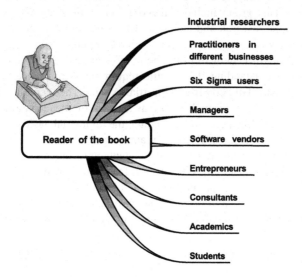

Fig. 0.4 Potential readers of the book

computational intelligence technologies in a language they can understand, and it will encourage them to find new applications in their businesses.

- *Six Sigma Users* – Six Sigma is a work process for developing high-quality solutions in industry. It has been accepted as a standard by the majority of global corporations. The users of Six Sigma are estimated to be in the tens of thousands of project leaders, called black belts, and hundreds of thousands of technical experts, called green belts. Usually they use classical statistics in their projects. Computational intelligence is a natural extension to Six Sigma in solving complex problems with a nonlinear nature, and both black and green belts can take advantage of that.

- *Managers* – Top-level and R&D managers will benefit from the book by understanding the mechanisms of value creation and the competitive advantages of computational intelligence. Middle- and low-level managers will find in the book a practical and nontechnical description of this emerging technology, which can make the organizations they lead more productive.

- *Software Vendors* – The group includes two types of vendors – of generic software for the development of computational intelligence systems, and of specialized software that uses appropriate computational intelligence techniques. Both groups will benefit from the book by better understanding the market potential for their products in this field.

- *Entrepreneurs* – This class consists of enthusiastic professionals who look to start new high-tech businesses, and venture capitalists who are searching for the Next Big Thing in technology investment. The book will give them substantial information about the nature of computational intelligence, its potential for value creation, and the current and future application areas. This will be a good basis for developing business plans and investment strategy analysis.

- *Academics* – This group includes the large class of academics who are not familiar with the research and technical details of the field and the small class of academics who are developing and moving computational intelligence ahead. The first group will benefit from the book by using it as an introduction to the field and by understanding the specific requirements for successful practical applications, defined directly from industrial experts. The second group will also benefit from the book through better awareness of the economic impact of computational intelligence, understanding the industrial needs, and learning about the details of successful practical applications.

- *Students* – Undergraduate and graduate students in technical, economics, medical, and even social disciplines can benefit from the book by understanding the advantages of computational intelligence and its potential for implementation in their specific fields. In addition, the book will help students to gain knowledge about the practical aspects of industrial research and the issues facing real-world applications.

How This Book Is Structured

The structure of the book with its organization in parts and chapters is given in Fig. 0.5.

Part I of the book is a condensed nontechnical introduction of the main technologies of computational intelligence. Chapter 1 clarifies the differences between

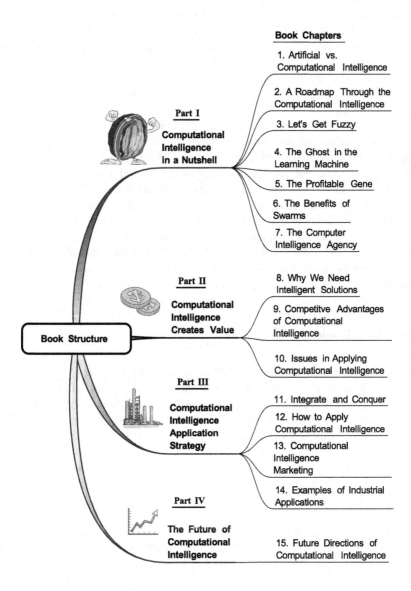

Book Chapters

1. Artificial vs. Computational Intelligence

2. A Roadmap Through the Computational Intelligence

Part I

Computational Intelligence in a Nutshell

3. Let's Get Fuzzy

4. The Ghost in the Learning Machine

5. The Profitable Gene

6. The Benefits of Swarms

7. The Computer Intelligence Agency

Part II

Computational Intelligence Creates Value

8. Why We Need Intelligent Solutions

9. Competitve Advantages of Computational Intelligence

Book Structure

10. Issues in Applying Computational Intelligence

Part III

Computational Intelligence Application Strategy

11. Integrate and Conquer

12. How to Apply Computational Intelligence

13. Computational Intelligence Marketing

14. Examples of Industrial Applications

Part IV

The Future of Computational Intelligence

15. Future Directions of Computational Intelligence

Fig. 0.5 Structure of the book

artificial intelligence and its current successor, while Chap. 2 presents an overview of the technologies and their key scientific principles. The key technologies are introduced in the next five chapters: Chap. 3 describes the main features of fuzzy systems; Chap. 4 introduces neural networks and support vector machines; Chap. 5 gives an overview of the different evolutionary computation techniques; Chap. 6 presents several methods based on swarm intelligence; and Chap. 7 describes the key capabilities of intelligent agents.

Part II focuses on the value creation potential of computational intelligence. It includes three chapters: Chap. 8 identifies the main application areas of computational intelligence; Chap. 9 defines the competitive advantages of this new emerging technology relative to the main, current research approaches in industry, such as first-principles modeling, statistics, heuristics, and classical optimization; finally the issues involved in applying computational intelligence in practice are discussed in Chap. 10.

Part III covers the most important topic of the book – defining an implementation strategy for successful real-world applications of computational intelligence. Chapter 11 emphasizes the critical importance of integrating different research approaches in industrial applications, and gives several examples of efficient integration; Chap. 12 presents the main steps of the application strategy and gives guidelines for the large group of Six Sigma users; Chap. 13 concentrates on the critical issue of computational intelligence marketing; and Chap. 14 gives specific examples of applications in manufacturing and new product design.

Finally, Part IV addresses the future directions of computational intelligence. Chapter 15 gives an introduction to the new technologies in computational intelligence, and looks ahead to the expected demands from industry.

What This Book Is NOT About

- *Detailed Theoretical Description of Computational Intelligence Approaches* – The book does not include a deep academic presentation of the different computational intelligence methods. The broad targeted audience requires descriptions that involve a minimal technical and mathematical burden. The reader who requires more detailed knowledge on any specific approach is forwarded to appropriate resources such as books, critical papers, and websites. The focus of the book is on the application issues of computational intelligence and all the methods are described and analyzed at a level of detail that enables their broad practical implementation.
- *Introduction of New Computational Intelligence Methods* – The book does not propose new computational intelligence methods or algorithms for the known approaches. The novelty in the book is on the application side of computational intelligence.
- *Software Manual of Computational Intelligence Approaches* – This is not an instruction manual for a particular software product – the interested reader is

directed to the corresponding websites. The author's purpose is to define a generic methodology for computational intelligence applications, independent of any specific software.

Features of the Book

The key features that differentiate this book from the other titles on computational intelligence are defined as:

1. *A Broader View of Computational Intelligence* – One of the main messages in the book is that focusing only on the technology and ignoring the other aspects of real-world applications is a recipe for failure. The winning application strategy is based on three key components – scientific methods, infrastructure, and people. We call it the Troika of Applied Research, shown in Fig. 0.6.

   The first component (People) represents the most important factor – the people involved in the whole implementation cycle, such as researchers, managers, programmers, and different types of final users. The second component (Methods), where most of the current analysis is focused, includes the theoretical basis of computational intelligence. The third component (Infrastructure) represents the necessary infrastructure for implementing the developed computational intelligence solution. It includes the required hardware, software, and all organizational work processes for development, deployment, and support. Promoting this broader view of computational intelligence, and especially clarifying the critical role of the human component for the success of practical applications, is the leading philosophy in the book.

2. *Balancing Scientific Purity with Marketing* – An inevitable requirement for broadening the audience of computational intelligence is changing the presentation language from technical jargon to nontechnical English. Unfortunately, some of the theoretical purity and technical details are lost in this translation.

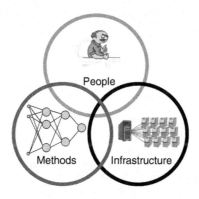

Fig. 0.6 The Troika of Applied Research

In principle, marketing-type language is much simpler and more assertive. The author is prepared for criticism from the research community but is firmly convinced of the benefits of changing the style to address nontechnical users.

3. *Emphasis on Visualization* – The third key feature of the book is the widespread use of different visualization tools, especially mind-maps[1] and clip art.[2] We strongly believe that nothing presents a concept better than a vivid visualization.

Mind-mapping (or concept mapping) involves writing down a central idea and thinking up new and related ideas that radiate out from the center.[3] By focusing on key ideas written down in your own words, and then looking for branches out and connections between the ideas, one can map knowledge in a manner that will help in understanding and remembering new information.

---

[1]The mind-maps in the book are based on the product ConceptDraw MindMap (http://www.conceptdraw.com/en/products/mindmap/main.php).

[2]The clip art in the book is based on the website http://www.clipart.com.

[3]A good starting book for developing mind-maps is: T. Buzan, *The Mind-map Book,* 3rd edition, BBC Active, 2003.

Acknowlegements

The author would like to acknowledge the contributions of the following colleagues from The Dow Chemical Company for introducing the technology and developing the computational intelligence capabilities in the company: Elsa Jordaan, Flor Castillo, Alex Kalos, Kip Mercure, Leo Chiang, and Jeff Sweeney; the former Dow employees Mark Kotanchek from Evolved Analytics and Wayne Zirk, as well as Ekaterina Vladislavleva from the University of Antwerp, Belgium.

Above all, the author would like to emphasize the enormous contribution of Guido Smits, the visionary guru who not only introduced most of the computational intelligence approaches into the company but has significantly improved some of them.

The author was lucky to have support from top visionary technical leaders and managers, such as Dana Gier, David West, Tim Rey, and Randy Collard, which was a critical factor in introducing, applying, and leveraging computational intelligence in the company.

Special thanks go to the Bulgarian cartoonist Stelian Sarev for his kind gesture to gift his funny caricatures in Chap. 13, and to John Koza for his kind permission to use the evolutionary computation icon.

The author would like to thank many people whose constructive comments substantially improved the final manuscript, including Flor Castillo, Elsa Jordaan, Dimitar Filev, Lubomir Hadjiski, Mincho Hadjiski, Alex Kalos, Mark Kotanchek, Bill Langdon, Kip Mercure, Adam Mollenkopf, Danil Prokhorov, Ray Schuette, Guido Smits, George Stanchev, Mikhail Stoyanov, David West, and Ali Zalzala. The author also highly appreciates the useful comments and suggestions of the two unknown reviewers. He is especially grateful for the full support and enthusiasm of Ronan Nugent, which was critical for the success of the whole project.

Lake Jackson Texas   Arthur K. Kordon
March 2009

Contents

Part II Computational Intelligence Creates Value

Part IV   The Future of Computational Intelligence

Part I
Computational Intelligence in a Nutshell

Chapter 1
Artificial vs. Computational Intelligence

There are three great events in history. One, the creation of the universe. Two, the appearance of life. The third one, which I think is equal in importance, is the appearance of Artificial Intelligence.

Edward Fredkin

Artificial Intelligence is not a science.

Maurizio Matteuzzi

Hardly any area of research has created such wide publicity and controversy as Artificial Intelligence (AI). To illustrate, the abbreviation AI is one of the few scientific buzzwords accepted in contemporary English language. Artificial intelligence has successfully penetrated Hollywood and even grabbed the attention of star director Steven Spielberg. His blockbuster movie *AI* spread the artistic vision of the concept to millions.

Considering the enormous popularity of AI, the reader may ask the logical question: Why would we need a different name? The first chapter objective is to answer this question by defining the fields of artificial and computational intelligence and clarifying their differences. In both cases, the focus will be on the application capabilities of these broad research areas. The chapter introduces the most important techniques and application areas of classical AI.[1] The key methods and application issues of computational intelligence are discussed in detail in the rest of the book. If a reader is not interested in the technical nature of AI, she/he can skip Sect. 1.1.

---

[1] the state of AI before the appearance of computational intelligence.

A.K. Kordon, *Applying Computational Intelligence*,
DOI 10.1007/978-3-540-69913-2_1, © Springer-Verlag Berlin Heidelberg 2010

1.1   Artificial Intelligence: The Pioneer

The enormous popularity of AI has its price. The perception of the field from different research communities is very diverse and the opinions about its scientific importance vary from one extreme of treating AI as low as contemporary *alchemy* to another extreme of glorifying its impact as one of the key events in history. The reason for this controversy is the complex nature of the key subject of AI – human intelligence. AI research efforts are focused on analyzing human intelligence using the methods of computer science, mathematics, and engineering. The classical definition of AI, according to *The Handbook of Artificial Intelligence* is as follows:

> Artificial Intelligence is the part of computer science concerned with designing intelligent computer systems, that is, systems that exhibit the characteristics we associate with intelligence in human behavior – understanding language, learning reasoning, solving problems, and so on.[2]

Meanwhile, human intelligence is a very active research area of several humanity-based scientific communities, such as neurophysiology, psychology, linguistics, sociology, and philosophy, to name a few. Unfortunately, the necessary collaboration between the computer science-based and humanity-based research camps is often replaced by the fierce exchange of "scientific artillery fire", which does not resolve the issues but increases the level of confusion. There is a common perception inside and outside the academic community that the research efforts on analyzing intelligence have not been handled in a very intelligent way.

1.1.1   Practical Definition of Applied Artificial Intelligence

As a result of the academic schisms, different definitions about the nature of human and artificial intelligence are fighting for acceptance from the research community at large. The interested reader could find them in the recommended references at the end of the chapter. Since our focus is on value creation, we will pay attention to those features of AI that are important from an implementation point of view.

The accepted birth of AI is the famous two-month workshop at Dartmouth College during the summer of 1956 when ten founding fathers defined the initial scientific directions in this new research area – imitating human reasoning, understanding language (machine translation), finding a General Problem Solver, etc. For many reasons, however, these generic directions did not produce the expected results in the next decade even with very generous funding from several governmental agencies in different countries. One of the lessons learned was that the initial strategy of neglecting domain knowledge (the so-called *weak methods* for building general-purpose solutions valid for any problems) failed. Research efforts have been

---

[2]A. Barr and E. Feigenbaum, *The Handbook of Artificial Intelligence*, Morgan Kaufmann, 1981.

Fig. 1.1 The icon of applied AI – "Put the expert in the computer"

redirected to developing reasoning mechanisms based on narrow areas of expertise. It turns out that this approach not only delivered successful scientific results in the early 1970s but opened the door to numerous practical applications in the 1980s.

From the practical point of view, the most valuable feature of human intelligence is the ability to make correct predictions about the future based on all available data from the changing environment. Usually, these predictions are specific and limited to some related area of knowledge. From that perspective, we define applied AI as *a system of methods and infrastructure to mimic human intelligence by representing available domain knowledge and inference mechanisms for solving domain-specific problems*. In a very elementary way we can describe this narrow-focused definition of AI as "Put the expert in the box (computer)", and we visualize it in Fig. 1.1.

In contrast to the initial generic-type of AI research, which had difficulties in identifying the potential for value creation, the defined applied AI has several sources for profit generation. The first one is improving the productivity of any process by operating continuously at the top experts' level. The second source of potential value creation is reducing the risks of decision-making through knowledge consolidation of the "best and the brightest" in the field. The third source of profit from using applied AI is by conserving the existing domain knowledge independently when the expert leaves the organization.

The definition of success for applied AI is a computerized solution that responds to different situations like the domain experts, imitates their reasoning and makes predictions which are indistinguishable from the experts. The practical AI dream is a machine clone of the domain-related "brain" of the expert.

1.1.2 Key Practical Artificial Intelligence Approaches

There are several methods to clone an expert "brain" and the most important are shown in Fig. 1.2. An expert system captures expert knowledge in the computer

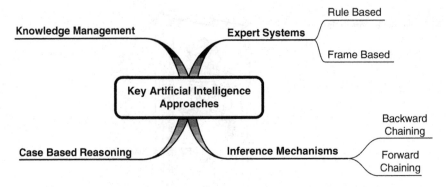

Fig. 1.2 Key approaches of applied AI

using rules and frames. Expert reasoning is mimicked by two key inference mechanisms – goal-driven (backward-chaining) and event-driven (forward chaining). If expert knowledge can be represented by different cases, it could be structured in a special type of indexed inference, called case-based reasoning. Knowledge management is a systematic way of knowledge acquisition, representation, and implementation. These key applied AI methods are discussed in more detail below.

1.1.2.1   Expert Systems (Rule-Based and Frame-Based)

Rule-Based Expert Systems

One possible way to represent human knowledge in a computer is by rules. A rule expresses a programmatic response to a set of conditions. A rule contains text and a set of attributes. When you create a new rule, you enter a two-part statement in its text. The first part, called the antecedent, tests for a condition. The second part, called the consequent, specifies the actions to take when the condition returns a value of true. This is an example of the text of a rule:

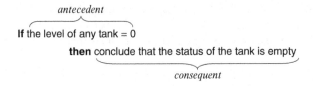

A rule-based expert system consists of a set of independent or related rules. The performance and the value creation of the rule-based expert system strongly depend on the quality of the defined rules. The domain expert plays a critical role in this process. Rule-based expert systems can only be as good as the domain experts they are programmed to mimic.

Frame-Based Expert Systems

Another way to represent domain knowledge is by frames. This method was introduced by one of the AI founding fathers, Marvin Minsky, in the 1970s and was pushed further by the emerging field of object-oriented programming. A frame is a type of knowledge representation in which a group of attributes describes a given object. Each attribute is stored in a slot which may contain default values. An example of a typical object in a manufacturing monitoring system is a process variable with its attributes. Object attributes include all necessary information for representing this entity, including the data server name, tag name, process variable value, update interval, validity interval of the data, history keeping specification, etc.

A unique and very powerful feature of frame-based expert systems is their ability to capture generic knowledge at a high abstract level by defining a hierarchy of classes. For example, the class hierarchy of a manufacturing monitoring system includes generic classes, such as process unit, process equipment, process variable, process model, etc. As generic-type objects, classes have attributes, which can be user-defined or inherited from the upper classes in the hierarchy. For example, the class process equipment includes inherited attributes, such as unit name and unit location from the upper class process unit, and user-defined attributes, such as dimensions, inlet process variables, outlet process variables, control loops, etc.

Classes may have associated methods, which define the operations characteristic of each class. Polymorphism allows methods representing generic operations to be implemented in class-specific ways. Code that invokes a method needs only to know the name of the object and the method's signature. For example, the process equipment class may include the methods for start-up and shut-down sequences.

The role of the domain-knowledge expert is to define the class hierarchy in the most effective way. The biggest advantage of frame-based expert systems is that once an object (or class of objects) is defined, the work is immediately reusable. Any object or group of objects can be cloned over and over again. Each cloned copy inherits all of the properties and behaviors assigned to the original object(s). It is possible to group objects, rules, and procedures into library modules that can be shared by other applications.

1.1.2.2 Inference Mechanisms

The next key topic related to applied AI is the nature of inference of the artificial expert. It addresses how the defined rules are selected and fired. At the core of this topic is the inference engine. An inference engine is a computer program which processes the contents of a knowledge base by applying search strategies to the contents and driving conclusions. The most important techniques used for inference are backward chaining and forward chaining.

1.1.2.3 Backward Chaining

The first inference mechanism, backward chaining, is based on goal-driven reasoning. Backward chaining begins by attempting to seek a value or values for a known goal (usually a possible solution, defined by the experts). This solution is usually a consequent in the rules. The process continues by seeking values on which the goal depends. If an appropriate rule is found and its antecedent section fits the available data, then it is assumed that the goal is proved and the rule is fired. This inference strategy is commonly used in diagnostic expert systems.

1.1.2.4 Forward Chaining

The second inference mechanism, forward chaining, is based on data-driven reasoning. Forward chaining begins by seeking eligible rules which are true for known data values. Processing continues by seeking additional rules which can be processed based on any additional data which has been inferred from a previously "fired" rule or a goal is reached. This inference strategy is commonly used in planning expert systems. It is a technique for gathering information and then using it for inference. The disadvantage is that this type of inference is not related to the defined goal. As a result, many rules could be fired that have no relationship with the objective.

1.1.2.5 Case-Based Reasoning

Reasoning by analogy is an important technique in human intelligence. Very often domain experts condense their knowledge on a set of past cases of previously solved problems. When a new problem appears, they try to find a solution by looking at how close it is to similar cases they have already handled. Case-based reasoning is an AI approach that mimics this behavior.

Presenting domain knowledge by cases has several advantages. First, case-based reasoning doesn't require a causal model or deep process understanding. Second, the technique gives shortcuts to reasoning, the capability to avoid past errors, and the capability to focus on the most important features of a problem. Third, the knowledge acquisition process is much easier and less expensive than the rule- or frame-based methods. In practice, cases require minimum debugging of the interactions between them, unlike rule-based systems.

Case-based reasoning incorporates a six-step process:

Step 1: Accepting a new experience and analyzing it to retrieve relevant cases from the domain experts.

Step 2: Selecting a set of best cases from which to define a solution or interpretation for the problem case.

Step 3: Deriving a solution or interpretation complete with supporting arguments or implementation details.

Step 4: Testing the solution or interpretation and assessing its strengths, weaknesses, generality, etc.

Step 5. Executing the solution or interpretation in practical situations and analyzing the feedback.

Step 6: Storing the newly solved or interpreted case into case memory and appropriately adjusting indices and other mechanisms.

When is it appropriate to apply case-based reasoning? One clear example is a system that is based on well-defined distinct entities (parts) supported by a rich history of cases. In this instance representing the domain knowledge by specific examples is a valid option that could significantly reduce the development cost relative to classical knowledge management. Another example is a system of a very difficult problem which requires very expensive development. In this case an approximate low-cost solution based on case-based reasoning is better than none at all.

1.1.2.6 Knowledge Management

Knowledge management is the process of building computer systems that require a significant amount of human expertise to solve a class of domain-specific problems. The key components involved in knowledge management are shown in Fig. 1.3 and described below:

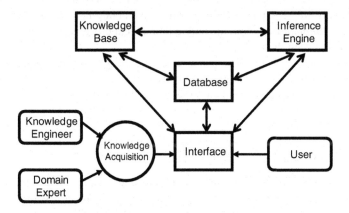

Fig. 1.3 Key components of knowledge management

- Knowledge base, consisting of a rule-based or frame-based expert system;
- Inference engine, which controls the artificial expert "thinking" and utilization of the knowledge;
- User interface, which organizes a (hopefully!) user-friendly dialog with the final non-expert type of user;
- Interface to a database, which contains the data necessary to support the knowledge base.
- Knowledge management is the process of effectively integrating these pieces.

The decisive procedure in knowledge management is the transfer of domain knowledge into the knowledge base. This is called "knowledge acquisition".

Knowledge acquisition and organization is the activity of gleaning and organizing knowledge from the domain expert in a form suitable for codification in an expert system. Many times this alone is of major benefit for knowledge consolidation. Frequently, during the knowledge acquisition, the experts will gain additional insight and understanding into their domain. Very often knowledge acquisition requires personal interviews of the experts, observing how they work, looking at expert journals, and "role playing" to capture knowledge. Where the knowledge largely consists of decisions or rules it is preferably organized into logic diagrams. If the knowledge largely consists of collections of objects or a data structure it can be adequately documented in frames.

Knowledge acquisition depends on the effective interaction between the two key actors – domain experts and knowledge engineers. The role of the domain expert has been discussed already. The character of knowledge engineer is discussed next.

A knowledge engineer is an individual who designs and builds knowledge-based systems. Knowledge engineers help domain experts map information into a form suitable for building a system. The knowledge engineer concentrates on the meaning of the data gathered, the logical interdependencies between facts and the schemes and inference rules that apply to the data. In order to be successful, the knowledge engineer has to have the following interpersonal skills:

- *Good and versatile communication skills* to interpret the whole variety of messages from the experts (verbal, written, even body language).
- *Tact and diplomacy* to avoid alienating domain experts and facilitating possible conflicts between them.
- *Empathy and patience* to establish a reliable team spirit with the experts. He or she needs to encourage without being patronizing, argue without appearing self-opinionated, and ask for clarification without appearing critical.
- *Persistence* to keep faith in the project despite knowledge gaps and inconsistencies.
- *Self-confidence* to lead the knowledge management process. He or she needs to behave as the "conductor" of the domain expert "choir".

The last (but not the least) component of knowledge management is the effective interface with the final user. Ease of use is vital for the final success of the applied expert system. A critical issue is the explanation capability of the captured expertise.

Since it is assumed that the average final user is not an expert, it is necessary to communicate the expert knowledge as clearly as possible. The best way to accomplish this is by combining the verbal and graphical presentation of the knowledge blocks and to make it understandable for the user.

1.1.3 Applied Artificial Intelligence Success Stories

The discussed approaches were the basis of the growing interest of industry in AI starting in the mid-1980s. Gradually, a mature software infrastructure for effective development and deployment of expert systems was developed by several vendors, which allowed effective real-world applications. The unique technical advantages of AI were defined and they created significant value in selected application areas. Applied AI gained credibility with numerous industrial applications in companies like the Ford Motor Company, DuPont, American Express, Dow Chemical, to name a few.

1.1.3.1 Integrated Software Infrastructure for Applied AI

The software infrastructure of applied AI has significantly evolved since the late 1950s. The first AI software tool was the language Lisp, developed by John McCarthy, followed two decades later by the programming language PROLOG, developed at the University of Marseilles. The real breakthrough for practical applications, however, was the development of expert system shells in the late 1980s. They usually consist of an inference engine, user interface facilities, and a knowledge base editor. The most popular are Level 5 Object, convenient for the development of small expert systems (with fewer than 100 rules), and Nexpert Object, convenient for large problems. One of the most advanced software tools which significantly increased productivity is Gensym Corporation's G2, which provides a complete, graphical development environment for modeling, designing, building, and deploying intelligent applications.

With G2, knowledge can be efficiently captured and applied by creating generic rules, procedures, formulas, and relationships that work across entire classes of objects. G2's structured natural language makes it easy to read, edit, and maintain the knowledge that is represented in G2 applications. This structured natural language does not require any programming skills to read, understand, or modify an application.

1.1.3.2 Technical Advantages of Applied AI

The developed methods and gradually improving infrastructure of applied AI since the late 1980s have opened the door for value creation by solving various industrial problems. In order to select the best opportunities for the technology, a clear identification of its technical advantages is needed. In this way the potential users and project stakeholders will be aware of the unique capabilities of applied AI, which will contribute to the effective solution of the specific problem. The key features that give a clear technical advantage to applied AI relative to the other known engineering approaches are described briefly below.

1.1.3.3 Domain Expertise Is Captured

The key strength of applied AI is in its capability to transfer the domain-specific knowledge of an expert into a user-friendly computer program. The expertise is represented by rules and frames and is convenient for the expert, the expert system developer, and the final user. Capturing domain knowledge into an expert system allows preserving trade secrets, distributing unique expertise, and increasing the efficiency of the decision-making process within the specific domain to the expert level.

1.1.3.4 Knowledge Is Presented in Natural Language

The ability of applied AI to represent and manipulate knowledge in natural language facilitates the introduction of this emerging technology to a broad audience of potential users. The captured knowledge can be understood by all stakeholders of the expert system – domain experts, knowledge engineers, and final users. This feature also reduces the development cost since the most time-consuming task of knowledge acquisition can be done on any word editor.

1.1.3.5 Rule Structure Is Uniform

The If-Then structure of the rules is universal; the syntax is self-documented and easy for interpretation and inference. This allows a uniform template for rule definition independent of the specific knowledge domain. As a result, the development and maintenance cost for defining specific rules is reduced.

1.1.3.6 Interpretive Capability

An important feature of applied AI, which appeals to the final user, is the explanation capability. The system behaves like human experts who explain the reasoning processes behind their recommendations. It is a very important capability in diagnostics-based expert systems and critical for health-based expert systems. Often the explanation capability is a decisive factor for the final user in accepting the expert system recommendations.

1.1.3.7 Separation of Knowledge from Inference Mechanisms

This feature of applied AI allows the development of a standard expert system shell that could be used for different domain-specific applications. As a result, the development productivity is significantly increased. In practice, this is the critical feature in applied AI infrastructure that opened the door for industrial applications.

1.1.3.8 Application Areas of AI

The discussed technical advantages of applied AI can be transferred into value in different ways. The common theme, however, is very simple: making profit from the top expertise, available all the time. The different forms of this common theme define the key application areas of AI, which are shown in Fig. 1.4 and described briefly below.

1.1.3.9 Advisory Systems

The most widespread use of expert systems is as a trusted artificial consultant on specific topics. The value is created by significantly reducing the labor cost of human experts. In most of the cases the benefits are not from eliminating the jobs of the top experts but from reducing their routine tasks and using their time for creating new expertise, which potentially will generate more value. Another benefit of advisory systems is that the final result is some recommendation to the user, i.e. the responsibility for the final action is transferred to her/him. This eliminates the potential for legal actions, which are one of the biggest issues in using expert systems products, especially related to medicine.

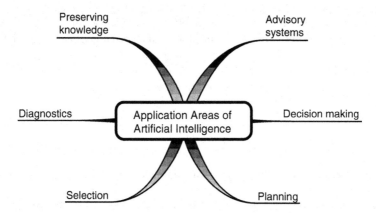

Fig. 1.4 A mind-map of the key application areas for value creation from AI

1.1.3.10   Decision-Making

Applied AI allows consolidating the existing knowledge, removing the inconsistencies, and refining the inference process of different experts. The final expert system has much better decision-making capabilities, which are beyond the competency of individual experts. Another important factor of the automated decision-making process is the elimination of purely human aspects, such as fatigue, stress, or any kind of strong emotions. Of special importance is the computer-driven decision-making during stressful events in real time, when the best decision must be taken and executed in a very short time. The value from proper handling of emergency events in a single manufacturing facility alone can be millions of dollars.

1.1.3.11   Planning

Planning a complex sequence of events dependent on multiple factors is an area where top human expertise is invaluable. Transferring this knowledge (or part of it) to a computer system allows continuous operation with minimum losses due to inefficient scheduling or inadequate response to current market demands. The bigger the enterprise, the bigger the economic impact of proper planning. For example, the impact of an AI-based manufacturing planning system at the Ford Motor Company (described in the next section) is billions of dollars.

1.1.3.12   Selection

Very often choosing a proper material, recipe, tool, location, etc. in conditions of interactive factors and conflicting objectives is not trivial and requires highly experienced specialists. Automating these types of activities with an expert system translates into significant reduction of wasted materials and energy. A typical example is the electrical connector assembly expert system, applied in the 1980s at the Boeing Company. It automatically selects the correct tools and materials for assembling electrical connectors, which is one of the time-consuming operations in aircraft manufacturing. As a result, the average assembly time has been reduced from 42 minutes to just 10 minutes with fewer errors.

1.1.3.13   Diagnostics

Capturing experts' knowledge on tracking complex fault detection and diagnostics is one of the key application areas of applied AI. The value is created by reducing the time for detecting the problem, by increasing the reliability of its early discovery, and, in many cases, by preventing a very costly or hazardous accident.

The range of implemented diagnostic expert systems is very broad – from medical diagnostics to space shuttle diagnostics.

1.1.3.14 Preserving Knowledge

Top experts are a hot commodity in any state of the job market. Losing a key source of knowledge under any circumstances (disease, accident, retirement, etc.) may have an immediate negative impact on the performance of a business. In some cases, the knowledge gap is irrecoverable. The best strategy is to capture the knowledge through an expert system. A critical point in this process, however, is to offer incentives to the top experts to encourage the transfer of knowledge.

1.1.3.15 Examples of Successful AI Real-World Applications

The benefits of applied AI have been actively explored by academia, industry, and government since the mid-1980s. From the thousands of known applications we'll focus on several milestone examples.

We'll begin with DENDRAL, which is accepted as the first successfully applied expert system.[3] It was developed in the early 1970s at Stanford University and supported by NASA. The original objective was to design a computer program to determine the molecular structure of Martian soil, based on the data from a mass spectrometer. There was no automated scientific algorithm to map the data into the correct molecular structure. However, DENDRAL was able to do so by using organized knowledge from experts. The domain experts significantly reduced the number of possible structures by looking for well-known patterns in the data and suggesting a few relevant solutions for consideration. At the end, the final program had achieved the level of performance of an experienced human chemist.

Another success story is MYCIN, also developed at Stanford University in 1972–1976. It began one of the most valuable application areas of AI – diagnostic systems. MYCIN was a rule-based expert system for the diagnosis of infectious blood diseases. It included 450 independent rules and provided a doctor with competent and user-friendly advice. The novelty was the introduction of certainty factors, given by the experts. The final product performed consistently better than junior doctors.

One of the first industrial success stories from the early 1980s was the R1 expert system for selecting components for VAX computers before they are shipped. It included 8000 rules; the development efforts were equivalent to 100 man years, and system maintenance required about 30 people. However, the Digital Equipment Corporation (DEC) claimed savings of about $20 million per year.

---

[3]Details of most of the described applications are given in: E. Feigenbaum, P. McCorduck, and H. Nii, *The Rise of the Expert Company*, Vintage Books, 1989.

Encouraged by these early successes, many companies in different industries began applying AI systems in the late 1980s. DuPont started a mass-scale campaign to introduce expert systems within the company, known as "Put Mike in the box". More than 3000 engineers learned how to capture their knowledge in a small rule-based expert system. As a result, about 800 systems have been deployed in the areas of diagnostics, selection, and planning, which together delivered profit in the range of $100 million by the early 1990s. British Petroleum developed an expert system with 2500 rules called GASOIL to help design refineries. The Campbell Soup Company improved control of their recalcitrant sterilizers by capturing the experts' knowledge. Dow Chemical has successfully applied several expert systems for production scheduling, materials selection, troubleshooting of complex processes, and alarm handling.

We'll finish the long list of successful industrial applications of AI with an impressive application at the Ford Motor Company.[4] At the core of the entire manufacturing process planning system in the company is an embedded AI component, known internally as the Direct Labor Management System. This smart system automatically reads and interprets process instructions and then uses the information to calculate the time needed to do the work. In addition, other factors, such as knowledge about ergonomics, are used by the AI system to prevent any potential ergonomic issues before beginning vehicle assembly.

The Direct Labor Management System is based on two methods – the Standard Language for knowledge syntax and the Description Logic for knowledge representation. The goal of Standard Language is to develop a clear and consistent means of communicating process instructions between various engineering functions in manufacturing. The use of Standard Language eliminates almost all ambiguity in process sheet instructions and creates a standard format for writing process sheets across the corporation. What is really amazing in this application is that the whole variety of engineering operations is represented by a vocabulary of 400 words only!

At the core of the decision-making process of the Direct Labor Management System is the knowledge base that utilizes a semantic network model to represent all of the automobile assembly planning information. The system has been in operation since the early 1990s and has demonstrated reliable maintenance performance even after thousands of changes in the knowledge base due to the introduction of new tools and parts, and new technologies, such as hybrid vehicles and satellite radio. Recently, an automatic machine translation system has been added for the translation of the instructions from English to German, Spanish, Portuguese, and Dutch. That allows global functioning of the AI system with multilingual users. Since this system is an integrated part of car manufacturing, its economic impact is estimated to be in the order of billions of dollars.

---

[4]N. Rychtyckyj, Intelligent Manufacturing Applications at Ford Motor Company, *Proceedings of NAFIPS 2005*, Ann Arbor, MI, 2005.

1.1.4 Applied Artificial Intelligence Issues

Despite the impressive and highly publicized implementation record, applied AI faced many challenges. The key issues of the three applied AI components – technical, infrastructural, and people-related are discussed below.

1.1.4.1 Technical Issues of Applied AI

Knowledge Consistency

One of the biggest technical flaws of expert systems is the lack of general and practical methodology to prove rules completeness and consistency. As a result, it is very difficult to identify incorrect knowledge. In principle, the expert system is based on the slogan "In Expert we trust". Knowledge consistency and correctness depends almost entirely on the quality and objectivity of the expert. The issue is even more complicated when knowledge from multiple experts has to be reconciled and aligned. Handling differences between experts is not trivial. Sometimes reaching consensus to define rules is not the best strategy, since one of the experts could be right and the others could be wrong on a specific topic. Another important issue is whether the rules interact properly with each other to achieve the best result. This issue can be especially troublesome in a large rule-based expert system.

Scale-up

Scaling-up rule-based knowledge with interdependent rules is extremely time consuming and costly. Even with the relatively small scale-up of increasing the number of rules from dozens to hundreds, knowledge management and especially knowledge acquisition becomes almost intractable. Unfortunately, very often the increased complexity in rule interaction appears during scale-up and cannot be captured during initial expert system design.

Fortunately, due to the object-oriented nature of frame-based expert system, their scale-up involves minimal development efforts.

Static Nature, No Learning

Once implemented, the expert system operates without changes in its rules or frames. The system has no built-in learning capabilities and cannot modify, add or delete rules automatically. This creates the danger of not recognizing expert system boundaries and responding inadequately to a changing environment. It is the responsibility of the knowledge engineer to introduce the changes after consulting with the domain experts. It is possible, however, that process knowledge under the new operating conditions is nonexistent and even the experts will need some learning before they can define the new rules.

Subjective Nature of Representing Intelligence

The performance of the expert system depends on the availability and professional level of domain experts. The "bag of tricks" they offer is their subjective heuristic model of the specific problem within their area of expertise. Often it is difficult to justify the defined rules of thumb either by physical interpretation or by statistical analysis of the available data. The subjective nature of represented human intelligence raises fundamental questions about the adequate responses of the applied AI system to the objective nature of the problem to which it is applied.

1.1.4.2 Infrastructure Issues of Applied AI

Limited Computer Capabilities in the Early 1980s

An obvious issue in applied AI, especially at the beginning of industrial applications in the early 1980s, was the limited capability of the available hardware and software infrastructure at that time. Most of the systems were implemented on minicomputers – such as the VAX - which significantly raised the hardware cost, into the range of hundreds of thousands dollars. Initially, the software infrastructure required specialized languages, such as Lisp or PROLOG, with steep learning curves due to their nontraditional style of programming. Even at the peak of its popularity in the late 1980s and early 1990s, applied AI had to deal with personal computers with limited computational power and network capabilities. The biggest progress was on the software front where the developed expert systems shells significantly improved implementation efforts.

High Total Cost of Ownership

The key bottleneck, however, was the slow development of applied AI due to the time-consuming nature of knowledge acquisition. The design of the classical applied AI stars (like DENDRAL) took 30 man years of development time. Even recently, with more advanced tools available (like G2), the development process takes several months, mostly because of knowledge acquisition.

Another component of the total-cost-of-ownership that was significantly underestimated was the maintenance cost. Without learning capabilities, supporting an expert system is very inefficient and expensive, especially if new interdependent rules have to be added. One hidden requirement for the maintenance of such a system is the continuous involvement of the knowledge engineer and the corresponding domain experts. Usually these are highly educated specialists (typically at a Ph.D. level) and it is very inefficient to spend their time on maintenance. Their long-term involvement is also an unrealistic assumption, especially for the domain experts. As a result, the average cost of ownership of an expert system at the peak of their popularity in the late 1980s and early 1990s was relatively high, in the range of $250,000 to $500,000 per year.[5]

1.1.4.3   People Issues of Applied AI

Knowledge Extraction

The key issue in resolving the bottleneck of expert systems – knowledge acquisition is the relationship of domain experts with the system. Good experts are too busy and sometimes difficult to deal with. They don't have real incentives to share their knowledge since most of them are at the top of their professional career. In addition, in the recent environment of constant lay-offs, there is a fear for their jobs if they transfer their expertise into a computer program. The recommended solution is in defining significant incentives in combination with high recognition of the value of the shared knowledge and clear commitment for job security from the management.

Incompetence Distribution

An unexpected and unexplored issue in applied AI is the bureaucratic influence of big hierarchical organizations on the quality of the defined rules. In some sense, it is an effect of the famous Peter Principle[6] on the nature of expertise with increased hierarchical levels. According to the Peter Principle, the technical expertise is located at the lowest hierarchical levels. It gradually evaporates in the higher levels and is replaced by administrative work and document generation. Unfortunately, in selecting domain experts as sources for knowledge acquisition, the Peter Principle effects have not been taken into account. As a result, some of the experts have been chosen based on their high hierarchical position and not on their technical brilliance. The deteriorating consequences of the defined rules from such Peter Principle-generated pseudo-experts are very dangerous for the credibility of the developed expert system. The real experts immediately catch the knowledge gaps and probably are not using the system. Challenging the rules, defined by their bosses, is often not a good career move. The final absurd result is that instead of enhancing top expertise, such bureaucratically generated expert systems tend to propagate incompetence.

---

[5]E. Feigenbaum, P. McCorduck, and H. Nii, *The Rise of the Expert Company*, Vintage Books, 1989.

[6]In a hierarchy every employee tends to rise to his level of incompetence.

Legal Issues

Another, initially underestimated, application issue of AI is the possible legal impact from the computerized expertise. It is of special importance in countries with high potential for lawsuits, such as the United States. One application area, which is very suitable for rule-based systems, is medical diagnostics. Due to the high risk of lawsuits, however, the field is almost frozen for mass-scale expert system applications.

1.1.5   *Artificial Intelligence Application Lessons*

Applied AI was one of the first emerging technologies that was introduced on a mass scale in the late 1980s and early 1990s. On one hand, this push created many successful industrial applications, which continue to deliver billions of dollars of value. On the other hand, the initial enthusiasm and technological faith have been replaced by disappointment and loss of credibility. Many lessons have been learned and the following is a short summary. It is not a surprise that most of them are not related to the technical issues of applied AI but to the human factor. Most of these lessons are valid for potential applications, based on computational intelligence as well.

1.1.5.1   *Application AI Lesson 1:* **Do not create unrealistic expectations**

Introducing AI as a voodoo technology, which will revolutionize science and industry, created an irrational exuberance and euphoria. Supported by the media and science fiction, it inflated the expectations of AI to enormous proportions. In many cases top management was convinced that applying AI would dramatically increase productivity in a short time. Unfortunately, the application issues of AI, especially related to the human factor, were significantly underestimated and not taken into account. The reality correction in the form of unsuccessfully applied AI systems as either technical failures or financial black holes eroded the credibility of the technology and ended management support.

1.1.5.2   *Application AI Lesson 2:* **Do not push new technologies by campaigns**

Introducing emergent technologies in mass-scale either by force from management or by mass-participation of enthusiastic but technically-ignorant users is not a very efficient strategy. In both cases, the new technology is artificially imposed on problems, when other simpler solutions may be better. As a result, the efficiency of most of the applications is low and that erodes the credibility of the technology. The idea that thousands of experts can capture their own knowledge and create expert systems by themselves has not been entirely successful. Just the opposite, it created the wrong impression that the role of the knowledge engineer could be delegated to the domain expert. It may work in very simple and trivial rule bases but in most cases professional knowledge management skills are needed.

1.1.5.3 *Application AI Lesson 3:* **Do not underestimate maintenance and support**

Maintenance and support cost is the most difficult component of the total- cost-of-ownership to evaluate. Unfortunately, in the case of applied AI it was significantly underestimated. The key neglected factor was the high cost of updating knowledge. In practice, it requires continuous involvement of the original development team of domain experts and knowledge engineers. In cases with complex and interdependent rules, knowledge base updating is very time-consuming. An additional issue could be the availability of the domain experts. If the domain experts have left the company, knowledge update will take much longer and be much costlier.

1.1.5.4 *Application AI Lesson 4:* **Clarify and demonstrate value as soon as possible**

As with the introduction of any new technology, the sooner we show the value, the better. The best strategy is to select a "low hanging fruit" application with low complexity and clearly defined deliverables. One little detail about value creation from expert systems needs to be clarified, though. There is a perception that cloning the domain expert's "brain" into a computer does not add new knowledge and therefore does not automatically create value. However, knowledge consolidation from multiple experts definitely creates value. The improvement in decision quality from using top expert rather than average expert recommendations is another source of value. An additional undisputed source of value from expert systems is their continuous accessibility and immediate response.

1.1.5.5 *Application AI Lesson 5:* **Develop a strategy for sustainable application growth**

Unfortunately, many AI applications were "one-trick ponies". Due to the difficulty in scale-up with complex interdependent rules, most of the applications didn't grow from their original implementations. The initial enthusiasm towards the technology did not mature into a line of successive applications. On top of that, most of the applications were not self-supportive, they experienced increasing maintenance costs and they lost original sources of expertise. As a result, management support gradually evaporated and with that the application record declined.

1.1.5.6   *Application AI Lesson 6:* **Link the success of the application with incentives to all stakeholders**

Applied AI challenged one of the most sensitive issues in humans – their intelligence. That created a different way of human-machine interaction than using devices without built-in brain power. In some sense, the human expert looks at "Mike in the box" as a competitor. We also observed a hidden form of rivalry between the expert system and the user who always tries to challenge the computer-generated recommendations. The best way to resolve this issue, and to improve the interaction, is by giving incentives to all stakeholders of the applied AI system. Above all, the incentives have to encourage the final user to support and not to sabotage the application.

Applied AI gave a new birth to the AI scientific field in the 1980s by opening the door to industry. It has generated numerous industrial applications, which created value in the range of billions of dollars. However, as a result of the combined effect of technical, infrastructural, and people-related issues, the efficiency and credibility of most of the applications have gradually deteriorated. The silent death of the inflexible rules, the ignorance towards the growing avalanche of data, and the fast-changing dynamics and complexity of global business have challenged the relevance of classical AI. The new reality requires more advanced features which add adaptability, learning capability, and generate novelty. Gradually the respected pioneer, applied AI, passed on the torch to its successor, applied computational intelligence.

1.2   Computational Intelligence: The Successor

Several research approaches, mostly born in the 1960s–1970s, progressively delivered the missing capabilities of classical AI. The critical learning features were made possible by a number of approaches from the emerging scientific field of machine learning, such as artificial neural networks, statistical learning theory, and recursive partitioning. The missing capabilities of representing ambiguity and mimicking the human thought process with vague information were addressed by a new type of logic – fuzzy logic. Impressive new capabilities to automatically discover novel solutions were developed by the fast growing research areas of evolutionary computation and swarm intelligence. Integrating all of these impressive features and related technologies created a new way of representing intelligence, called computational intelligence. This chapter focuses on defining this new research area from a practical perspective, identifying its key approaches, and clarifying the differences with its predecessor – applied AI.

1.2.1 Practical Definition of Applied Computational Intelligence

One of the most accepted definitions of computational intelligence by the scientific community is as follows:[7]

> Computational intelligence is a methodology involving computing that exhibits an ability to learn and/or to deal with new situations, such that the system is perceived to possess one or more attributes of reason, such as generalization, discovery, association and abstraction.
>
> Silicon-based computational intelligence systems usually comprise hybrids of paradigms such as artificial neural networks, fuzzy systems, and evolutionary algorithms, augmented with knowledge elements, and are often designed to mimic one or more aspects of carbon-based biological intelligence.

The academic definition of computational intelligence emphasizes two key elements. The first element is that the technology is based on new learning capabilities which allow mimicking at more abstract levels of thinking like generalization, discovery, and association. The second element in the definition clarifies that at the core of computational intelligence is the integration of machine learning, fuzzy logic, and evolutionary computation.

Having in mind the academic definition and the focus on value creation by profitable practical implementations, we define applied CI in the following way:

> Applied computational intelligence is a system of methods and infrastructure that enhance human intelligence by learning and discovering new patterns, relationships, and structures in complex dynamic environments for solving practical problems.

In contrast to applied AI, which is visually represented by putting the expert in the computer (see Fig. 1.1), applied computational intelligence is visualized by an effective collaboration between the human and the computer, which leads to enhanced human intelligence, as shown in Fig. 1.5.

Three key factors contributed to the development of computational intelligence: (1) the unresolved issues of classical AI; (2) the fast progression of computational power; and (3) the booming role of data. The role of the first factor we discussed

Fig. 1.5 The icon of applied computational intelligence

---

[7]From the document: http://www.computelligence.org/download/citutorial.pdf

already. The importance of the second factor is related to the high computational power, required by most of the machine learning and especially evolutionary computation algorithms. Without sufficient computational resources these methods cannot operate and be useful for solving real-world problems. The third factor emphasizes the enormous influence of data on contemporary thinking. The tendency towards data-driven decisions, pushed enormously by the Internet, will grow further with new wireless technologies. Computational intelligence methods benefit from the current data explosion and will be the engine to transfer data into knowledge and ultimately into value.

The driving force for value creation from computational intelligence, according to the definition, is enhanced human intelligence as a result of the continuous adaptation and automatic discovery of new features in the applied system. The knowledge gained may deliver multiple profit-chains from new product discovery to almost perfect optimization of very complex industrial processes under significant dynamic changes. One of the objectives of the book is to identify and explore these enormous opportunities for value creation from computational intelligence.

1.2.2   Key Computational Intelligence Approaches

We'll begin our journey through computational intelligence with a short introduction of the main approaches that comprise this emerging technology. The mind-map of the approaches is shown in Fig. 1.6 and the short descriptions follow. Each method is also visually represented by a corresponding icon, which is used in most

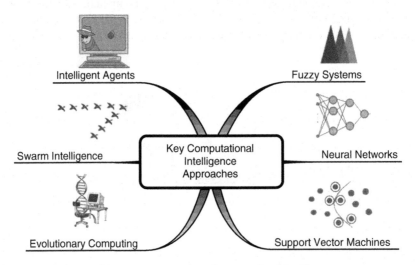

Fig. 1.6 A mind-map with the core approaches of computational intelligence

of the figures in the book. The detailed analysis of each of the technologies is given in Chaps. 3–7.

1.2.2.1 Fuzzy Systems

A fuzzy system is the component of computational intelligence that emulates the imprecise nature of human cognition. It mimics the approximate reasoning of humans by representing vague terms in a quantitative way. This allows inferring numerically in the computer with a different type of logic, called fuzzy logic, which is much closer to the real world than the classical crisp logic. In this way the computer "knows" the meaning of vague terms, such as "slightly better" or "not very high" and can use them in calculating the logical solution. Contrary to common perception, the final results from fuzzy logic are not fuzzy at all. The delivered answers are based on exact numerical calculations.

1.2.2.2 Neural Networks

The learning capabilities of computational intelligence are based on two entirely different methods – artificial neural networks and support vector machines. Artificial neural networks (or simply neural networks) are inspired by the capabilities of the brain to process information. A neural network consists of a number of nodes, called neurons, which are a simple mathematical model of the real biological neurons in the brain. The neurons are connected by links, and each link has a numerical weight associated with it. The learned patterns in the biological neurons are memorized by the strength of their synaptic links. In a similar way, the learning knowledge in the artificial neural network can be represented by the numerical weights of their mathematical links. In the same way as biological neurons learn new patterns by readjusting the synapse strength based on positive or negative experience, artificial neural networks learn by readjustment of the numerical weights based on a defined fitness function.

1.2.2.3 Support Vector Machines

Support vector machines (SVM) deliver learning capabilities derived from the mathematical analysis of statistical learning theory. In many practical problems in engineering and statistics, learning is the process of estimating an unknown relationship or structure of a system using a limited number of observations. Statistical learning theory gives the mathematical conditions for design of such an empirical learning machine, which derives solutions with optimal balance between accurately representing the existing data and dealing with unknown data. One of the key advantages of this approach is that the learning results have optimal complexity for the given learning data set and have some generalization capability.

In support vector machines the learned knowledge is represented by the most informative data points, called support vectors.

1.2.2.4 Evolutionary Computation

Evolutionary computation automatically generates solutions of a given problem with defined fitness by simulating natural evolution in a computer. Some of the generated solutions are with entirely new features, i.e. the technology is capable of creating novelty. In a simulated evolution, it is assumed that a fitness function is defined in advance. The process begins with creation in the computer of a random population of artificial individuals, such as mathematical expressions, binary strings, symbols, structures, etc. In each phase of simulated evolution, a new population is created by genetic computer operations, such as mutation, crossover, copying, etc. As in natural evolution, only the best and the brightest survive and are selected for the next phase. Due to the random nature of simulated evolution it is repeated several times before selecting the final solutions. Very often the constant fight for high fitness during simulated evolution delivers solutions beyond the existing knowledge of the explored problem.

1.2.2.5 Swarm Intelligence

Swarm intelligence explores the advantages of the collective behavior of an artificial flock of computer entities by mimicking the social interactions of animal and human societies. A clear example is the performance of a flock of birds. Of special interest is the behavior of ants, termites, and bees. The approach is a new type of dynamic learning, based on continuous social interchange between the individuals. As a result, swarm intelligence delivers new ways to optimize and classify complex systems in real time. This capability of computational intelligence is of special importance for industrial applications in the area of scheduling and control in dynamic environments.

1.2.2.6 Intelligent Agents

Intelligent agents are artificial entities that have several intelligent features, such as being autonomous, responding adequately to changes in their environment, persistently pursuing goals, and being flexible, robust, and social by interacting with other agents. Of special importance is the interactive capability of the intelligent agents since it mimics human interaction types, such as negotiation, coordination, cooperation, and teamwork. We can look at this technology as a modern version of AI, where knowledge presentation is enhanced with learning and social interaction. In the current environment of global wired and wireless networks intelligent agents may play the role of a universal carrier of distributed artificial intelligence.

1.3 Key Differences Between AI and CI

1.3.1 Key Technical Differences

There are many technical differences between applied AI and applied computational intelligence. The top three are shown in the mind-map in Fig. 1.7 and discussed below.

1.3.1.1 Key Difference #1 – On the main source of representing intelligence

While applied AI is based on representing the knowledge of domain experts in a computer, applied computational intelligence extracts the knowledge from the available data. The main source in the first case is the human and in the second case is the data. However, it is the human who recognizes, interprets, and uses the knowledge.

1.3.1.2 Key Difference #2 – On the mechanisms of processing intelligence

At the core of classical AI are symbolic reasoning methods[8] while computational intelligence is based on numerical methods.

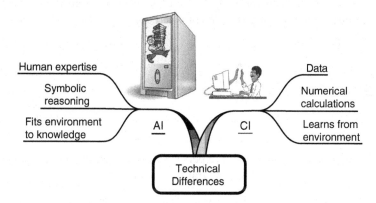

Fig. 1.7 Main technical differences between applied AI and applied computational intelligence

---

[8]In the case of symbolic reasoning the inference is based on symbols, which represent different types of knowledge, such as facts, concepts, and rules.

1.3.1.3   Key Difference #3 – On the interactions with the environment

Applied AI tries to fit the environment to the known solutions, represented by static knowledge bases. In contrast, applied computational intelligence uses every opportunity to learn from the environment and create new knowledge.

1.3.2   The Ultimate Difference

In January 2001 Professor Jay Liebowitz published an exciting editorial in the respected international journal *Expert Systems with Applications*.[9] The title of the editorial is very intriguing: "If you are a dog lover, build expert systems; if you are a cat lover, build neural networks"; see Fig. 1.8.

Professor Liebowitz's arguments for this astonishing "scientific" discovery are as follows.

Since expert systems require a lot of human interaction and devotion, it is preferable if the developers have good people skills. This correlates with the dog lovers, who like companionship, expressiveness, and empathy. Cat lovers, according to Professor Liebowitz, are typically independent types who don't cry for attention. This correlates to the black-box nature of neural networks, which are trained with patterns and require little human intervention and interaction. Neural networks almost seem sterile with respect to expert systems – almost independent like cats. If we look at expert systems as a symbol of applied AI and neural networks as a symbol of applied computational intelligence, Professor Liebowitz's criterion defines the Ultimate Difference between these two applied research fields (see Fig. 1.9).

The proposed hypothesis for the Ultimate Difference between AI and computational intelligence has at least one correct data point. The author is a devoted cat lover.

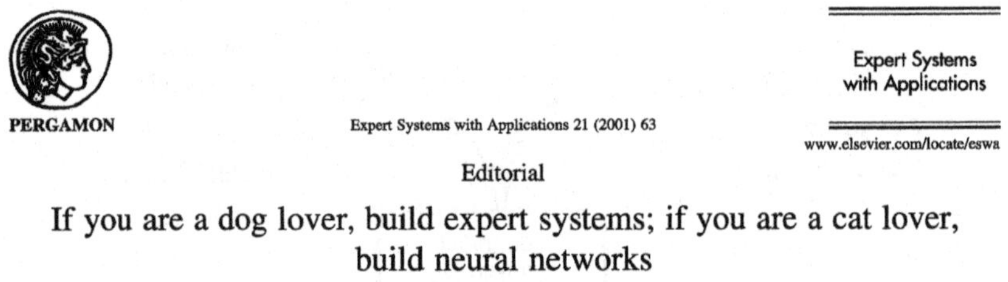

PERGAMON   Expert Systems with Applications 21 (2001) 63

Expert Systems
with Applications

www.elsevier.com/locate/eswa

Editorial

If you are a dog lover, build expert systems; if you are a cat lover, build neural networks

Fig. 1.8 The historic editorial of Prof. Liebowitz defining the ultimate difference between AI and computational intelligence

---

[9]J. Liebowitz, If you are a dog lover, build expert systems; if you are a cat lover, build neural networks, *Expert Systems with Applications*, *21*, pp. 63, 2001.

Fig. 1.9 The Ultimate Difference between applied AI and applied computational intelligence

1.3.3   An Integrated View

Recently, another research direction has emerged for AI which takes the view that intelligence is concerned mainly with rational actions. The artifact that delivers this behavior is the intelligent agent. Building such a computer creature requires a synergy between the approaches of both classical AI and computational intelligence. Probably, this integrated approach will be the next big advance in the field. The synergetic nature of intelligent agents is discussed in Chap. 7.

1.4   Summary

Key messages:

Applied AI mimics human intelligence by representing available expert knowledge and inference mechanisms for solving domain-specific problems.

Applied AI is based on expert systems, inference mechanisms, case-based reasoning, and knowledge management.

Applied AI generated billions of dollars of value in different industries in the areas of advisory systems, decision-making, planning, selection, diagnostics, and knowledge preservation.

Applied AI has significant issues with knowledge consistency, scale-up, lack of learning capabilities, difficult maintenance, and high total-cost-of-ownership.

Applied computational intelligence enhances human intelligence by learning and discovering new patterns, relationships, and structures in complex dynamic environments for solving practical problems.

While applied AI is static and is based on human expertise and symbolic reasoning, applied computational intelligence learns from the changing environment and is based on data and numerical methods.

The Bottom Line

Applied AI exhausted its value creation potential and needs to pass on the torch to its successor – applied computational intelligence.

Suggested Reading

This book presents an interesting discussion of the practical interpretation of intelligence from the founder of Palm Computing:
J. Hawkins, *On Intelligence*, Owl Books, NY, 2005.

An enthusiastic book about the key artificial intelligence applications in the 1980s:
E. Feigenbaum, P. McCorduck, and H. Nii, *The Rise of the Expert Company*, Vintage Books, 1989.

A book with hot philosophical discussions about the nature of artificial intelligence:
S. Franchi and G. Güzeldere, *Mechanical Bodies, Computational Minds: Artificial Intelligence from Automata to Cyborgs*, MIT Press, 2005.

The official AI Bible:
A. Barr and E. Feigenbaum, *The Handbook of Artificial Intelligence*, Morgan Kaufmann, 1981.

These are the author's favorite books on the state of the art of AI:
G. Luger, *Artificial Intelligence: Structures and Strategies for Complex Problem Solving*, 6th edition, Addison-Wesley, 2008.
S. Russell and P. Norvig, *Artificial Intelligence: a Modern Approach*, 2nd edition, Prentice Hall, 2003.

The details for the example with the AI system at the Ford Motor Company:
N. Rychtyckyj, Intelligent Manufacturing Applications at Ford Motor CompanyFord Motor Company, published at *NAFIPS 2005*, Ann Arbor, MI, 2005.

A recent survey of the state of the art of expert system methodologies and applications:
S. Liao, Expert Systems Methodologies and Applications – a decade review from 1995 to 2004, *Expert Systems with Applications*, *28*, pp. 93–103, 2005.

The famous editorial with the Ultimate Difference between AI and computational intelligence:
J. Liebowitz, If you are a dog lover, build expert systems; if you are a cat lover, build neural network, *Expert Systems with Applications*, *21*, p. 63, 2001

Chapter 2
A Roadmap Through the Computational Intelligence Maze

Signposts can turn a highway into labyrinth.
Stanisław Jerzy Lec

The objective of the first part of the book is to give the reader a condensed overview of the main computational intelligence methods, such as fuzzy systems, neural networks, support vector machines, evolutionary computation, swarm intelligence, and intelligent agents. The emphasis is on their potential for value creation. The journey begins in this chapter with a generic explanation of the most important features of the different approaches and their application potential. The focus is on answering the following three questions:

Question#1
What are the top three strengths and weaknesses of each approach?
Question#2
What are the scientific principles on which computational intelligence methods are based?
Question#3
How are the different methods related to the key computational intelligence application areas?

A more detailed description of the specific methods as well as an assessment of their application potential is given in Chaps. 3–7.

2.1 Strengths and Weaknesses of CI Approaches

One of the unique features of computational intelligence is the diverse nature and characteristics of the composed approaches. Unfortunately, it creates confusion to potential users. The broad diversity of methods creates the perception of a steep and time-consuming learning curve. Very often it leads to reducing the attention to one or two approaches from the package. In the last option, however, the potential for value creation is not explored sufficiently since often the benefits are in methods integration.

The first step into the computational intelligence maze is identifying the distinct advantages and potential issues of each approach. The discussed strengths and weaknesses are based not only on purely scientific advantages/disadvantages but on assessments of the potential for value creation as well. The good news is that some of the limitations can be overcome by integration with other approaches. These synergetic opportunities are clearly emphasized.

2.1.1 Strengths and Weaknesses of Fuzzy Systems

A fuzzy system deals with the imprecise nature of human cognition. It mimics the approximate reasoning of humans by representing vague terms in a quantitative way. The top three advantages and issues with fuzzy systems are presented in the mind-map in Fig. 2.1 and discussed briefly below.

Main advantages of applied fuzzy systems:

- *Capture Linguistic Ambiguity* – In contrast to other approaches, fuzzy systems can handle the vagueness of human expressions and build models based on linguistic descriptions, such as: "decrease the power slightly", "move around the corner", "unsatisfactory performance", etc. By replacing the classical crisp logic with the nuances of fuzzy logic, one can quantify linguistic expressions with the granularity of the available knowledge. For example, the phrase "decrease the power slightly" can be mathematically represented by a specific table (called a membership function, described in Chap. 3). The table includes the concrete numbers of declining power until it is completely turned off. The numerical range or the granularity of the linguistic meaning of "slightly decreased power" is problem specific and depends on the available expertise.
- *Computing with Words* – Translating vague linguistic expressions into the numerical world allows us to perform precise computer calculations in a similar way as with other quantitative modeling techniques. Another unique feature of fuzzy systems is presenting the results in a natural language as rules. Therefore

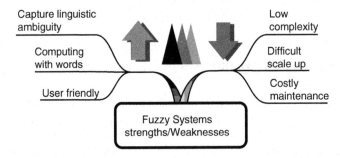

Fig. 2.1 Key strengths and weaknesses in applying fuzzy systems

the dialog with the user and the expert is entirely based on a natural language and no knowledge of the modeling techniques is needed. In many practical cases, using a verbal model of a problem is the only available option. Having the opportunity to use this knowledge in quantitative fuzzy systems is a big advantage of computational intelligence. Often, it is also economically inefficient to justify very high numerical modeling costs and the low fuzzy systems development cost is a viable alternative.

- *User-Friendly* – The interactions in natural language with users and experts significantly reduce the requirements for fuzzy systems development and use. All key components of the fuzzy system – the language description of the system, the definition of the membership functions by the experts, and the recommended solutions as verbal interpretable rules – are easy to understand and implement without special training. Developing a prototype is also relatively fast and easy.

Let's now focus on the dark side of applied fuzzy systems and discuss their top three issues:

- *Low Complexity* – All the discussed advantages of fuzzy systems are practically limited to no more than 10 variables or 50 rules. Beyond that point the interpretability of the system is lost and the efforts for describing the system, defining the linguistic variables, and understanding the recommended actions become too high. Unfortunately, many systems, especially in manufacturing are with higher dimensionality than the practical limits of fuzzy systems. One possible solution to this problem is to rely on other methods to reduce the complexity by extracting the important variables and records. Using statistical methods, neural networks, and support vector machines can significantly reduce the dimensionality of the problem. An example of an integrated technology which accomplishes this goal is given in Chap. 11.
- *Difficult Scale-up* – Often the quantification of linguistic variables is local, i.e. it is valid only for a specific case. For example, the membership function, representing the phrase "slightly decreased power" can be significantly different even between two similar pumps. As a result, scaling-up fuzzy systems is challenging and time-consuming. It is especially true in the case of many variables described by very complex membership functions. A possible solution to overcome partially this issue is by using evolutionary computing for automatically defining the membership functions and the rules.
- *Costly Maintenance* – The static nature of fuzzy systems and the local meaning of the defined membership functions and rules significantly increases the maintenance cost. Even minor changes in any component of the system may require a sequence of maintenance procedures, such as parameter retuning or complete redesign. One way to overcome this limitation and reduce maintenance cost is by combining the advantages of fuzzy systems with the learning capability of neural networks. The developed neuro-fuzzy systems respond to process changes by updating the membership functions from process data.

2.1.2 Strengths and Weaknesses of Neural Networks

The key feature of neural networks is their ability to learn from examples through repeated adjustments of their parameters (weights). The top three advantages and issues with neural networks are presented in the mind-map in Fig. 2.2 and discussed briefly below.

Key advantages of applied neural networks:

- *Learn from Data* – Neural networks are the most famous and broadly applied machine learning approach. They are capable of defining unknown patterns and dependencies from available data. The process of learning is relatively fast and does not require deep fundamental knowledge of the subject. The learning algorithms are universal and applicable for a broad range of practical problems. The required tuning parameters are few and relatively easy to understand and select. In cases of learning long-term patterns from many variables, neural networks outperform humans.
- *Universal Approximators* – It is theoretically proven that neural networks can approximate any continuous nonlinear mathematical function, provided that there is sufficient data and number of neurons. If there is a nonlinear dependence in a given data set, it can be captured by neural networks. The practical importance of this theoretical result is significant. For example, in the case of successful discovery of patterns or relationships, the model development process can be reliably initiated by different methods, including neural networks. On the contrary, in the case of neural network failure, it is doubtful that another empirical modeling technique will succeed.
- *Fast Development* – The neural networks based modeling process is very fast relative to most of the known approaches, such as first-principles modeling, fuzzy systems, or evolutionary computation. However, insufficient quantity and low quality of the available historical data may significantly slow down model development and increase cost. In the case of high-quality data with broad ranges of process variables, useful neural network models can be derived in a couple of hours even for problems with dozens of variables and thousand of records.

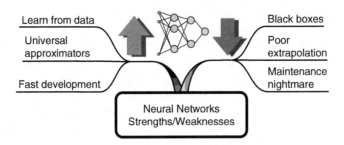

Fig. 2.2 Key strengths and weaknesses in applying neural networks

Key disadvantages of applied neural networks:

- *Black-Box Models* – Many users view neural networks as magic pieces of software that represent unknown patterns or relationships in the data. The difficult interpretation of the magic, however, creates a problem. The purely mathematical description of even simple neural networks is not easy to understand. A black-box links the input parameters with the outputs and does not give any insight into the nature of the relationships. As a result, black-boxes are not well accepted by the majority of users, especially in manufacturing, having in mind the big responsibility in controlling the plants. A potential solution to the problem is by adding interpretability of neural networks with fuzzy systems.

- *Poor Extrapolation* – The excellent approximation capabilities of neural networks within the range of model development data are not valid when the model operates in unknown process conditions. It is true that empirical models cannot guarantee reliable predictions outside the initial model development range, defined by available process data. However, the various empirical modeling methods deliver different levels of degrading performance in unknown process conditions. Unfortunately, neural networks are very sensitive to unknown process changes. The model quality significantly deteriorates even in minor deviations ($< 10\%$ outside the model development range). A potential solution for improving the extrapolation performance of neural networks is by using evolutionary computation for optimal structure selection.

- *Maintenance Nightmare* – The combination of poor extrapolation and black-box models significantly complicates neural network maintenance. The majority of industrial processes experience changes in their operating conditions beyond 10% during a typical business cycle. As a result, the performance of the deployed neural network models degrades, and triggers frequent model retuning, even complete redesign. Since the maintenance and support of neural networks requires special training, this inevitably increases maintenance cost.

2.1.3   Strengths and Weaknesses of Support Vector Machines

In engineering, learning is the process of estimating an unknown relationship or structure using a limited number of observations. Support vector machines (SVM) derive solutions with optimal balance between accurately representing the existing data and dealing with unknown data. In contrast to neural networks, the results from statistical machine learning have optimal complexity for the given learning data set and may have some generalization capability if proper parameters are selected. Support vector machines represent the knowledge learned by the most informative data points, called support vectors. The top three advantages and issues with support vector machines are presented in the mind-map in Fig. 2.3 and discussed briefly below.

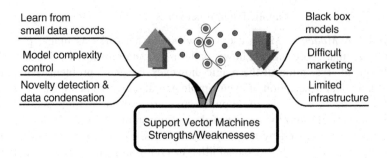

Fig. 2.3 Key strengths and weaknesses in applying support vector machines

Key advantages of applied support vector machines:

- *Learn from Small Data Records* – In contrast to neural networks, which require significant amounts of data for model development, support vector machines can derive solutions from a small number of records. This capability is critical in new product development where each data point could be very costly. Another application area is microarray analysis in biotechnology.
- *Model Complexity Control* – Support vector machines allow explicit control over complexity of the derived models by tuning some parameters. In addition, statistical learning theory, on which support vector machines are based, defines a direct measure of model complexity. According to this theory, an optimal complexity exists. It is based on the best balance between the interpolation capability of a model on known data and its generalization abilities on unknown data. As a result, the derived support vector machines models with optimal complexity have the potential for reliable operation during process changes
- *Novelty Detection and Data Condensation* – At the foundation of support vector machine models are the information-rich data points, called support vectors. This feature gives two unique opportunities. The first opportunity is to compare the informational content of each new data point with respect to the existing support vectors and to define novelty if some criteria are met. This capability is of critical importance for detecting outliers in on-line model operation. The second opportunity of applying the support vector machines method is to condense a large number of data records into a small number of support vectors with high information content. Very often data condensation could be significant and high-quality models could be built with only 10–15% of the original data records.

Main disadvantages of applied support vector machines:

- *Black-Box Models* – For a nonexpert, interpretability of support vector machines models is an even bigger challenge than understanding neural networks. In addition, developing the models requires more specialized parameter selection, including complex mathematical transformations, called kernels. The prerequisite for understanding the latter is some specialized knowledge of math.

The main flaw, however, is the black-box nature of the models with the same negative response from the users, already discussed. Even the better generalization capability of support vector machines relative to neural networks cannot change this attitude. One option of improvement is to add interpretability with fuzzy systems. Another option is to avoid direct implementation of support vector machines as final models but to use some of their advantages, such as data condensation, in model development of methods with higher interpretability, like genetic programming (to be described in Chap. 5). The last option is discussed in Chap. 11.

- *Difficult Marketing* – Explaining support vector machines and their basis, statistical learning theory, includes very heavy math and is a challenge even to an experienced research audience. In addition, the black-box nature of the models and the limited software options create significant obstacles in marketing the approach to a broad class of users. SVM is the most difficult computational intelligence approach to sell. The gap between the technical capabilities of this method and its application potential is big and needs to be filled with adequate marketing. A first step to fill this gap is the condensed presentation of the method in Chap. 4.

- *Limited Infrastructure* – The difficulties in popular nontechnical explanations of support vector machines is one of the issues in propagating the approach to a broader application audience. Another big issue is the lack of user-friendly software. Most of the existing software is developed in academia. While doing a great job computationally, playing with the algorithms, it is not ready for real-world applications. For this purpose, a professional vendor who delivers user-friendliness, integration with the other methods, and professional support is needed. In addition to the professional software deficit, a critical mass of support vector machines users is needed to kick off of this promising technology in industry.

2.1.4 Strengths and Weaknesses of Evolutionary Computation

Evolutionary computation creates novelty by automatically generated solutions of a given problem with defined fitness. The top three advantages and issues with evolutionary computation are presented in the mind-map in Fig. 2.4 and discussed briefly below.

Key advantages of applied evolutionary computation:

- *Novelty Generation* – The most admirable feature of evolutionary computation is its unlimited potential to generate novel solutions for almost any practical problem. When the fitness function is known in advance and can be quantified, the novel solutions are selected after a prescribed number of evolutionary phases. In the case of unknown and purely qualitative fitness, such as fashion design, the novel solutions are interactively selected by the user in each phase of the simulated evolution.

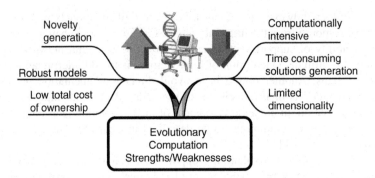

Fig. 2.4 Key strengths and weaknesses in applying evolutionary computation

- *Robust Models* – If complexity has to be taken into account, evolutionary computation can generate nonblack-box models with optimal balance between accuracy and complexity. This is a best-case scenario for the final user, since she/he can implement simple empirical models which can be interpreted and are robust towards minor process changes. Several examples of robust models generated by evolutionary computation are given in different chapters in the book.
- *Low Total-Cost-of-Ownership* – A very important feature of evolutionary computation is that the total development, implementation, and maintenance cost is often at an economic optimum. The development cost is minimal since the robust models are generated automatically with minimal human intervention. In the case of a multiobjective fitness function, the model selection is fast and evaluating the best trade-off solutions between the different criteria takes a very short time. Since the selected empirical solutions are explicit mathematical functions, they can be implemented in any software environment, including Excel. As a result, the implementation cost is minimal. The robustness of the selected models also reduces the need for model readjustment or redesign during minor process changes. This minimizes the third component of the total-cost-of-ownership – maintenance cost.

Main disadvantages of applied evolutionary computation:

- *Computationally Intensive* – Simulated evolution requires substantial number-crunching power. In cases of using other modeling simulation packages in the evolutionary process (for example, simulation of electronic circuits), even computer clusters are needed. Fortunately, the continuous growth of computational power, according to Moore's law,[1] gradually resolves this issue. Recently, the majority of evolutionary computation applications can be developed on a standard PC. Additional gains in productivity are made by improved algorithms.

---

[1]Moore's law states that the number of transistors included in an integrated circuit doubles approximately every two years.

The third way to reduce computational time is shrinking down the data dimensionality. For example, only the most informative data can be selected for the simulated evolution by other methods like neural networks and support vector machines.

- *Time-Consuming Solution Generation* – An inevitable effect of the computationally intensive simulated evolution is the slow model generation. Dependent on the dimensionality and the nature of the application, it may take hours, even days. However, the slow speed of model generation does not significantly raise the development cost. The lion's share of this cost is taken by the model developer's time which is relatively low and limited to a couple of hours for model selection and validation.

- *Limited Dimensionality* – Evolutionary computation is not very efficient when the search space is tremendously large. It is preferable to limit the number of variables to 100 and the number of records to 10,000 in order to get results in an acceptable time. It is strongly recommended to reduce dimensionality before using evolutionary computation. An example of using neural networks for variable selection and support vector machines for data record reduction, in combination with model generation using evolutionary computation, is given in Chap. 11.

2.1.5 Strengths and Weaknesses of Swarm Intelligence

Swarm intelligence offers new ways of optimizing and classifying complex systems in real time by mimicking social interactions in animal and human societies. The top three advantages and issues with swarm intelligence are presented in the mind-map in Fig. 2.5 and discussed briefly below.

Main advantages of applied swarm intelligence:

- *Complex Systems Optimization* - Swarm intelligence algorithms are complementary to classical optimizers and are very useful in cases with static and dynamic combinatorial optimization problems, distributed systems, and very noisy and

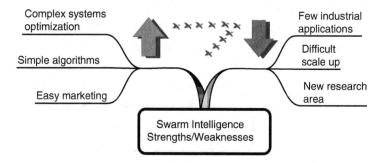

Fig. 2.5 Key strengths and weaknesses in applying swarm intelligence

complex search space. In all of these cases classical optimizers cannot be efficiently applied.

- *Simple Algorithms* – In contrast to classical optimizers, most swarm intelligence algorithms are simple and easy to understand and implement. However, they still need the setting of several tuning parameters, most of them problem specific.
- *Easy Marketing* – The principles of swarm intelligence systems are well understood by potential users. Most of them are willing to take the risk in applying the approach.

Key disadvantages of applied swarm intelligence:

- *Few* Industrial *Applications* – The record of impressive industrial applications based on swarm intelligence is still very short. However, the situation can change very quickly with successful marketing and identifying proper classes of applications.
- *Difficult Scale-up* – Applying swarm intelligence solutions to similar problems on larger scale is not reproducible. This increases the development cost and reduces productivity.
- *New Research Area* – Relative to most other computational intelligence approaches, swarm intelligence has a short research history and many theoretical issues. However, the practical appeal of this approach is very high and its application potential can be exploited even in this early phase of scientific development.

2.1.6 Strengths and Weaknesses of Intelligent Agents

Intelligent agents or agent-based models can generate emerging behavior by mimicking human interactions, such as negotiation, coordination, cooperation, and teamwork. The top three advantages and issues with intelligent agents are presented in the mind-map in Fig. 2.6 and discussed briefly below.

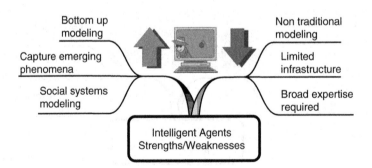

Fig. 2.6 Key strengths and weaknesses in applying intelligent agents

Key advantages of applied intelligent agents:

- *Bottom-up Modeling* – In contrast to first-principles modeling, based on the Laws of Nature, or empirical modeling, based on different techniques of data analysis, intelligent agents are based on the emerging behavior of its constituents – the different classes of agents. At the foundation' of agent-based modeling is the hypothesis that even simple characterization of the key agents can generate complex behavior of the system. Such types of systems cannot be described by mathematical methods. A classical example is analyzing the complex behavior of supply chain systems based on a bottom-up description of the key types of agents – customer, retailer, wholesaler, distributor, and factory.
- *Capture Emerging Phenomena* – A unique feature of intelligent agents is generating totally unexpected phenomena during the modeling process. The source of these emerging patterns is agents' interaction. One impressive advantage of agent-based modeling is that the emerging phenomena can be captured with relatively simple representations of the key classes of agents. This feature has tremendous application potential, especially in the growing area of business-related modeling.
- *Social Systems Modeling* – Intelligent agents are one of the few available methods to simulate social systems. It is our strong belief that the social response simulations will be the most significant addition to the current modeling techniques in industry. The fast market penetration in the global world with diverse cultures of potential customers requires modeling efforts for accurate assessment of customers' acceptance of future products. Intelligent agents can help to fill this gap.

Key disadvantages of applied intelligent agents:

- *Nontraditional Modeling* – Intelligent agents modeling is quite unusual from the mathematical or statistical perspective. One cannot expect results as tangible models like equations or defined patterns and relationships. Usually the results from the modeling are visual and the new emergent phenomena are represented by several graphs. In some cases it is possible that the simulation outcome is verbally condensed in a set of rules or a document. However, the qualitative results may confuse some of the users with a classical modeling mindset and the credibility of the recommendations could be questioned. Validating agent-based models is also difficult since the knowledge of emerging phenomena is nonexistent.
- *Limited Infrastructure* – The available software for development and implementation of intelligent agents is very limited and requires sophisticated programming skills. The methodology for agents-based modeling is still in its infancy.[2] Even the notion of intelligent agents is different in the various research communities. As a result, the level of users' confusion is relatively high.

---

[2]The book *Managing Business Complexity* by M. North and C. Macal, Oxford University Press, 2007, is one of the first sources offering a practical intelligent agents modeling methodology.

Significant marketing efforts are needed to promote the technology for industrial applications.

- *Broad Expertise Required* – The key for success in intelligent agents modeling is the adequate representation of the behavior of the main agent classes. This requires participation of the best experts in defining and refining the classes and analyzing modeling results. The process could be time-consuming, especially when diverse expertise has to be defined and organized. Analyzing modeling results and capturing emergent behavior is also nontrivial and requires high expertise.

2.2   Key Scientific Principles of Computational Intelligence

The next step in the computational intelligence roadmap is to give the reader an overview about the technical basis of the field. The starting point is the mind-map in Fig. 2.7.

At the foundation of computational intelligence are three scientific disciplines: bio-inspired computing, machine learning, and computer science. While being very diverse in their principles, all of these disciplines contribute to the development of computer-based solutions which enhance human intelligence. The specific scientific principles are discussed next.

2.2.1   Bio-inspired Computing

It is not a surprise that one of the key founding disciplines of computational intelligence is related to the sources of intelligence in nature, i.e. biology. Bio-inspired computing uses methods, mechanisms, and features from biology to develop novel computer systems for solving complex problems. The motivation of bio-inspired computing is discovering new techniques with competitive advantage over the traditional approaches for a given class of problems. In the case of computational intelligence, biologically motivated computing has driven the development of three methods - neural networks, evolutionary computation, and swarm intelligence from their biological metaphors, respectively, brain functioning, evolutionary biology, and social interaction.

- *Brain Functioning* – It is natural to look first at the brain as the key source in understanding intelligence. One of the most popular computational intelligence methods – neural networks – is directly inspired by the brain's functionality. In the same way as neurons are the basic units for information processing in the brain, their simplified mathematical models are the main calculation units of artificial neural networks. The principles of neuron organization in the brain, the

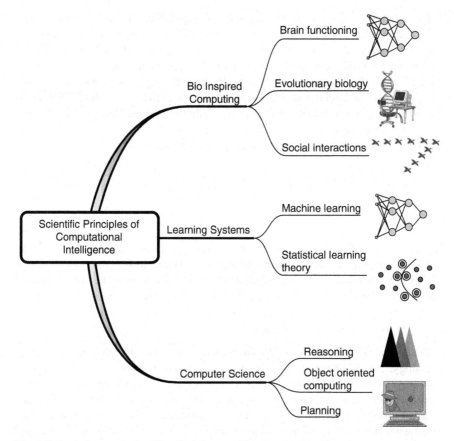

Fig. 2.7 A mind-map of scientific principles on which computational intelligence is based

information flows, and the learning mechanisms are used in neural network design.

- *Evolutionary Biology* – Darwin's "dangerous idea" about natural evolution is the other key source of imitating nature in understanding and simulating intelligence. The important step is to look at natural evolution as an abstract algorithm which can be used in a computer for creating new artifacts or improving the properties of existing artifacts. Evolutionary computation represents in various forms the three major processes in the theory of natural evolution: (1) existence of a population of individuals that reproduce with inheritance, (2) adaptation to environmental changes by variation in different properties in individuals, and (3) natural selection where the "best and the brightest" individuals survive and the losers perish. At the center of evolutionary computation is the hypothesis that the simulated evolution will generate individuals which will improve their performance in a similar way as biological species did in millions of years of natural evolution.

- *Social Interactions* – The third key source of mimicking biology for understanding and simulation of intelligence is the social life of different biological species, such as insects, birds, fish, humans, etc. One way is reproducing the collective behavior of some specific forms of social coordination, like finding food or handling the dead bodies in ant colonies in computer algorithms. Another way of exploring social interactions is by representing the abilities of humans to process knowledge through sharing information. In both cases, unique algorithms for complex optimization have been developed even by very simple models of collective behavior of both biological species and humans.

2.2.2   Learning Systems

The capability to learn about the changing environment and adapt accordingly is one of the distinct features of human intelligence. Numerous research disciplines, such as neuroscience, cognitive psychology, and different branches of computer science, are exploring this complex phenomenon. We'll focus our attention on two of them – machine learning and statistical learning theory – which have direct impact on computational intelligence.

- *Machine Learning* – The approach enables computers to learn by example, analogy, and experience. As a result, the overall performance of a system improves over time and the system adapts to changes. A broad range of computer-driven learning techniques have been defined and explored. Some of them, called supervised learning, require an artificial teacher who gives examples that navigates the machine tuition process. In many cases the computerized learning process does not have the luxury of having a teacher and can rely only on analysis of existing data (unsupervised learning). Sometimes the machine can learn in a hard way, by trial-and-error, using reinforcement learning.
- *Statistical Learning Theory* – It is very important in practical applications to have the capability to learn from a small number of data samples. Statistical learning theory gives the theoretical basis for extracting patterns and relationships from a few data points. At the core of this learning machine is the balance between performance and complexity of the derived solutions. Statistical learning theory describes two key ways of predictive learning, which are very important from the application point of view. The first approach, which prevails in industry, is the inductive-deductive method. In it, a model is developed from available data, as a result of inductive learning (i.e. progressing from particular cases to general relationships, captured in the model). Then, the derived model is applied for prediction on new data using deduction (i.e., progressing from the generic knowledge of the model to the specific calculation of the prediction). The second approach, which has big application potential, is the transductive learning method. It avoids the generalization step of building a model from the data and gives direct predictions from available data.

2.2.3 Computer Science

Since computers are the natural environment of applied computational intelligence, its links to different disciplines in the broad research area of computer science is expected. We'll focus on three disciplines: reasoning, object-oriented computing, and planning due to their relation to fuzzy systems and intelligent agents.

- *Reasoning* – The issue of mimicking the way humans think and make conclusions is central to both artificial and computational intelligence. Several methods for describing logical inference, such as propositional logic, first-order logic, probabilistic logic, and fuzzy logic, are used. The key assumption is rational behavior which will lead to sound reasoning and logical inference. Unfortunately, this assumption is not entirely correct in observing social systems.
- *Object-Oriented Computing* – The fast development of object-oriented methodologies and programming languages in the last 10 years offers very important features for the design and implementation of distributed intelligent systems, especially intelligent agents. Of special importance for agent functionality are the features of inheritance, encapsulation, message passing, and polymorphism.
- *Planning* – One of the most interesting and useful features of intelligent agents is based on their ability to automatically design plans for specific tasks. Several methods, such as conditional planning, reactive planning, execution monitoring, and re-planning are used.

2.3 Key Application Areas of Computational Intelligence

The objective of the final third step in the roadmap through the computational intelligence maze is to help the reader to match the key approaches with the application areas with most perspective. The mind-maps representing these links are given in Fig. 2.8 for manufacturing applications and in Fig. 2.9 for business applications. A short description of how the related methods are used in each application area is given below. A more detailed analysis of the application potential of computational intelligence is given in Chap. 8.

2.3.1 Invention of New Products

A typical list of new product invention tasks is: identification of customer and business requirements for new products; product characterization; new product optimal design; and design validation on a pilot scale. Two computational intelligence approaches – evolutionary computation and support vector machines – are especially appropriate in solving problems related to discovery of new products. Evolutionary computation generates automatically or interactively novelty on almost any new idea that can be represented by structural relationships and

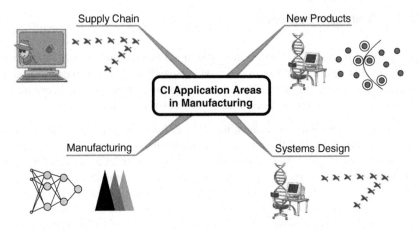

Fig. 2.8 A mind-map of the key computational intelligence application areas in manufacturing and new products and related methods

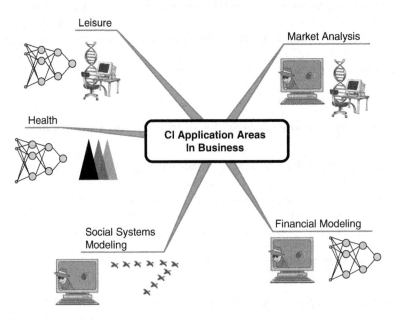

Fig. 2.9 A mind-map of the key computational intelligence application areas in business

supported by data. Support vector machines are capable of delivering models from very few data records, which is the norm in new product development. This allows the use of models in making critical decisions at relatively early implementation phases of the invention. In addition, the development cost is reduced due to the smaller number of expensive experiments and more reliable scale-up.

2.3.2 Systems Design

Systems design includes activities like structure generation, optimization, and implementation. Evolutionary computation, especially genetic programming, is very efficient in generating structures of different physical systems. Typical examples are automatically generated electronic circuits and optical systems.[3] Other evolutionary computation approaches, such as genetic algorithms and evolutionary strategies, have been used in numerous industrial applications for design optimization of car parts, airplane wings, optimal detergent formulations, etc. Recently, swarm intelligence and especially particle swarm optimizers offer new capabilities for optimal design in industrial problems with high noise level.

2.3.3 Manufacturing

Advanced manufacturing is based on good planning, comprehensive measurements, process optimization and control, and effective operating discipline. This is the application area where all of the computational intelligence methods have demonstrated impact. For example, intelligent agents can be used for planning. Neural networks, genetic programming, and support vector machines are used for inferential sensors and pattern recognition. Evolutionary computation and swarm intelligence are widely implemented for process optimization. The list of industrial neural networks-based and fuzzy logic-based control systems is long and includes chemical reactors, refineries, cement plants, etc. Effective operating discipline systems can be built, combining pattern recognition capabilities of neural networks with fuzzy logic representations of operators' knowledge and decision-making abilities of intelligent agents.

2.3.4 Supply Chain

The main activity with greatest impact in the supply chain is optimal scheduling. It includes actions like vehicle loading and routing, carrier selection, replenishment time minimization, management of work-in-progress inventory, etc. Two computational intelligence methods – evolutionary computation and swarm intelligence – broaden the optimal scheduling capabilities in these problems with high dimensionality and noisy search landscapes with multiple optima. For example, Southwest Airlines saved $10 million/year from a Cargo Routing Optimization

---

[3]J. Koza, *et al. Genetic Programming IV. Routine Human-Competitive Machine Intelligence,* Kluwer, 2003.

solution designed using evolutionary computation.[4] Procter & Gamble saved over $300 million each year thanks to a Supply Network Optimization system based on evolutionary computation and swarm intelligence. Another important supply chain application, based on computational intelligence, is optimal scheduling of oil rigs at the right places at the right times, which maximizes production and reserves while minimizing cost.

The third useful computational intelligence technology for this type of application – intelligent agents – is appropriate in predicting customer demand, which is a key factor in successful supply chain planning.

The application areas related to business applications are shown in the mind-map in Fig. 2.9.

2.3.5   Market Analysis

Analysis of current and future markets is one of the key application areas where computational intelligence can make a difference and demonstrate value potential. Typical activities include, but are not limited to, customer relation management, customer loyalty analysis, and customer behavior pattern detection. Several computational intelligence methods, such as neural networks, evolutionary computation, and fuzzy systems give unique capabilities for market analysis. For example, self-organizing maps can automatically detect clusters of new customers. Evolutionary computation also generates predictive models for customer relation management. Fuzzy logic allows quantification of customer behavior, which is critical for all data analysis numerical methods. Some specific examples of the impact of using computational intelligence in market analysis include: a 30% system-wide inventory reduction at Procter & Gamble as a result of customer attention analysis and a system for customer behavior prediction with a loyalty program and marketing promotion optimization.[5]

Intelligent agents can also be used in market analysis by simulating customers' responses to new products and price fluctuations.

2.3.6   Financial Modeling

Financial modeling includes computation of corporate finance problems, business planning, financial securities portfolio solutions, option pricing, and various economic scenarios. The appropriate computational intelligence techniques for this application area include neural networks, evolutionary computation, and intelligent agents. Neural networks and genetic programming allow the use of nonlinear

---

[4]http://www.nutechsolutions.com/pdf/SO_Corporate_Overview.pdf

[5]http://www.nutechsolutions.com

forecasting models for various financial indicators derived from available data of past performance, while intelligent agents can generate emergent behavior based on economic agents' responses. An additional benefit of evolutionary computing in financial modeling is the broad area of optimizing different portfolios.

2.3.7 Modeling Social Behavior

One growing trend in industrial modeling is improving business decisions through simulating social systems. Three computational intelligence techniques are critical in this process. Intelligent agents are the most appropriate method to represent different types of behavior of social agents. Swarm intelligence has the capability to model social behavior based on different social interactive mechanisms. Fuzzy systems allow the transfer of the qualitative knowledge of humans into the numerical world. An example is the social simulation of the effect of small-scale experimental drug discovery in contrast to large-scale R&D organization at Ely Lilly. According to the simulation, based on intelligent agents and swarm intelligence, the drug discovery time can be reduced from 40 months to 12 months and the cost reduction is from roughly $25 million to the order of $2.7 million.[6]

2.3.8 Health

It is no secret that the growing healthcare-related needs will define more and more opportunities for creating value. Computational intelligence can be a significant player in this process in the areas of medical diagnosis and modeling, health monitoring, and building personal health advisory systems. For example, medical diagnosis can benefit significantly from the pattern recognition capabilities of neural networks. Developing medical models is extremely difficult and requires integration of empirical relationships, derived by support vector machines and evolutionary computation, with qualitative rules of thumb, defined by physicians and captured by fuzzy logic. The final products from medical diagnosis and modeling will be personal health monitoring and advisory systems, based on these computational intelligence technologies.

2.3.9 Leisure

The application area of leisure-related industry is expected to grow with the coming army of well-educated computer-savvy baby boomers having one key objective in

---

[6]http://www.icosystem.com/releases/ico-2007-02-26.htm

their retirement years – how to spend their free time with more exciting activities. Some of them could be interested in the capabilities of evolutionary art, even in playing intelligent games. An interesting opportunity is designing new types of games as mental gymnastics to counteract the degrading cognitive capabilities with aging. Three computational intelligence technologies can be used in this process – intelligent agents, evolutionary computing, and neural networks. Usually, evolving neural networks are at the basis of the learning capabilities of designed games. Intelligent agents offer a proper framework to define the characters of the game.

2.4   Summary

Key messages:

Understanding the distinct advantages and limitations in applying each specific computational intelligence method is the first step in entering the field.

The main scientific principles of computational intelligence are based on bio-inspired computing, machine learning, and computer science.

The diverse scientific principles require a broader scope of scientific knowledge for understanding the technical details of different methods.

The next critical step in entering the field of applied computational intelligence is understanding the application areas of each specific method.

The Bottom Line

Computational intelligence is based on diverse scientific principles and covers broad application areas in almost all business activities.

Suggested Reading

The following two books give detailed technical descriptions of the key computational intelligence techniques:

A. Engelbrecht, *Computational Intelligence: An Introduction*, 2nd edition, Wiley, 2007.

A. Konar, *Computational Intelligence: Principles, Techniques, and Applications*, Springer, 2005.

The best references for the scientific principles of bio-inspired computational intelligence method are:

L. de Castro, *Fundamentals of Natural Computing*, Chapman & Hall, 2006.

R. Eberhart and Y. Shi, Computational *Intelligence: Concept to Implementations*, Elsevier, 2007.

Chapter 3
Let's Get Fuzzy

*Everything is vague to a degree you do not realize till you
have tried to make it precise.*

Bertrand Russell

We begin our journey through the computational intelligence maze by analyzing the value creation capabilities of fuzzy systems. The reason is the popularity of this approach relative to other computational intelligence methods. Unfortunately, the "celebrity" status is often combined with misconceptions and lack of understanding of fuzzy systems principles. One of the main objectives of this chapter is to reduce the confusion and to defuzzify the key terms and methods. Fuzzy logic does not mean vague answers. Fuzzy sets and fuzzy logic allow reasoning about grey areas using simple numerical schemes. The only limitation on the precision of fuzzy logic models is the trade-off between increased accuracy and high cost. The basic assumption of fuzzy systems is the existence of an effective human solution to a problem, which is close to industrial reality but often is the basis of sharp academic criticism.

This chapter addresses the value creation capabilities of fuzzy systems in the following manner. Firstly, the main concepts and methods are discussed in a popular nonmathematical way. Secondly, attention is focused on the key benefits and issues of applied fuzzy systems. Thirdly, the topic of applying fuzzy systems, as well as the best known application areas, is discussed and illustrated with specific industrial applications. Examples of fuzzy system marketing as well as links to several related resources on the Internet are given at the end of the chapter.

The same format will be used for presenting the other key approaches of applied computational intelligence in the first part of the book.

3.1 Fuzzy Systems in a Nutshell

A fuzzy system is a qualitative modeling scheme describing system behavior using a natural language and non-crisp fuzzy logic. We suggest practitioners focus on the following key topics, related to fuzzy systems, shown in Fig. 3.1.

A.K. Kordon, *Applying Computational Intelligence*,
DOI 10.1007/978-3-540-69913-2_3, © Springer-Verlag Berlin Heidelberg 2010

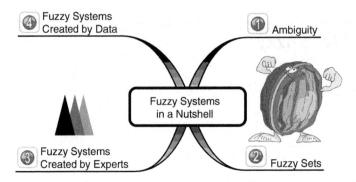

Fig. 3.1 Key topics related to fuzzy systems

It is not a surprise that we begin the introduction to the fuzzy systems world by discussing their most unique feature – dealing with ambiguity. The next step includes clarifying the notion of fuzzy sets, which are at the basis of translating vague concepts into corresponding levels of degree, and finally into numbers. The key topics in this section are the descriptions of the two main types of fuzzy systems – classical fuzzy system based on expert knowledge and the more modern data-based fuzzy systems where the knowledge is extracted from the data and refined by the experts.

3.1.1   Dealing With Ambiguity

Handling imprecise and vague information is typical for human intelligence. Unfortunately, reproducing this activity in the numerical universe of computers is not trivial. Fuzzy systems have the unique capability to represent vague concepts as a level of degree and to build the bridge to computer calculations. At a fundamental level, expressing the semantic ambiguity of natural language with mathematical means allows us to analyze scientifically subjective intelligence. More importantly, this capability has tremendous value creation opportunities, which have been explored and used by different industries at the very early development phases of the field.

It is accepted that fuzzy systems were born with the famous paper of Prof. Lotfi Zadeh with the provocative title "Fuzzy Sets" in the 1965 issue of *Information and Control* journal.[1] The new ideas, however, were furiously challenged by part of the academic community, especially in the United States.[2] Fortunately, the value

---

[1]The paper is available at http://www-bisc.eecs.berkeley.edu/Zadeh-1965.pdf

[2]The interesting history of fuzzy logic is described in the book of R. Seising, *The Fuzzification of Systems: The Genesis of Fuzzy Set Theory and Its Initial Applications: Developments up to the 1970s,* Springer, 2007.

Fig. 3.2 An interesting example of fuzzy math[3]

creation potential of the approach was detected before settling the academic wars. Japanese industry took the lead and several big companies like Matsushita, Sony, Hitachi, and Nissan, have developed impressive applications in various areas since the late 1970s. As a result, industrial support played a critical role for the survival of this emerging research field. The history of fuzzy systems shows that when successful industrial applications speak even arrogant academic gods listen.

Before moving to the key mechanisms of representing ambiguity by fuzzy systems we would like to clarify the meaning and reduce the confusion with the famous phrase "fuzzy math". Fuzzy math has nothing in common with fuzzy systems. It is a synonym of messy or wrong calculations. Fuzzy math is well illustrated by the example shown in Fig. 3.2.

Two basic concepts of *graduation* and *granulation* are at the core of understanding how fuzzy systems deal with ambiguity. On the one hand, in fuzzy systems everything is allowed to be *graduated*. On the other hand, in fuzzy systems, everything is allowed to be *granulated*, with a granule being a clump of attribute-values which capture the essence of an entity. In this way the ambiguity of human intelligence can be represented by linguistic variables, which may be viewed as granulated variables whose granular values are linguistic labels of granules.

The philosophical basis of *graduation* and *granulation* is the Principle of Incompatibility, defined by Prof. Zadeh:

The Principle of Incompatibility

As the complexity of a system increases, our ability to make precise and yet significant statements about its behavior diminishes until a threshold is reached beyond which precision and significance (or relevance) become almost exclusive characteristics.[4]

---

[3]From the website: www.DirtyButton.com

[4]Lotfi Zadeh, Outline of a new approach to the analysis of complex systems and decision processes, *IEEE Trans. Syst. Man and Cybern.*, vol. SMC-3, no. 1., pp. 28–44, 1973.

Briefly: high precision is incompatible with high complexity. The solution, according to Prof. Zadeh, is by using the unique capabilities of fuzzy logic. And it has been proven in many areas of science and in industry.

The main objective of applied fuzzy systems is transferring ambiguity in natural language into value. Lawyers are already doing this very effectively earning tremendous amounts of money.

3.1.2 Fuzzy Sets

Fuzzy sets are the key for understanding fuzzy logic and fuzzy systems. We'll illustrate the concept and the difference between crisp sets and fuzzy sets with an example. Imagine that we have a bowl full of white and black cherries in a different ratio. Let's analyze how we can answer the question: Is this a bowl of white cherries?

In the case of a crisp logic the potential answers are limited to two crisp values - Yes or No, i.e. the only adequate cases assume that the bowl does contain either only white cherries or only black cherries. One way to represent visually this crisp set is by the degree of membership, which in this case is zero when the bowl does not contain white cherries and one when the bowl is totally full with white cherries (see Fig. 3.3).

In real life, however, the bowl may contain many different ratios between black and white cherries. The distinct ratios can be linguistically represented by specific fuzzy phrases, such as Slightly, Sort Of, Mostly, etc. The ratio between black and white cherries of each phrase can be quantified by its degree of membership, which goes from zero (no membership, i.e. no white cherries in the bowl) to one (complete membership, i.e. bowl totally full with white cherries). The degree of membership is also known as the membership or truth function since it establishes a one-to-one correspondence between an element of the fuzzy set (the ratio between the black and the white cherries in a bowl) and a truth value indicating its degree of

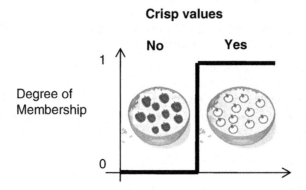

Fig. 3.3 Crisp values for the answer to the question "Is this a bowl of white cherries?"

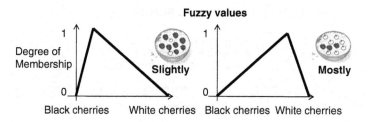

Fig. 3.4 Fuzzy sets Slightly and Mostly for the answer to the question "Is this a bowl of white cherries?"

membership in the set. An example of a simple triangular membership function for two possible fuzzy sets, Slightly and Mostly, is shown in Fig. 3.4.

This example illustrates clearly the notion of the fuzzy set, defined by its membership function. A membership function is a curve that defines how each point in the input space of the linguistic variable is mapped to a membership value between 0 and 1. The values of the membership function measure the degree to which objects satisfy imprecisely defined properties. Membership functions can be of different shapes – triangular, trapezoidal, sigmoid, etc. As shown in this example, fuzzy sets are clearly (crispy) defined by the membership function and there is nothing ambiguous about the definition itself.

3.1.3 Fuzzy Systems Created By Experts

There are two main sources for creating of fuzzy systems – experts and data. We'll call them expert-based and data-based fuzzy systems. In expert-based fuzzy systems, the fuzzy rules are usually defined by the experts while in data-based fuzzy systems, the rules are extracted from clusters in the data. The first class is discussed in the current section and the second class is the topic of the next section. A typical structure of an expert-based fuzzy system is shown in Fig. 3.5.

We begin with clarifying several key terms, such as linguistic variable, fuzzy proposition, fuzzy inference, fuzzification, and defuzzification.

The unique feature of fuzzy systems is their ability to manipulate linguistic variables. A linguistic variable is the name of a fuzzy set directly representing a specific region of the membership function. For example, the phrase "Slightly" in Fig. 3.4 is a fuzzy set of the linguistic variable "a bowl of white cherries".

The inference part of fuzzy systems is based on fuzzy propositions. In general, this is a conditional proposition with the general form,

If *w* is *Z*,
    then *x* is *Y*

Where w and x are model scalar values and Z and Y are linguistic variables. The proposition following the *if* term is the antecedent or predicate and can be any fuzzy proposition. The proposition following the *then* term is the consequent and is also a

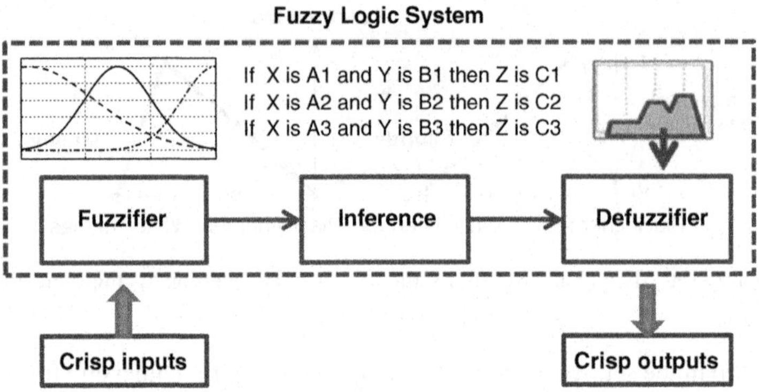

Fuzzy Logic System

If X is A1 and Y is B1 then Z is C1
If X is A2 and Y is B2 then Z is C2
If X is A3 and Y is B3 then Z is C3

Fuzzifier → Inference → Defuzzifier

Crisp inputs   Crisp outputs

Fig. 3.5 Key blocks of an expert-based fuzzy systems

fuzzy proposition. However, the consequent is correlated with the truth of the antecedent, i.e. x is member of Y to the degree that w is a member of Z.

Example:
    if Reactor Temperature is low,
      then Bottom Pressure is high,
    i.e. Bottom Pressure is a member of the linguistic variable "low" to the degree that Reactor Temperature is a member of the linguistic variable "high".

Fuzzy inference is a method that interprets the input values and based on a set of rules assigns values to the output. Fuzzification includes the process of finding the membership value of a number in a fuzzy set.

Aggregating two or more fuzzy output sets yields a new final fuzzy output set, which is converted into a crisp result by the deffuzifier. Defuzzification is the process of selecting a single value from a fuzzy set. It is based on two basic mechanisms - center of gravity and maxima. The center of gravity method finds a balance point of a property, such as the weight area of each fuzzy set. The other basic mechanism, maximum possibility, searches for the highest peak of the fuzzy set.

We'll illustrate the functionality of expert-based fuzzy systems with an example of a very simple system for process monitoring. It represents the big application area of capturing the best process operators' knowledge in monitoring and control of industrial processes.

The objective of the fuzzy system is to monitor the top pressure of a chemical reactor based on two input variables – reactor temperature and caustic flow to the reactor. Process operators use three linguistic levels for each input and output variables – low, good, stable or normal, and high. Their unique knowledge of the process is captured by membership functions for each variable at three linguistic levels. For example, the reactor temperature behavior is represented by the three sigmoid functions, shown in Fig. 3.6 where the y-axis represents the membership function value between 0 and 1 and the x-axis represents the physical measurements of the temperature in degrees C between 100°C and 250°C, which is the full range of change of the temperature.

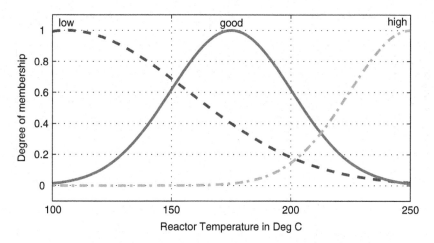

Fig. 3.6 Membership function plots of reactor temperature input variable

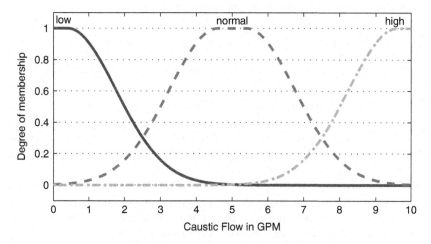

Fig. 3.7 Membership function plots of caustic flow input variable

The low reactor temperature linguistic variable represents the range below 150°C with definite inclusion of the temperatures around 100°C. The broad range of the good reactor temperature linguistic variable covers the temperatures between 150°C and 200°C with the highest level of membership around 175°C. The high reactor temperature linguistic variable represents the range above 200°C with strongest influence of the temperatures around 250°C.

The membership function for the caustic flow, shown in Fig. 3.7 is organized in a similar way but the shape of the sigmoid functions for low, normal, and high flow is different. In the case of the output variable – the reactor top pressure – process operators prefer to represent the low, stable, and high pressure by triangular membership functions (see Fig. 3.8).

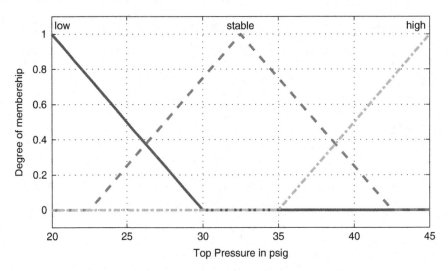

Fig. 3.8 Membership function plots of top pressure output variable

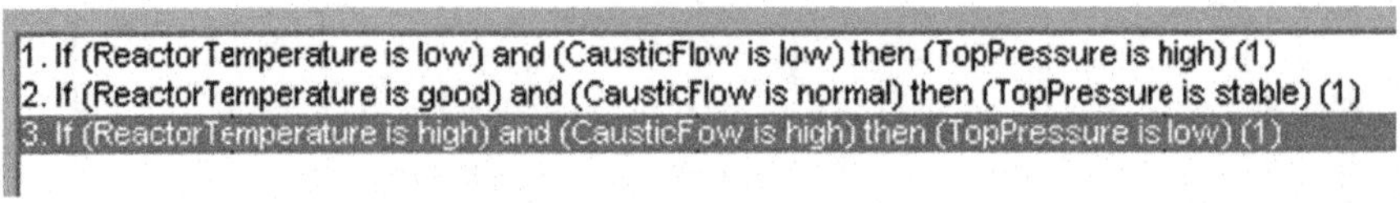

1. If (ReactorTemperature is low) and (CausticFlow is low) then (TopPressure is high) (1)
2. If (ReactorTemperature is good) and (CausticFlow is normal) then (TopPressure is stable) (1)
3. If (ReactorTemperature is high) and (CausticFow is high) then (TopPressure is low) (1)

Fig. 3.9 Fuzzy rule base for top pressure defined by process operators

The other way of capturing operators' knowledge is through rules based on the defined linguistic variables. For example, the monitoring process of the top pressure is represented by three fuzzy rules, shown in Fig. 3.9.

One of the advantages of fuzzy rules over classical rule-based systems is the quantitative nature of the decision space. In the case of our example, the decision space can be visualized as a response surface of the top pressure in the full ranges of change of the two input variables, reactor pressure and caustic flow, shown in Fig. 3.10.

As is clearly seen, the qualitative description, captured in the fuzzy rules in Fig. 3.9 is translated by the membership functions of the linguistic variables into an entirely quantitative response surface.

3.1.4 Fuzzy Systems Created by Data

The other main type of fuzzy system is based on fuzzy clusters, derived from available data. Its key advantage is the reduced role of experts in defining the system. As a result, the development cost could be significantly lower. The main blocks of a data-driven fuzzy system are shown in Fig. 3.11 and are described below.

Data-driven fuzzy systems assume availability of high quality data from the key process inputs and outputs. In order to help the discovery of a fuzzy rule, it is

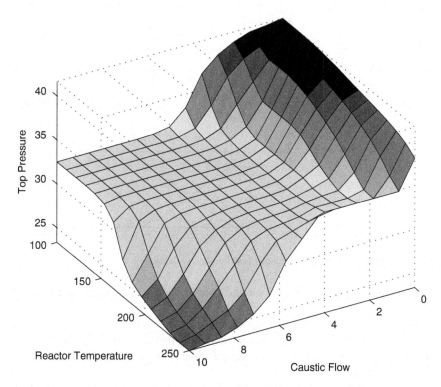

Fig. 3.10 Top pressure fuzzy system response surface based on reaction temperature and caustic flow

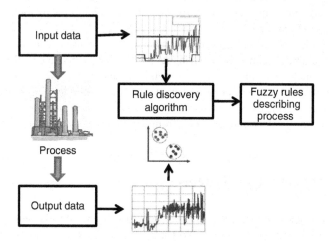

Fig. 3.11 Key blocks of a fuzzy inference systems based on rule discovery from data

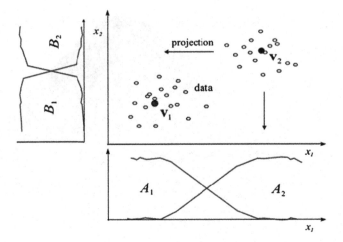

Fig. 3.12 Defining membership functions based on fuzzy clusters

recommended to collect data with the broadest possible ranges of identified variables.

The idea behind data-driven fuzzy systems is to define fuzzy rules and membership functions by finding clusters in data, assuming that each recognized cluster is a rule. The cluster is projected on a single dimension to project the degrees of membership of the data.

The rule discovery algorithm is based on different types of fuzzy clustering algorithms, which partition the data into a given number of clusters with fuzzy boundaries. The principle of these algorithms is minimizing some distance of a data point to a cluster candidate and the membership of the data point to the selected proto-cluster. As a result of applying the fuzzy clustering algorithm, the data are divided into the discovered number of clusters. Each cluster can be represented by a rule. The corresponding fuzzy sets in the rule are identified by projecting the clusters into each variable.

A very simple example of fuzzy clustering based on two clusters with centers V_1 and V_2 and two variables x_1 and x_2 is shown in Fig. 3.12. In this particular case the projected membership functions A_1 and A_2 for x_1 as well as B_1 and B_2 for x_2 are well defined.

Unfortunately, it is more common practice to have overlapping clusters with blurred edges. The rule discovery in this case requires additional techniques for rule merging and, above all, some feedback from the experts.

3.2 Benefits of Fuzzy Systems

One of the key benefits of fuzzy systems is their focus on what the system should do rather than trying to model how it works. However, it assumes sufficient expert knowledge even in the case of data-driven fuzzy rule discovery for the formulation

Fig. 3.13 Key benefits from applying fuzzy systems

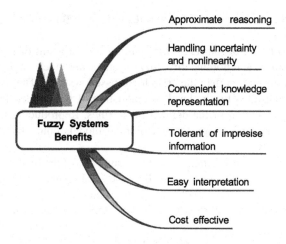

of the membership functions, the rule base, and the defuzzification. It is an adequate approach if the process is very complex and mathematical modeling is a very expensive alternative.

The main benefits from applying fuzzy systems are shown in the mind-map in Fig. 3.13 and discussed briefly below.

- *Approximate Reasoning* – Fuzzy systems convert complex problems into simpler problems using approximate reasoning. The system is described by fuzzy rules and membership functions using natural language and linguistic variables. It is easy for experts to design and even maintain the system. Their unique condensed system description is a form of generalization.
- *Handling Uncertainty and Nonlinearity* – Fuzzy systems can effectively represent both the uncertainty and nonlinearity of complex systems. Very often it is technically challenging to define mathematically uncertainty and nonlinearity of a complex industrial system and especially their dynamics. Usually the uncertainty of the system is captured by the design of the constituent fuzzy sets and the nonlinearity is captured by the defined fuzzy rules.
- *Convenient Knowledge Representation* – In fuzzy systems the knowledge is distributed between the rules and the membership functions. Usually the rules are very general and the specific information is captured by the membership functions. This allows decoupling of the potential problems with tuning the membership functions first. A comparison with classical expert systems shows that the number of rules in a fuzzy system is significantly lower. In fuzzy systems, knowledge acquisition is easier, more reliable, and less ambiguous than in classical expert systems. In fact, crisp logic-based classical expert systems are a special case of fuzzy reasoning. An observed advantage of fuzzy expert systems from numerous practical applications is

that the number of fuzzy rules is 10 to 100 times less than the number of crisp rules.[5]

- *Tolerant of Imprecise Information* – As a result of handling ambiguity, uncertainty, and nonlinearity, fuzzy systems are much more robust towards imprecise information than the crisp logic-based systems. The tolerance to imprecision is controlled by the level of graduation of the defined fuzzy sets and the size of granulation of the defined fuzzy rules. Reproducing this level of tolerance by either first-principles or empirical models could be extremely inefficient and costly.

- *Easy Interpretation* – The two main parts of the fuzzy system - the fuzzy sets and rules – are interpretable by a large audience of potential users. Defining or refining the membership functions and the fuzzy rules by the experts does not require special training and skills. Due to the high interpretability, the designed fuzzy systems are communicated to the final user with very low effort. Their maintenance also benefits from the user-friendly nature of the models.

- *Cost Effective* – Fuzzy systems are cost effective for a wide range of applications, especially when low hardware and software resources are needed, as in home appliances. The development, implementation and support cost is low relative to most of the other computational intelligence technologies.

3.3 Fuzzy Systems Issues

In addition to the well-known fuzzy systems issues already discussed in the previous chapter (low complexity of the models, difficult scale-up, and costly maintenance in changing operating conditions) we'll focus on two important problems: membership function tuning and feasibility of defuzzification results.

As the system complexity increases, defining and keeping track of the rules becomes more and more difficult. An open issue is to determine a sufficient set of rules that adequately describes the system. In addition, tuning the membership functions and adjusting the defined rules with the available data can be very time consuming and costly. Beyond 20 rules the development and maintenance cost becomes high.

Another well-known issue of fuzzy systems is the heuristic nature of the algorithms used for defuzzification and rule evaluation. In principle, heuristic algorithms do not guarantee feasible solutions in all possible operating conditions. As a result, undesirable and confusing results can be produced, especially in conditions even slightly different from the defined.

---

[5]T. Terano, K. Asai, M. Sugeno, *Applied Fuzzy Systems*, Academic Press, 1994.

3.4   How to Apply Fuzzy Systems

As the key objective of applied fuzzy systems is to transfer ambiguity into value, the main application effort is the effective translation of the vague information from the natural language of the experts into the precise and highly interpretable language of fuzzy sets and fuzzy rules. Experts play the key role in this process even in the case of data-driven fuzzy systems where the rules are automatically discovered by the clustering algorithms but must be interpreted and blessed by the experts.

3.4.1   When Do We Need Fuzzy Systems?

Here is a shortlist:

- Vague knowledge can be included in the solution.
- The solution is interpretable in a form of linguistic rules, i.e. we want to learn about our data/problem.
- The solution is easy to implement, use, and understand.
- Interpretation is as important as performance.

3.4.2   Applying Expert-Based Fuzzy Systems

There is no fixed order for the design of a fuzzy system. An attempt to define an application sequence for classical expert-based systems is given in Fig. 3.14.

Probably 80% of the application success depends on the efficiency and quality of knowledge acquisition. This is the process of extraction of useful information from the experts, data sets, known documents and commonsense reasoning applied to a specific objective. It includes interviewing the experts and defining key features of the fuzzy system, such as: identifying input and output variables, separate crisp and fuzzy variables, formulating the proto-rules using the defined variables, ranking the rules according to their importance, identifying operational constraints, and defining expected performance.

As a result of knowledge acquisition, the structure of the fuzzy system is defined by its functional and operational characteristics, key inputs and outputs and performance metric. The next phase of the application sequence is applying the defined structure into a specific software tool, such as the Fuzzy Logic toolbox in MATLAB. The development process includes designing the membership functions, defining the rules, and the corresponding defuzzification methods. The aggregated model is simulated and validated with independent data in several iterations until the defined performance is achieved mostly by tuning the membership functions. A run time version of the model can be applied in a separate software environment, such as Excel.

Fig. 3.14 Key steps of applying expert-based fuzzy systems

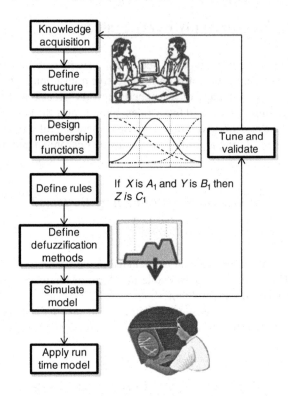

3.4.3 Applying Data-Based Fuzzy Systems

If the success of expert-based fuzzy systems depends mostly on the quality of knowledge acquisition, the success of data-based fuzzy systems is heavily linked to the quality of the available data. The key steps for applying this type of fuzzy system are shown in Fig. 3.15.

The defined structure at the beginning of the application includes mostly the data-related issues for selection of process inputs and outputs from which it is expected to find potential rules. Data collection is the critical part in the whole process and could be a significant obstacle if the data are with very narrow ranges and the process behavior cannot be represented adequately. The chance for appropriate fuzzy rule discovery is very low in this case.

The data processing part includes discovery of proto-clusters from the data and definition of the corresponding rules. The most interesting step of the design is the decision about the size of the granule of the proto-clusters. In principle, the broader the cluster space, the more generic the defined rule. However, some important nonlinear behavior of the process could be lost. It is recommended that the proper size of the fuzzy clusters be decided with the domain experts. The final result of the development process is a fuzzy system model based on the generalized rules. There is no difference in the run-time application between both approaches.

Fig. 3.15 Key steps of
applying data-based fuzzy
systems

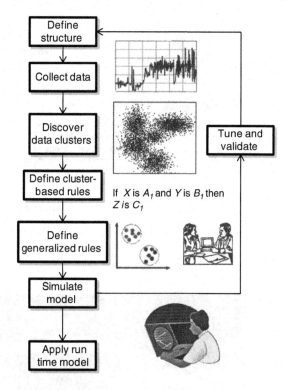

3.5   Typical Applications of Fuzzy Systems

Fuzzy systems cover very broad application areas from the subway system in Japan to the popular rice cooker. The key application areas are systematized by the mind-map in Fig. 3.16. We'll give a representative list for the most impressive applications in each area and the reader can find the details from the key websites or by Googling on the Web. Two important applications – the Sendai subway operation by Hitachi and the fully automatic washing machine of Matsushita Electric – will be described in more detail.

- *Transport Systems* – One of the industrial applications that made fuzzy systems famous was the predictive fuzzy controller for automatic train operation in the Sendai subway system in Japan, implemented by Hitachi in July 1987 (see Fig. 3.17).[6] Train operation involves control that begins acceleration at a departure signal, regulates train speed so that it does not exceed the control speed, and stops the train at a fixed position at the next station that requires a stop. Analyzing the performance of the most experienced train operators has shown

---

[6]T. Terano, K. Asai, M. Sugeno, *Applied Fuzzy Systems*, Academic Press, 1994.

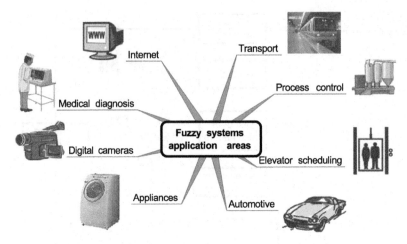

Fig. 3.16 Key applications of fuzzy systems

Fig. 3.17 Sendai subway operation where a predictive fuzzy controller was applied for automatic train operation[7]

the critical role of linguistic factors in their decision-making. For example, for stopping control, they carry out quality operations by considering the number of passengers, the strength of the brakes, and whether they can stop and maintain passenger comfort at the time of acceleration and deceleration.

---

[7]http://osamuabe.ld.infoseek.co.jp/subway/maincity/sendai/sendai.htm

Operators consider fuzzy control objectives like "stop just fine", "accurate stop", and "passenger comfort" as they drive trains.

The developed fuzzy predictive controller is based on six ambiguous objectives, defined as fuzzy sets: safety, comfort, energy savings, speed, elapsed time, and stopping precision. The designed fuzzy sets use linguistic variables like "good", "bad", "very good", "very bad", and "intermediate value" and trapezoidal, triangular, and sigmoid shapes for the membership functions. What is amazing in this pioneering application of fuzzy systems is that the tremendous complexity of the subway train control is captured by nine rules only! The subway train moves so smoothly that standing passengers don't even need to hold on to poles or straps. A fish tank could travel the entire 8.4 mile route, stop at all 16 stations, and never even spill a drop of water.

- *Industrial Process Control* – One of the first famous process control application was a cement kiln control built in Denmark in the early 1980s. Other typical applications are heat exchanger control, activated sludge wastewater treatment process control, water purification plant control, polyethylene control by Hoechst, etc.

- *Elevator Scheduling* – Companies such as Otis, Hitachi, and Mitsubishi have developed smart elevators – devices that employ fuzzy logic to adjust to the level of traffic. By knowing how many passengers are currently in each elevator, which floors have passengers waiting to board, and the locations of all of the elevators currently in use, an optimum strategy of operation can be dynamically achieved with fuzzy systems.

- *Automotive Industry* – Fuzzy systems have been used by key car manufacturers, such as Nissan and Subaru, for cruise control, automatic transmissions, antiskid steering systems, and anti-lock brake systems.

- *Appliances* – We'll illustrate the application potential of fuzzy systems in the big market of home appliances with the fully automatic washing machine, based on fuzzy logic, implemented by Matsushita Electric.[8] It triggered a flood of similar applications in other home appliances, such as vacuum cleaners, air conditioners, microwave ovens, and the famous rice cooker. The biggest problem in washing machines is that it is difficult to obtain the optimal relationship between the level of dirtiness and washing time. It is nearly impossible to collect detailed experimental data for all kinds of dirtiness and statistically relate them to washing time.

Due to the high development cost, developing a first-principles model is economically unacceptable. It turns out that the skilled launderers' know-how can be captured in rules and membership functions in a fuzzy system.

The system is based on three key parameters: (1) the time until the transmittance reaches saturation, Ts (which is related to the quality of dirt); (2) the output level of saturation, Vs (which is related to the amount of dirt), and the remaining

---

[8]N. Wakami, H. Nomura, S. Araki, Fuzzy logic for home appliances, *in Fuzzy Logic and Neural Network Handbook*, C. Chen (Editor), pp. 21.1–21.22, McGraw-Hill, 1996.

wash time Wt. The membership functions for these three parameters are triangle and trapezoid-based and defined by skilled launderers. The rules are like these two examples:

Rule 1
If the output level of saturation Vs is "Low" and the saturation time Ts is "Long" **Then** the wash time Wt is "Very Long"

Rule 2
If the output level of saturation Vs is "High" and the saturation time Ts is "Short" **Then** the wash time Wt is "Very Short"

By using a simple fuzzy system, the complex nonlinear relationship between the degree of dirtiness and the washing time can be expressed by only six rules and three membership functions which can be very effectively implemented on any hardware environment. The savings of both energy and time due to reduced excessive or inadequate washing are enormous.

- *Digital Cameras* – This is one of the early application areas of fuzzy systems. For example, Canon has come out with something called fuzzy focus. Traditional self-focusing cameras would simply bounce an infrared (or ultrasound) signal of whatever single object was at the dead center of the field of view and then use that information to determine the distance. But if there were two or more objects present, such auto-focusing cameras could get confused. Canon solved the problem by allowing the camera to consider multiple targets and, utilizing fuzzy logic, incorporates them all. In 1990 Matsushita introduced a fuzzy camcorder that automatically reduces the amount of jitter caused by hand-held operation. Matsushita used fuzzy logic to develop what it called a digital image stabilizer. The stabilizer compares each pair of successive frames to see how much they have shifted, and then adjusts them accordingly.
- *Medical Diagnosis* – Typical applications in this area are: fuzzy logic control of inspired oxygen, fuzzy control of the depth of anesthesia in a patient during surgery, medical decisions based of fuzzy logic in intensive care, coronary heart disease risk assessment based on fuzzy rules, etc.
- *Internet* – Some examples of fuzzy logic implementation for improvement in Web search engines, which is one of the biggest application areas in the future, are: scalable Web usage mining, concept-based search engines, fuzzy modeling of the Web user, fuzzy query and search at British Telecom, opinion mining based on a fuzzy thesaurus by Clairvoyance Corporation, etc.

3.6   Fuzzy Systems Marketing

Fuzzy systems require relatively low marketing efforts since the approach is easy to communicate. In addition, the list of impressive industrial applications in leading companies is long. In order to help the marketing efforts we'll give examples of two

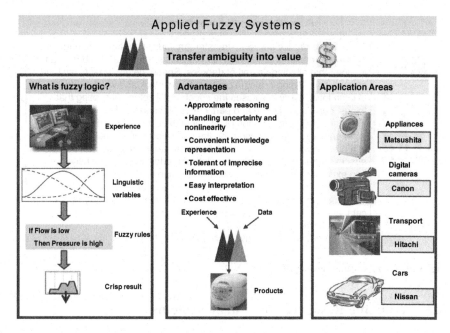

Fig. 3.18 Fuzzy systems marketing slide

documents – marketing slide and elevator pitch for each discussed computational intelligence approach. The fuzzy systems marketing slide is shown in Fig. 3.18.

The purpose of the marketing slide is to deliver a very condensed message in one page/PowerPoint slide about the technology. On the top of the slide is the key slogan that captures the essence of the value creation capability of the technology. In the case of applied fuzzy systems the slogan is "Transfer ambiguity into value".

The marketing slide is divided into three sections. The left section demonstrates in a very simplified and graphical way how the technology works. In the case of fuzzy logic, the emphasis is on the translation of human experience via linguistic variables and fuzzy rules into crisp results in the computer. The middle section of the slide advertises the advantages of the approach and visualizes the key uniqueness of fuzzy systems – product design based on effective use of expertise and data. The right section focuses attention on the well-known application areas and gives references to famous companies, which have applied the approach or have developed specialized fuzzy system products.

This organization of the marketing slide allows a nontechnical person to have a rough idea about the approach, its unique capabilities, and application record in around 30 seconds.

An example of an elevator pitch for presenting the approach to top managers is given on Fig. 3.19.

The timing of the pitch is around 60 seconds. It is up to the reader to convince the boss about the undisputed value of fuzzy systems.

Fuzzy Systems Elevator Pitch

Fuzzy systems, like lawyers, transfer ambiguity in natural language and data interpretation into value. Fuzzy logic is not fuzzy and is based on the idea that all things admit of degrees. Speed, height, beauty, attitude all come in a sliding scale. In contrast to the black and white world of crisp logic, fuzzy logic uses the spectrum of colors, accepting that things can be partly true and partly false at the same time. This allows very effective representation of available experience by defining rules in natural language and codifying the used words like "small variation" in a precise way in the computer. Fuzzy systems are especially effective when developing mathematical models is very expensive and expertise on a specific problem is available.

Fuzzy systems are easy to develop, implement, and maintain. There are numerous applications in appliances, digital cameras, subway operations, cars, and industrial process control by companies like Matsushita, Hitachi, Nissan, Subaru, etc.

It's an emerging technology where the sky is the limit.

Fig. 3.19. Fuzzy systems elevator pitch

3.7 Available Resources for Fuzzy Systems

One of the issues with computational intelligence is that due to the high dynamics of the field, the information becomes obsolete very soon. The best advice in the Age of Google is to search and find the latest updates. The recommended generic resources are http://www.scholarpedia.org/ and http://www.wikipedia.org/ which include many useful links on all computational intelligence technologies. The key resources for each computational intelligence method given in the book, will reflect the state of the art at the time of writing the final phase of this manuscript (March 2009).

3.7.1 Key Websites

Our recommended list includes the following sites:

Berkeley Initiative of Soft Computing:
http://www-bisc.cs.berkeley.edu

North American Fuzzy Information Processing Society (NAFIPS)
http://nafips.ece.ualberta.ca/

International Fuzzy Systems Association (IFSA)
http://www.cmplx.cse.nagoya-u.ac.jp/~ifsa/

A good general source about fuzzy systems with many links:
http://www.cse.dmu.ac.uk/~rij/general.html

A good introduction to fuzzy systems:
http://blog.peltarion.com/2006/10/25/fuzzy-math-part-1-the-theory

3.7.2 Selected Software

The links to the two dominant software tools for fuzzy systems development – the toolboxes from Wolfram and Mathworks – are given below:

The Fuzzy Logic toolbox for Mathematica:
http://www.wolfram.com/products/applications/fuzzylogic/

The Fuzzy Logic toolbox for MATLAB:
http://www.mathworks.com/products/fuzzylogic/

3.8 Summary

Key messages:

Ambiguity is captured by fuzzy sets and fuzzy rules.

Fuzzy systems are not fuzzy.

Fuzzy systems are created by two sources - the domain experts and the data. In expert-based fuzzy systems the fuzzy rules are defined by experts while in data-based fuzzy systems the rules are extracted by clusters in the data.

In fuzzy systems the knowledge is distributed between the rules and the membership functions. Usually the rules are very general and the specific information is captured by the membership functions.

Fuzzy systems can effectively represent both the uncertainty and nonlinearity of even a complex system. Usually the uncertainty of the system is captured by the design of the constituent fuzzy sets and the nonlinearity is captured by the defined fuzzy rules.

Fuzzy systems have been successfully applied in transportation, industrial process control, cars, digital cameras, and many appliances like washing machines, microwave ovens, vacuum cleaners, and the famous rice cooker.

The Bottom Line

Applied fuzzy systems have the capability to transfer ambiguity in natural language and data interpretation into value.

Suggested Reading

The following books give detailed technical descriptions of the key fuzzy systems techniques:

R. Berkan and S. Trubatch, *Fuzzy Systems Design Principles: Building Fuzzy IF-THEN Rule*Rule *Bases*, IEEE Press, 1997.

T. Terano, K. Asai, M. Sugeno, *Applied Fuzzy Systems*, Academic Press, 1994.

R. Yager and D. Filev, *Essentials of Fuzzy Modeling and Control*, Wiley, 1994.

The state of the art of fuzzy systems is given in the book:

W. Pedrycz and F. Gomide, *Fuzzy Systems Engineering: Toward Human-Centric Computing*, Wiley-IEEE Press, 2007.

Chapter 4
Machine Learning: The Ghost in the Learning Machine

Learning without thought is useless, thought without learning is dangerous.

Confucius

Since ancient time learning has played a significant role in building the basis of human intelligence. The tendency for learning with the increased dynamics and complexity of the global economy is growing. If in the past most of the learning efforts were concentrated in high-school and college years, in the 21st century learning becomes a continuous process during the whole working career. Applied computational intelligence can play a significant role in enhancing the learning capabilities of humans through its unique abilities to discover new patterns and dependencies.

The broad research discipline that delivers these capabilities is called machine learning, with the key objective to automatically extract information from data by computational and statistical methods. At a general level, machine learning is based on two types of inference: *inductive–deductive* and *transductive*, shown in Fig. 4.1.

Inductive machine learning methods extract rules and patterns out of data sets. It is a process of generalization from particular cases represented by training data to newly discovered patterns or dependencies (captured as an estimated model). In the *deduction* step of inference, the derived model is used to apply the generalized knowledge for specific predictions. The level of generalization is limited to *a priori* defined assumptions, for example, low and high limits of input variables. The majority of applied machine intelligence systems are based on *inductive–deductive* inference. However, for many practical situations there is no need to go to the expensive process of building a generalized model. The most appropriate solution in this case is *transductive* inference where the result does not require explicitly constructing a function. Instead, the predictions of the new outputs are based on available training and test data. From a cost perspective, it is obviously a much cheaper solution.

Machine learning algorithms are classified based on the desired outcome of the algorithm. The most frequently used algorithm types include supervised learning, unsupervised learning, and reinforcement learning:

A.K. Kordon, *Applying Computational Intelligence*,
DOI 10.1007/978-3-540-69913-2_4, © Springer-Verlag Berlin Heidelberg 2010

Fig. 4.1 Types of machine
inference: induction,
deduction, and transduction

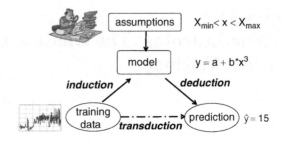

$X_{min} < x < X_{max}$

$y = a + b*x^3$

$\hat{y} = 15$

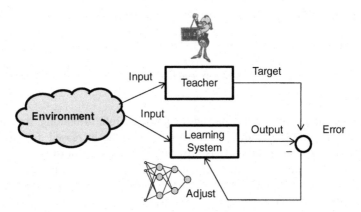

Fig. 4.2 Supervised machine learning

- *Supervised Learning* – In which the algorithm generates a function that maps inputs to desired outputs. The key assumption is the existence of a "teacher" who provides the knowledge about the environment by delivering training input-target samples (see Fig. 4.2). The parameters of the learning system are adjusted by the error between the target and the actual response (output). Supervised learning is the key method for the most popular neural networks - multilayer perceptrons and for support vector machines.[1]
- *Unsupervised Learning* – The algorithm models a set of inputs since target examples are not available or do not exist at all. Unsupervised learning does not need a teacher and requires the learner to find patterns based on self-organization (see Fig. 4.3). The learning system is self-adjusted by the discovered structure in the input data. A typical case of unsupervised learning is the self-organized maps using neural networks.
- *Reinforcement Learning* – The algorithm is based on the idea of how to learn by interacting with an environment and adapting the behavior to maximize an *objective function* specific to this environment (see Fig. 4.4). The learning

---

[1]Multilayer perceptrons and support vector machines will be described in detail in subsequent sections.

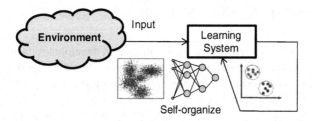

Fig. 4.3 Unsupervised machine learning

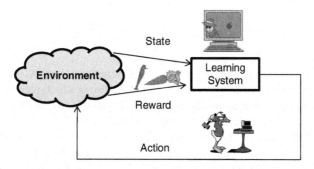

Fig. 4.4 Reinforcement machine learning

mechanism is based on trial-and-error of actions and evaluating the reward. Every action has some impact on the environment, and the environment provides carrot-and-stick type of feedback that guides the learning algorithm. The aim is to find the *optimal behavior*: the one whose actions maximize long-term reinforcement. Reinforcement learning is often used in intelligent agents.

The objective of applied machine learning is to create value through automatic discovery of new patterns and relationships by learning the behavior of a dynamic complex environment. The focus of this chapter is on two of the most widely applied machine learning techniques – neural networks and support vector machines. Several other approaches, such as recursive partitioning (or decision trees), Bayesian learning, instance-based learning, and analytical learning are not discussed since their value creation potential is significantly lower.[2] The only exceptions are decision trees, which are used in data mining applications and are part of most available data mining software packages.

---

[2]A good survey of all machine learning methods is the book by T. Mitchell, *Machine Learning*, McGraw-Hill, 1997.

4.1 Neural Networks in a Nutshell

It is no surprise that the first source for inspiration in developing machine learning algorithms is the human learning machine - the brain. Neural networks are models that attempt to mimic some of the basic information processing methods found in the brain. The field has grown tremendously since the first paper of McCulloch and Pitts in 1943 with thousands of papers and significant value creation by numerous applications.[3] We suggest practitioners focus on the following key topics shown in Fig. 4.5.

4.1.1 Biological Neurons and Neural Networks

Biological neurons are the building blocks of the nervous system. At one end of the biological neuron are tree-like structures called dendrites which receive incoming signals from other neurons across junctions called synapses. At the other end of the neuron is a single filament leading out of the cell body, called an axon. The structure of the biological neuron is shown in Fig. 4.6.

Biological neurons communicate with each other by means of electrical signals. They are connected to each other through their dendrites and axons. One can define dendrites as the inputs to the neuron and the axon as the output. An axon carries information through a series of action potentials that depend "on the neuron's potential". Each biological neuron collects the overall electrical stimulus from its incoming signals (inputs) and if the potential is beyond a certain threshold, produces an output signal of constant magnitude to other neurons. Biological neurons can be treated as very simple processing units that weight the incoming signals and

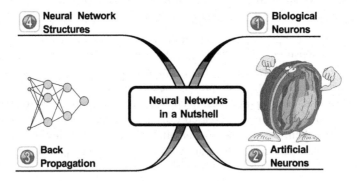

Fig. 4.5 Key topics related to neural networks

[3]W. McCulloch and W. Pitts, A logical calculus of ideas immanent in nervous activity, *Bulletin of Mathematical BioPhysics*, 5, p. 115, 1943.

activate the neurons beyond a critical threshold. The processing capabilities of biological neurons are very limited. It is the interaction of many biological neurons combined into large networks that give the human brain its ability to learn, reason, generalize, and recognize patterns. The possibilities of interconnecting the multiple inputs and outputs of biological neurons are practically infinite. The typical number of connections (synapses) with other neurons of nerve cells in the brain is between 1000 and 10,000. It is estimated that there are approximately 10 billion neurons in the human cortex, and 60 trillion synapses.[4]

Research efforts on analyzing the mechanisms of biological learning show that synapses play a significant role in this process. For example, synaptic activity facilitates the ability of biological neurons to communicate. Thus a high degree of activity between two neurons at one time, captured by their synapses, could facilitate their ability to communicate at a later time. Another discovery from biological learning with significant effect in artificial neural network development is that learning strengthens synaptic connections.

4.1.2 Artificial Neurons and Neural Networks

Let's represent the biological neuron structure, shown in Fig. 4.6 in a systematic view, shown in Fig. 4.7, where the synaptic strength is represented by the connection weights w_{ij} and visualized by the widths of the arrows. This schematic

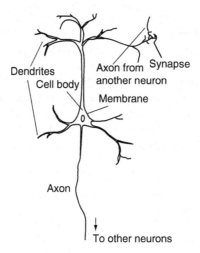

Fig. 4.6 Biological neuron structure[5]

---

[4]S. Haykin, *Neural Networks: A Comprehensive Foundation*, 2nd edition, Prentice Hall, 1999.
[5]http://www.learnartificialneuralnetworks.com

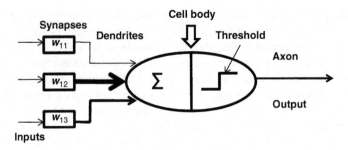

Fig. 4.7 A schematic view of the biological neuron

Fig. 4.8 Artificial neuron structure

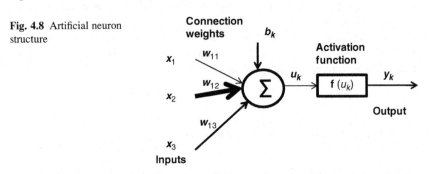

representation is at the basis of defining the artificial neuron structure, shown in Fig. 4.8.

An artificial neuron, or processing element, emulates the axons and dendrites of its biological counterpart by connections and emulates the synapses by assigning a certain weight or strength to these connections. A processing element has many inputs and one output. Each of the inputs x_j is multiplied by their corresponding weights w_{ij}. The sum of these weighted inputs u_k is then used as the input of an activation function $\mathbf{f}(u_k)$, which generates the output y_k for that specific processing element. The bias term b_k, also called a "threshold term", serves as a zero adjustment for the overall inputs to the artificial neuron. Excitatory and inhibitory connections are represented as positive or negative connection weights, respectively.

Artificial neurons use an activation function to calculate their activation level as a function of total inputs. The most distinguished feature between existing artificial neurons is based on the selected activation function. The main popular options are based on a sigmoid function and Radial Basis Function (RBF). A typical shape of a sigmoid function is shown in Fig. 4.9.

The important feature of this activation function is its continuous threshold curve that allows computing derivatives. The last property is critical for the prevailing learning algorithm of artificial neural networks - back propagation.

A typical shape of a two-dimensional radial basis activation function is shown in Fig. 4.10.

Fig. 4.9 A sigmoid activation function for artificial neurons

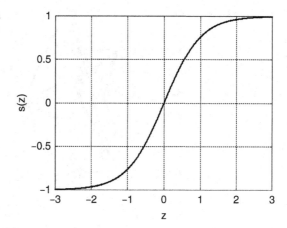

Fig. 4.10 A two-dimensional radial basis activation function for artificial neurons

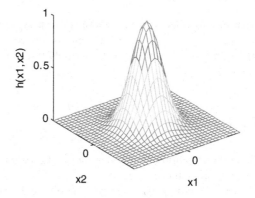

A key feature of a radial basis activation function is its ability to yield its maximum value only for inputs near its center, and to be nearly zero for all other inputs. The center and width of the RBF may be varied to achieve different artificial neuron behavior.

An artificial neural network consists of a collection of processing elements, organized in a network structure. The output of one processing element can be the input of another processing element. Numerous ways to connect the artificial neurons into different network architectures exist. The most popular and widely applied neural network structure is the multilayer perceptron. It consists of three types of layers (input, hidden, and output), shown in Fig. 4.11.

The input layer connects the incoming patterns and distributes these signals to the hidden layer. The hidden layer is the key part of the neural network since its neurons capture the features hidden in the input patterns. The learned patterns are represented by the weights of the neurons and visualized in Fig. 4.11 by the widths of the arrows. These weights are used by the output layer to calculate the prediction.

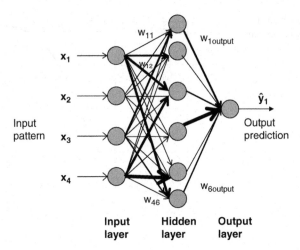

Fig. 4.11 Structure of a multilayer perceptron with one hidden layer

Artificial neural networks with this architecture may have several hidden layers with different numbers of processing elements as well as outputs in the output layer. The majority of applied neural networks use only three layers, which are similar to those shown in Fig. 4.11.

4.1.3   Back-propagation

The key function of neural network-learning is based on representing the desired behavior by adjusting the connection weights. Back-propagation, developed by Paul Werbos in the 1970s, is the most popular among the known methods for neural network-based machine learning.[6] The principle behind back-propagation is very simple and is described next.

The performance of the neural network is usually measured by some metric of the error between the desired and the actual output at any given instant. The goal of the back-propagation algorithm is to minimize this error metric in the following way.

First, the weights of the neural network are initialized, usually with random values. However, in some cases it is also possible to use prior information to make an informed guess of the initial weights. The next step is to calculate the outputs y and the associated error metric for that set of weights. Then the derivatives of the error metric are calculated with respect to each of the weights. If increasing a weight results in a larger error then the weight is adjusted by reducing its value. Alternatively, if increasing a weight results in a smaller error then this weight is corrected by increasing its value. Since there are two sets of weights, a set corresponding to the

---

[6]P. Werbos, *Beyond Regression: New Tools for Prediction and Analysis in the Behavioral Sciences*, Ph.D. Thesis, Harvard University, Cambridge, MA, 1974.

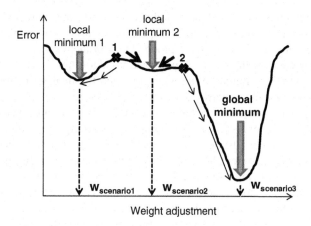

Fig. 4.12 Weight adjustment based on back-propagation of the error

connections between the input and the hidden layes (w_{ij} in Fig. 4.11), and another set between the output and the hidden layer ($w_{ioutput}$ in Fig. 4.11), the error calculated at the output layer is propagated backwards in order to also adjust the first set of weights – hence the name back-propagation. After all the weights have been adjusted in this manner, the whole process starts over again and continues until the error metric meets some predetermined threshold.

The basic back-propagation algorithm generally leads to a minimum error. Problems can occur, however, when local minima are present (which is generally the case). Depending on the starting point in weight space the basic back-propagation algorithm can get stuck in local minima. The effect is illustrated in Fig. 4.12 where the x-axis represents the adjustment of a specific weight and the y-axis is the error between the target and the predicted neural network output. Let's assume the existence of two local and one global minimum in this search space. In the original back-propagation algorithm, the weight adjustment could be significantly different depending on the initial random values. In the given example, three possible outcomes are possible. If the initial weight value is at position 1 (shown with the x sign in Fig. 4.12), the direction of the gradient (the first derivative of the error with respect to the weights, shown as arrows in Fig. 4.12) is pushing the solution to local minimum 1. The result of this local optimization of the error is the weight adjustment $w_{scenario1}$. The second scenario is based on even smaller differences in the initial weights at position 1 or 2. In both cases, the gradient will direct the solution toward local minimum 2 despite the fact that the error is higher than in local minimum 1. The adjusted weight according to this back-propagation optimization is $w_{scenario2}$.

Minor changes in the initialization values at position 2, however, could deliver significantly different results, according to the third possible scenario. The gradient of the weights steers toward the global minimum and the weight adjustment of $w_{scenario3}$ is obviously the preferable solution. The problem is that the probability for these three scenarios to happen is equal. As a result, we may have completely different weight adjustments and the neural network performance will vary significantly. It is strongly recommended to repeat the neural network development process several times to increase the reproducibility of the results.

Another problem with the back-propagation algorithm is its slow speed. Thousands of iterations might be necessary to reach the final set of weights that minimize the selected error metric.

Fortunately, these well-known flaws of back-propagation are overcome with adding tuning parameters, such as momentum and learning rate. The momentum term causes the weight adjustment to continue in the same general direction. This, in combination with different learning rates, speeds up the learning process and minimizes the risk of being entrapped in local minima. As a result, the error back-propagation process is more efficient and in control. It has to be taken into account, however, that biological neurons do not work backwards to adjust the strength of their synapse, i.e. back-propagation is not a process that emulates the brain activity.

To propagate or not to propagate? That is the most important neural networks question.

4.1.4   Neural Network Structures

Among the many available structures of neural networks[7] we'll focus our attention on three: multilayer perceptrons, self-organized maps, and recurrent neural networks. They represent the key capabilities of neural networks – to learn respectively functional relationships, unknown patterns, and dynamic sequences.

The first type of neural network structure is the multilayer perceptron, which was introduced in Sect. 4.1.2 and illustrated by Fig. 4.11. One of the most important properties of this architecture is its ability to be a universal approximator, i.e. to represent any nonlinear dependence in the available data. The capacity of the nonlinear fit depends on the number of neurons in the hidden layer. The majority of applied neural networks is based on multilayer perceptrons.

The second type of neural network structure is self-organizing maps, invented by the Finnish professor Teuvo Kohonen, and is based on the associative neural properties of the brain. The topology of the Kohonen Self-Organizing Maps (SOM) network is shown in Fig. 4.13.

This neural network contains two layers of nodes - an input layer and a mapping (output) layer in the shape of a two-dimensional grid. The input layer acts as a distributor. The number of nodes in the input layer is equal to the number of features or attributes associated with the input. The Kohonen network is fully connected, i.e. every mapping node is connected to every input node. The mapping nodes are initialized with random numbers. Each actual input is compared with each node on

---

[7]For detailed description, see S. Haykin, *Neural Networks: A Comprehensive Foundation*, 2nd edition, Prentice Hall, 1999.

Fig. 4.13 Structure of a Kohonen self-organized map network

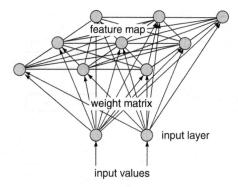

Fig. 4.14 A self-organized map for three iris flowers: Iris-Setosa, Iris-Versicolor and Iris-Virginica

the mapping grid. The "winning" mapping node is defined as the one with the smallest error. The most popular metric in this case is the Euclidean distance between the mapping node vector and the input vector. The input thus maps to a given mapping node. The value of the mapping node vector is then adjusted to reduce the Euclidean distance and all of the neighboring nodes of the winning node are adjusted proportionally. In this way, the multidimensional (in terms of features) input nodes are mapped to a two-dimensional output grid, called feature map. The final result is a spatial organization of the input data organized into clusters.

An example of a self-organized map for clustering of three iris flowers: Iris-Setosa, Iris-Versicolor and Iris-Virginica, is shown in Fig. 4.14.

The grid in the feature map is based on 100 neurons and the training data includes 50 samples of the three Iris flowers. The three distinct classes in the map are clearly shown in Fig. 4.14.

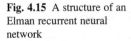 **Fig. 4.15** A structure of an Elman recurrent neural network

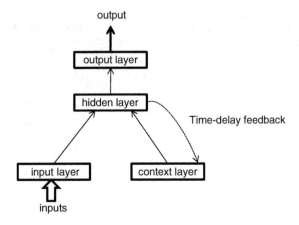

The third type of neural network structure is a recurrent neural network, which is able to represent dynamic systems by learning time sequences. One key requirement for capturing process dynamics is the availability of short-term and long-term memory. Long-term memory is captured into the neural network by the weights of the static neural network while the short-term memory is represented by using time delays at the input layer of the neural network.

A common feature in most recurrent neural networks is the appearance of the so-called context layer, which provides the recurrence. At current time t the context units receive some signals resulting from the state of the network at the previous time sample t–1. The state of the neural network at a certain moment in time thus depends on previous states due to the ability of the context units to remember certain aspects of the past. As a result, the network can recognize time sequences and capture process dynamics. An example of one of the most popular architectures of recurrent neural network - the Elman networks is shown in Fig. 4.15.

In this configuration the input layer is divided into two parts: one part that receives the regular input signals (which could be delayed in time) and another part made up of context units, which receive as inputs the activation signals of the hidden layer from the previous time step. One of the advantages of the Elman recurrent neural network is that the feedback connections are not modifiable so the network can be trained using conventional training methods, such as back-propagation.

4.2   Support Vector Machines in a Nutshell

An implicit assumption in developing and using neural networks is the availability of sufficient data to satisfy the "hunger" of the back-propagation learning algorithm. Unfortunately, in many practical situations we don't have this luxury and here is the place for another machine learning method, called support vector

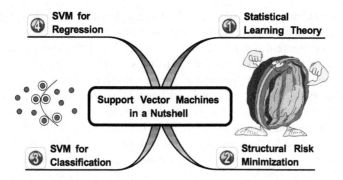

Fig. 4.16 Key topics related to support vector machines

machines (SVM). It is based on the solid theoretical ground of statistical learning theory which can handle effectively statistical estimation with small samples. The initial theoretical work of two Russian scientists – Vapnik and Chervonenkis from the late 1960s – grew into a complete theoretical framework in the 1990s[8] and has triggered numerous real-world applications since the early 2000s. SVMs are one of the fastest growing approaches of computational intelligence in terms of both publications and real-world applications.

We recommend practitioners focus on the following main topics, related to support vector machines, which are shown in Fig. 4.16 and discussed below.

We need to warn nontechnical readers that explaining SVMs is challenging and requires a mathematical background.

4.2.1 Statistical Learning Theory

The key topic of statistical learning theory is defining and estimating the capacity of the machine to learn effectively from finite data. Excellent memory is not an asset when it comes to learning from limited data. A machine with too much capacity to learn is like a botanist with a photographic memory who, when presented with a new tree, concludes that it is not a tree because it has a different number of leaves from anything she or he has seen before; a machine with too little capacity to learn is like the botanist's lazy brother, who declares that if it's green, it's a tree. Defining the right learning capacity of a model is similar to developing efficient cognitive activities by proper education. In the same way as the mind was spelled out as the "ghost" in the human "machine",[9] model capacity could be defined as the "ghost" in the computer learning machine.

---

[8]V. Vapnik, *Statistical Learning Theory*, Wiley, 1998.

[9]A. Koestler, *The Ghost in the Machine*, Arkana, London, 1989.

Statistical learning theory addresses the issue of learning machine capacity by defining a quantitative measure of complexity, called Vapnik-Chervonenkis (VC) dimension. The theoretical derivation and interpretation of this measure requires a high level of mathematical knowledge.[10] The important information from a practical point of view is that VC dimension can be calculated for most of the known analytical functions. For learning machines linear in parameters the VC dimension is given by the number of weights, i.e., by the number of "free parameters". For learning machines nonlinear in parameters, the calculation of the VC dimension is not trivial and could be done by simulations. A finite VC dimension of a model guarantees generalization capabilities beyond the range of the training data. Infinite VC dimension is a clear indicator that it is impossible to learn with the selected functions. The nature of VC dimension is further clarified by the Structural Risk Minimization Principle, which is at the basis of statistical learning theory.

4.2.2 Structural Risk Minimization

The Structural Minimization Principle (SRM) defines a trade-off between the quality of the approximation of the given learning data and the generalization ability (ability to predict beyond the range of learning or training data) of the learning machine. The idea of the Structural Minimization Principle is illustrated in Fig. 4.17, where the y-axis represents the error or the risk of prediction and the x-axis represents model complexity.

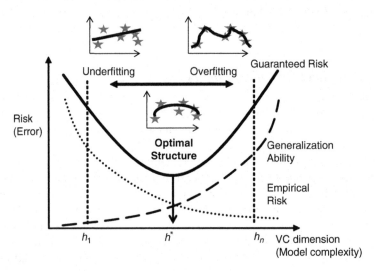

Fig. 4.17 The Structural Risk Minimization Principle

---

[10]Curious readers may look for details in the book of V. Vapnik, *The Nature of Statistical Learning Theory*, 2nd edition, Springer, 2000.

In this particular case, model complexity, defined by its VC dimension, is equal to the polynomial order from h_1 to h_n. The approximation capability of the set of functions is shown by the empirical risk (error) which declines with increased complexity (polynomial order). If the learning machine uses a too high complexity (for example, a polynomial of order 10), the learning ability may be good, but the generalization ability is not. The learning machine will overfit the data to the right of the optimal complexity with VC dimension of h^* in Fig. 4.17.

On the other hand, when the learning machine uses too little complexity (for example, a first-order polynomial), it may have good generalization ability, but not impressive learning ability. This underfitting of the learning machine corresponds with the region to the left of the optimal complexity. The optimal complexity of the learning machine is the set of approximating functions with lowest VC dimension *AND* lowest training error (for example, a third-order polynomial).

There are two approaches to implementing the SRM inductive principle in learning machines:

1. Keep the generalization ability (which depends on VC dimension) fixed and minimize the empirical risk.
2. Keep the empirical risk fixed and minimize the generalization ability.

Neural network algorithms implement the first approach, since the number of neurons in the hidden layer is defined a priori and therefore the complexity of the structure is kept fixed.

The second approach is implemented by the support vector machines (SVM) method where the empirical risk is either chosen to be zero or set to an *a priori* level and the complexity of the structure is optimized. In both cases, the SRM principle pushes toward an optimal model structure that should match the learning machine capacity with training data complexity.

4.2.3 Support Vector Machines for Classification

The key notion of SVMs is the support vector. In classification, the support vectors are the vectors that lie on the margin (i.e. the largest distance between two closest vectors to either side of a hyperplane) and they represent the input data that are the most difficult to classify. Above all, the generalization ability of the optimal separating hyperplane is directly related to the number of support vectors, i.e. the separating hyper-plane with minimum complexity has the maximal margin. From the mathematical point of view, the support vectors correspond to the positive Lagrangian multipliers from the solution of a quadratic programming (QP) optimization problem. Only these vectors and their weights are then used to define the decision rule or model. Therefore this learning machine is called the support vector machine. It is important to emphasize that, in contrast to neural networks, the defined support vectors are derived by global optimization and there is no issue in generating inefficient solutions due to local minima.

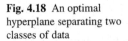

Fig. 4.18 An optimal
hyperplane separating two
classes of data

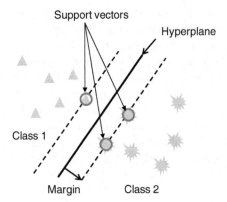

An example of support vectors for classification of two distinct classes of data is shown in Fig. 4.18. The data points from the two classes are represented by stars and triangles. The support vectors that define the hyperplane or the decision function with the maximal margin of separation are encircled. It is clear from Fig. 4.18 that these three support vectors are the critical data points that contain the significant information to define the hyperplane and separate reliably the two classes. The rest of the data points are irrelevant for the purpose of class separation even if their number is huge.

There is a social analogy that may help in understanding the notion of support vectors.[11] The political realities in the USA are such that there are hard-core supporters of both dominant political parties – the Democratic Party and the Republican Party – who don't change their backing easily. During presidential elections, approximately 40% of the electorate votes for any Democratic candidate and approximately the same percentage votes for any Republican candidate, independently of her or his platform. The result from the elections depends on the 20% of the undecided voters. These voters are the analogy of social "support vectors".

In general, the SVMs map the input data into a higher-dimensional feature space (see Fig. 4.19).

The logic behind this operation is the possibility for a linear solution in the feature space, which can solve the difficult nonlinear classification problem in the input space. The mapping can be done nonlinearly and the transformation function is chosen *a priori*. Usually it is a mathematical function, called the kernel (shown as Φ in Fig. 4.19) that satisfies some mathematical conditions given in the SVM literature and the selection is problem-specific. Typical cases of kernels are the radial basis function (see Fig. 4.10) and the polynomial kernel. In the feature space the SVM finally constructs an optimal approximating function that is linear in its parameters. In the classification case, the function is called a decision function or

---

[11]V. Cherkassky and F. Mulier, *Learning from Data: Concepts, Theory, and Methods*, 2nd edition, Wiley, 2007.

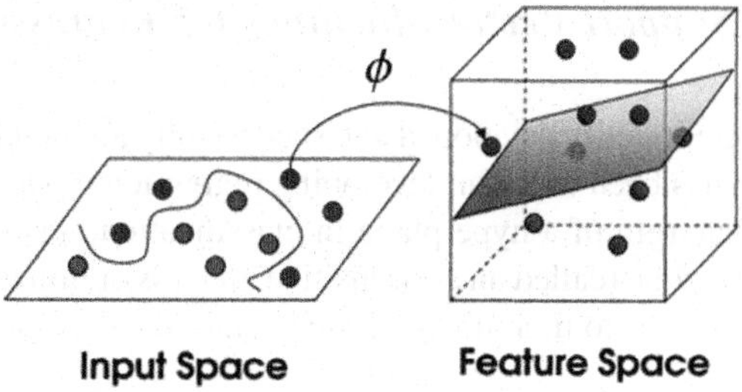

Fig. 4.19 Mapping the input
space into a high-dimensional
feature space

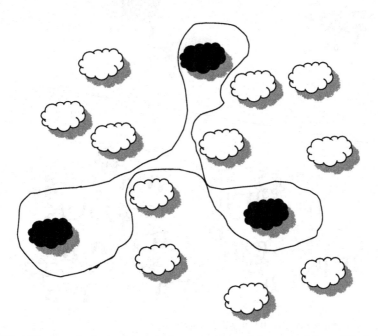

Input Space   **Feature Space**

Fig. 4.20 A 2D view of a flock of sheep and a separation curve between white and black sheep

hyperplane, and the optimal function is called an optimal separating hyperplane. In
the regression case, it is called an approximating function or in statistical terms a
hypothesis.

The mapping into high-dimensional feature space can be illustrated with the
following example. Imagine that a flock of sheep is observed from space. The 2D
view of the flock with a potential divisive curve separating white from black sheep
is shown in Fig. 4.20.

The 2D separation curve is very complex and most of the classification methods
fail. However, increasing the dimensionality to 3D may result in finding a simple
linear classifier, shown in Fig. 4.21.

4.2.4 *Support Vector Machines for Regression*

SVMs were initially developed and successfully applied for classification. The same approach has been used for developing regression models. In contrast to finding the maximal margin of a hyperplane in classification, a specific loss function with an insensitive zone (called an ε-insensitive zone) is optimized for SVMs for regression. The objective is to fit a tube with radius ε to the data (see Fig. 4.22).

Fig. 4.21 A 3D linear classifier for the flock of sheep

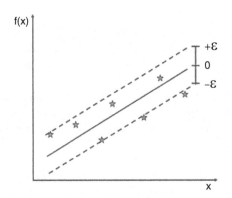

Fig. 4.22 The ε-insensitive zone for support vector regression

It is assumed that any approximation error smaller than ε is due to noise. It is accepted that the method is insensitive to errors inside this zone (visualized by the tube). The data points (vectors) that have approximation errors outside, but are close to the insensitive zone, are the most difficult to predict. Usually the optimization algorithm picks them as support vectors. Therefore, the support vectors in the regression case are those vectors that lie outside the insensitive zone, including the outliers, and as such contain the most important information from the data.

One impressive feature of the regression models based on SVMs is their improved extrapolation capabilities relative to those of neural networks. The best effect is achieved by combining the advantages of global and local kernels. It is observed that a global kernel (like a polynomial kernel) shows better extrapolation abilities at lower orders, but requires higher orders for good interpolation. On the other hand, a local kernel (like the radial basis function kernel) has good interpolation abilities, but fails to provide longer-range extrapolation.

There are several ways of mixing kernels. One possibility is to use a convex combination of the two kernels K_{poly} and K_{rbf}, for example

$$K_{mix} = \rho\, K_{poly} + (1 - \rho)\, K_{rbf},$$

where the optimal mixing coefficient ρ has to be determined. The results from an investigation of different data sets show that only a "pinch" of a RBF kernel ($1 - \rho = 0.01$–0.05) needs to be added to the polynomial kernel to obtain a combination of good interpolation and extrapolation in broad operating conditions.[12]

4.3 Benefits of Machine Learning

The key benefit of applied machine learning – creating value by automatically learning new patterns and dependencies from a complex environment – is delivered in different ways by its key approaches – neural networks and SVMs. First, we'll clarify the similarities and differences between these two methods. Second, we'll focus on the specific key benefits of each method.

4.3.1 Comparison Between Neural Networks and SVM

Independently of the different origins of neural networks and SVMs, both methods have several similar features. For example, both methods learn from data, their

---

[12]See details in G. Smits and E. Jordaan, Using mixtures of polynomial and RBF kernels for support vector regression, *Proceedings of WCCI 2002*, Honolulu, HI, IEEE Press, pp. 2785–2790, 2002.

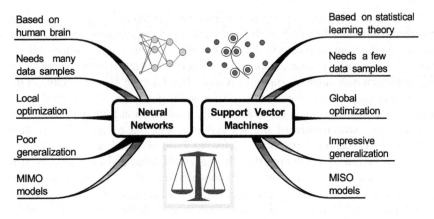

Based on
human brain

Needs many
data samples

Local
optimization

Poor
generalization

MIMO
models

Based on statistical
learning theory

Needs a few
data samples

Global
optimization

Impressive
generalization

MISO
models

Neural Networks

Support Vector Machines

Fig. 4.23 Key differences between neural networks and SVMs

models are black-boxes with the capability of universal approximation of any function to any desired degree of accuracy. The derived models are with comparable structures as well. For example, the number of support vectors plays a similar role to the number of neurons in the hidden layer of a multilayer perceptron.

The differences, however, are more substantial than the similarities. The key differences are shown in the mind-map in Fig. 4.23 and discussed next.

4.3.1.1   Key Difference #1 – On the method's basis

While neural networks are based on mimicking the human brain, SVMs are based on statistical learning theory, which gives the method a more solid theoretical foundation.

4.3.1.2   Key Difference #2 – On the necessary data for model development

While neural networks need a large data set for model building, SVMs are capable of deriving models from data sets with a small number of records even when the number of variables is larger than the number of observations. This capability is critical in biotechnology for classification of microarrays.

4.3.1.3   Key Difference #3 – On the optimization type

While neural networks, based on back-propagation, could get stuck in local minima, the SVMs solutions are always based on global optimization.

4.3.1.4 Key Difference #4 – On the generalization capability

While neural networks have very poor performance even slightly outside the training range, SVMs, due to the explicit complexity control, have much better generalization capability, especially in the case of properly selected mixed kernels.

4.3.1.5 Key Difference #5 – On the number of model outputs

While neural network models contain multiple inputs and multiple outputs (MIMO), SVM models are limited to multiple inputs and a single output (MISO).

4.3.2 Benefits of Neural Networks

The key benefits of neural networks are captured in the mind-map in Fig. 4.24 and discussed below.

- *Adaptive Learning* – Neural networks have the capability to learn how to perform certain tasks and adapt themselves by changing the network parameters in a surrounding environment. The requirements for successful adaptive learning are as follows: choosing an appropriate architecture, selecting an effective learning algorithm, and supporting model building with representative training, validation, and test data sets.
- *Self-organization* – In the case of unsupervised learning, a neural network is capable of creating its own representations of the available data. The data are automatically structured in clusters by the learning algorithms. As a result, new unknown patterns can be discovered.
- *Universal Approximators* – As we already mentioned, neural networks can represent nonlinear behavior at any desired degree of accuracy. One practical

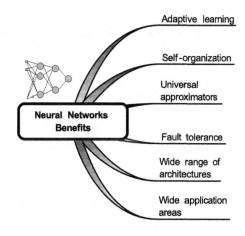

Fig. 4.24 Key benefits from neural networks

advantage of this property is a fast test of the hypothesis of an eventual functional input-output dependence in a given data set. If it is impossible to build a model with a neural network, there is very little hope that such a model could be developed by any other empirical method.

- *Fault Tolerance* – Since the information in the neural network is distributed in many process elements (neurons), the overall performance of the network does not degrade drastically when the information of some node is lost or some connections are damaged. In this case the neural network will repair itself by adjusting the connection weights according to the new data.

- *Wide Range of Architectures* – Neural networks are one of the few research approaches that offer high flexibility in the design of different architectures with its basic component - the neuron. In combination with the variety of learning algorithms, such as back-propagation, competitive learning, Hebbian learning, and Hopfield learning, the possibilities for potential solutions are almost infinite.[13] The key architectures from an application point of view are multilayer perceptrons (delivering nonlinear functional relationships), self-organizing maps (generating new patterns), and recurrent networks (capturing system dynamics).

- *Wide Application Areas* – The unique learning capabilities of neural networks combined with the wide range of available architectures allow for very broad application opportunities. The main application areas are in building nonlinear empirical models, classification, finding new clusters in data, and forecasting. The application topics vary from process monitoring and control of complex manufacturing plants to on-line fraud detection in almost any big bank.

4.3.3 Benefits of Support Vector Machines

The key benefits of SVMs are represented in the mind-map in Fig. 4.25 and discussed below.

- *Solid Theory* – One of the very important benefits of SVMs is that the solid theoretical basis of statistical learning theory is closer to the reality of practical applications. Modern problems are high-dimensional and the linear paradigm needs an exponentially increasing number of terms with the increased dimensionality of the inputs (the so-called effect of "the curse of dimensionality"). As a result, the high-dimensional spaces are terrifyingly empty and only the learning capability from sparse data, offered by statistical learning theory (on which SVMs are based), can derive a viable empirical solution.

---

[13]A survey with a good explanation of neural network learning methods can be found in the book: M. Negnevitsky, *Artificial Intelligence: A Guide to Intelligent Systems*, Addison-Wesley, 2002.

Fig. 4.25 Key benefits from
support vector machines

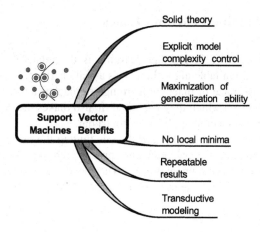

- *Explicit Model Complexity Control* – SVMs provides a method to control complexity independently of dimensionality of the problem. As was clearly shown in the classification example in Fig. 4.18, only three support vectors (i.e., the right complexity) are sufficient to separate two classes with maximal margin. In the same way, the modeler has full control over model complexity of any classification or regression solutions by selecting the relative number of support vectors. Obviously, the best solution is based on the minimal number of support vectors.
- *Maximization of Generalization Ability* – In contrast to neural networks, where the weights are adjusted only by the empirical risk, i.e. by the capability to interpolate training data, SVMs are designed by keeping the empirical risk fixed and maximizing the generalization ability (see Fig. 4.17). As a result, the performance of the derived models is with the best possible generalization abilities, i.e., they do not deteriorate dangerously under new operating conditions like the neural network models.
- *No Local Minima* – SVMs are derived based on global optimization methods, as linear programming (LP) or quadratic programming (QP) and are not stuck in local minima, as the typical back-propagation-based neural networks. The devastation effect of significantly different models, based on local minima and clearly illustrated in Fig. 4.12 is avoided in SVMs and this is a significant advantage.
- *Repeatable Results* – SVMs model development is not based on random initialization as is the case with most of neural network learning algorithms. This guarantees repeatable results.
- *Transductive Modeling* – SVMs is one of the few available approaches that allows reliable solutions of predicting outputs without building a model. The solution is defined by available data points only. This is currently an intensive research area with big practical potential.

4.4   Machine Learning Issues

An obvious issue in machine learning is the relatively large number and wide diversity of available algorithms. It is a real challenge to a practitioner to navigate through the maze of machine learning approaches and to understand the different mechanisms on which they are based. It also makes the task of selecting the proper method for a specific application nontrivial.

A common issue in all algorithms is finding the proper capacity of the specific learning machine. Very often this process requires more delicate tuning and a longer model development process. The specific issues of the most popular machine learning approaches, neural networks and SVMs, are discussed next.

4.4.1   Neural Networks Issues

In addition to the discussed weaknesses in Chap. 2, namely black-box models, poor extrapolation, and maintenance nightmare, neural networks have the following limitations:

- In case of high dimensionality (number of inputs > 50 and number of records >10,000), the convergence to the optimal solution could be slow and the training times could be long.
- It is difficult to introduce existing knowledge in the model, especially if it is qualitative.
- Large amounts of training, validation, and test data are required for model development. It is also necessary to collect data with the broadest possible ranges to avoid the poor generalization capability of neural networks.
- Results are sensitive to the choice of initial connection weights (see Fig. 4.12).
- There is no guarantee of optimal results due to the possibility of being stuck in local minima. As a result, the training may "bias" the neural network toward some operating conditions and deteriorate the performance in others.
- Model development requires substantial intervention of the modeler and can be time-consuming. A specialized knowledge of neural networks is assumed for model development and support.

4.4.2   Support Vector Machines Issues

In addition to the discussed weaknesses in Chap. 2, namely black-box models, difficult marketing, and limited infrastructure, SVMs possess the following issues:

- Very demanding requirements for practitioners who must understand the extremely mathematical and complex nature of the approach. Knowledge of statistics and optimization is also required.

- SVMs model development is very sensitive to the selection of a "good" kernel function. Since the kernel is problem-dependent, the final user must have some knowledge of SVMs.
- The application record of SVMs is relatively short. The experience in large-scale industrial applications as well as in model support is limited.

4.5 How to Apply Machine Learning Systems

Applying machine learning is significantly different from implementing fuzzy systems. Since the automatic learning is based on data, the data quality and information content is critical. The famous Garbage-In-Garbage-Out (GIGO) effect is in full power. As a result, the disappointment in some practical applications due to low quality data can be very big.

4.5.1 When Do We Need Machine Learning?

Most machine learning applications are focused on four areas - function approximation, classification, unsupervised clustering, and forecasting, shown in Fig. 4.26.

One of the most widely used capabilities of machine learning algorithms is building empirical models. Both neural networks and SVMs are universal approximators and can fit arbitrary complex nonlinear functions to multidimensional data. The derived data-driven models have significantly lower cost than first-principles models and could be used for prediction, optimization, and control.

Another big application area is classification. Reliable classification is important in applications like fraud detection, medical diagnosis, and bioinformatics. Unsupervised clustering capabilities of neural networks are very useful in situations when the naturally formed clusters in the data are unknown in advance. Self-organized

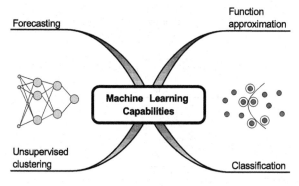

Fig. 4.26 Key machine learning application areas

maps define automatically those clusters and simultaneously reveal the spatial relationships in them.

Forecasting is based on the ability of neural networks to learn time sequences through some specific architecture, such as recurrent networks. Economic forecasts are a must in any planning activities of big enterprises and many used tools are based on machine learning algorithms.

4.5.2 Applying Machine Learning Systems

In spite of the wide diversity of available machine learning algorithms, it is possible to define a generic application sequence, shown in Fig. 4.27.

The first step in the machine learning application sequence is defining the fitness function, which is different for the various application areas. In case of function approximation, the fitness could be the prediction error between the model and the measured value. In case of classification, the fitness could be the classification error between defined classes. In case of forecasting, the fitness could be the error between the forecast in a selected time horizon and the real value. The values in both fitnesses are based on historic data.

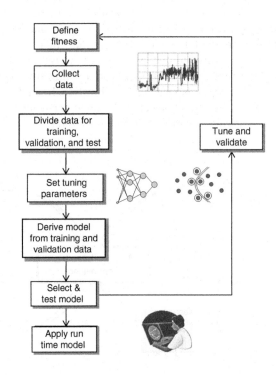

Fig. 4.27 Key steps in applying machine learning systems

Since machine learning algorithms are based on data, data collection and pre-processing is of ultimate importance for the application success. Very often this step occupies 80% of the application efforts. It is also necessary to reduce the dimensionality of the data as much as possible. It is strongly recommended to apply variable selection and minimize the number of inputs by preliminary statistical analysis and physical considerations. Usually human learning includes activities like attending classes, preparing homework, and finally, passing exams. The training is performed during the classes; the absorbed knowledge is validated in the homework and tested at the exam. It is recommended that machine learning applications follow this pattern. For that purpose, the available data are divided into three categories. The first category includes the training data with which the model is developed by learning its corresponding parameters (connection weights in the case of neural networks) and builds its approximation capability. Then the performance of the trained model has to be proved in the "homework" – how the model predicts on an independent validation data set (the second category).

In the case of neural networks, this process includes changing the number of hidden layer neurons. Finally, the selected model with optimal capacity, i.e. minimal validation error, has to pass the final exam on another data set, called the test data. The purpose of this third category of data is to prove the model performance on conditions different from those used during model development (where both training and validation data have been used). The division between the three data categories is problem dependent. The ideal distribution is 50% for training data, 30% for validation data and 20% for test data. Most machine learning software packages are doing the data division automatically by random selection based on the defined percentages. For time-series data, however, the division into these three classes is not trivial and has to be done completely differently. Usually the validation and test data are selected at the end of the time series.

Running machine learning algorithms requires selection of the corresponding tuning parameters. For example, neural networks with multilayer perceptron architecture require selection of the following parameters: number of hidden layers, number of neurons in the hidden layers, type of activation function, stopping criteria (number of epochs or steps when the training should be stopped), momentum, learning speed, and type of weight initialization. In most of the available packages the majority of these tuning parameters have universal default values and the only tuning parameter is the number of neurons in the hidden layers.

SVMs require several tuning parameters, controlled by the size of the tube ε (see Fig. 4.20), such as: type of kernel (linear, polynomial, RBF, sigmoid, etc.), percentage of support vectors used for model building, complexity parameter C (explained later), type of optimizer (linear or quadratic), and type of loss function (linear or quadratic).

The final model is selected after several iterations by changing the appropriate tuning parameters and analyzing the performance on the validation and test data sets.

Once the neural networks or the SVMs have been trained, it is said to have learned the input-output mapping function implicitly. One can now fix the weights and use a run-time version of the model.

4.5.3 Applying Neural Networks: An Example

The generic machine learning application sequence will be illustrated with an example of developing empirical models that infer expensive process measurements from available low cost measurements, such as temperature, pressure, or flow meters. The full details of the development are given in Chap. 11. In this section we illustrate the development of a neural networks-based model.

The developed application is in the class of function approximation. Among the many possible fitness metrics, related to the error between the predicted and measured values, the final user selected $1-R^2$ (R^2 is one of the key statistical measures of model performance. In the case of perfect match between predictions and measurements, R^2 is one. In the opposite case of lack of any relationship, R^2 is zero).

The objective of model development in this example is the prediction of emissions from a chemical reactor based on four process variables. One of the input variables is the rate, which is assumed to be correlated with the emissions in a nonlinear fashion. Data collection was based on large changes in operating conditions, including cases with very low rates (related to product loss) and took several weeks. The collected data were pre-processed and finally separated into 251 records for training and 115 records for validation and test. The range of the emission variable in the validation/test data is ~20% outside the range of the training data, which is a good test of the generalization capability of the model.

The neural network model included four inputs, one hidden layer, and one output – the predicted emissions. The key tuning parameter in this case is the number of hidden neurons, which was changed from 1 to 50. The results from the neural network models with minimal, optimal, and maximal number of neurons are shown in Fig. 4.28 and 4.29.

The three subplots in Fig. 4.28 show the evolution of the neural network performance with increased capacity on training data. It is obvious that the neural networks improve their approximation ability from an R^2 of 0.89 with one neuron in the hidden layer to an almost perfect fit of R^2 of 0.97 with 50 neurons in the hidden layer.

However, the generalization performance on the validation/test data, shown on Fig. 4.29, has a different behavior. The evolution of the training and test errors (in this case represented by $1-R^2$) with increased model capacity (number of neurons in the hidden layer) is shown in Fig. 4.30.

As was expected, while the training error continuously decreases with increased neural network complexity, the test error increases drastically beyond some capacity. The selected neural net model was with the minimal test error (in this case, a structure with 20 neurons in the hidden layer). The best performance of the selected structure with 20 neurons on test data is clearly seen in Fig. 4.29 when compared with the structures with one neuron and with 50 neurons. It has the highest R^2 and doesn't predict negative values in the low

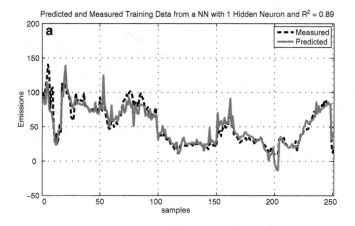

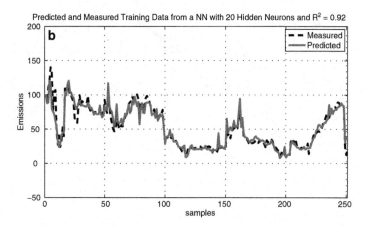

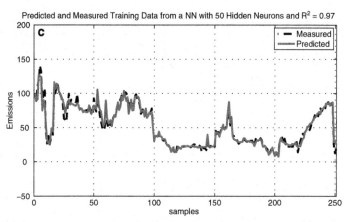

Fig. 4.28 Performance of a neural network model for emissions estimation with different capacity (number of neurons in the hidden layer) on training data

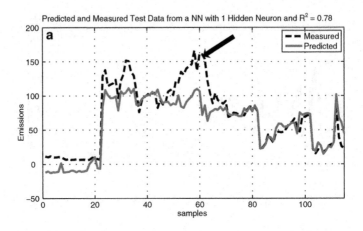

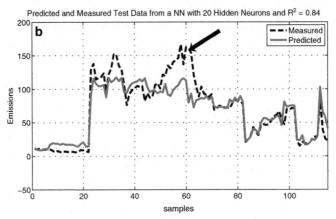

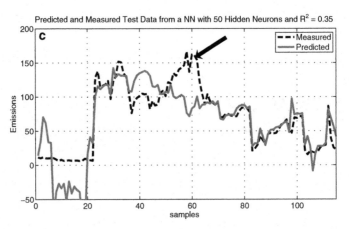

Fig. 4.29 Performance of a neural network model for emissions estimation with different capacity (number of neurons in the hidden layer) on test data

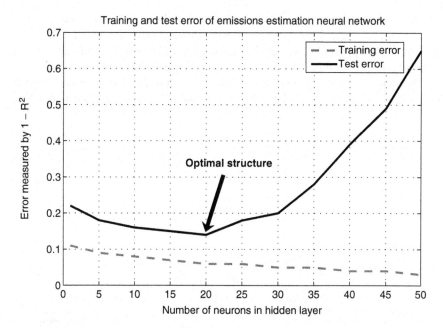

Fig. 4.30 Selecting the optimal structure of a neural network for emissions estimation based on minimal test data error

rate operating mode (the first 20 samples). However, due to the limited extrapolation capabilities of neural networks, it cannot predict accurately in the interesting range of high emissions above 140 (shown by the arrow in samples 55–63). The acceptable performance of the neural network model proves the hypothesis that it is possible to estimate the important emissions variable from the selected four process measurements. The relatively high complexity of the optimal structure (20 neurons in the hidden layer) indicates the nonlinear nature of the dependence.

4.5.4 Applying Support Vector Machines: An Example

The same data sets were used for development of a SVMs model for regression. Since the objective is building a model with good generalization ability, a SVMs model with mixed kernels of polynomials and RBF was selected due to its known extrapolation properties. The other selected tuning parameters were: complexity parameter C = 100,000, optimization method: quadratic programming; loss function: linear ε-sensitive. The system was optimized with the following ranges of the tuning parameters: polynomial kernel order from one to five, RBF kernel width from 0.2 to 0.7, ratio between polynomial and RBF kernels between 0.8 and 0.95, percentage of support vectors between 10 and 90%.

The best performance on test data was achieved by a SVMs model based on a mixture of a third-order polynomial and RBF with a width of 0.5 with a ratio of 0.9 by using 74 support vectors (29.4% of training data). The performance of the selected model on training and test data sets is shown in Fig. 4.31.

The approximation quality of the SVMs is similar to the neural network models (see subplot a) in Fig. 4.31 where the 74 used support vectors are encircled). The

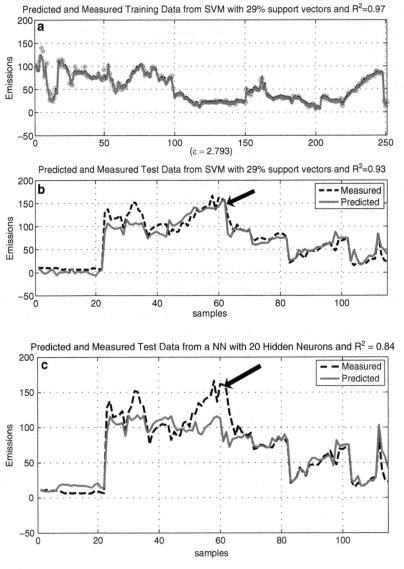

Fig. 4.31 Performance of the optimal SVM model for emissions prediction and comparison with the best neural network model

improved generalization capability of the SVMs model on the test data (subplot b) are clearly seen in comparison with the optimal neural net model (subplot c). In contrast to the optimal neural network model, the optimal SVMs model predicts with acceptable accuracy the important high emissions range above 140 (shown with the arrow in Fig. 4.31 in samples 55–63).

4.6 Typical Machine Learning Applications

Among all computational intelligence approaches machine learning has the lion's share of real-world applications. Most of it is based on the popularity of neural networks which have penetrated to almost every business. At the other extreme are the SVMs, which still struggle for mass-scale industrial acceptance.

4.6.1 Typical Applications of Neural Networks

Neural networks have been applied in thousands of industrial applications in the last 20 years. Among the interesting cases are: using an electronic nose for classification of honey, applying neural networks for predicting beer flavors from chemical analysis, and using an on-boat navigator to make better decisions in real time during competitions. The selected neural networks application areas are shown in Fig. 4.32 and discussed briefly next.

- *Marketing* – Neural networks are used in a wide range of marketing applications by companies like Procter & Gamble, EDS, Sears, Experian, etc. Typical applications include: targeting customers by learning critical attributes of

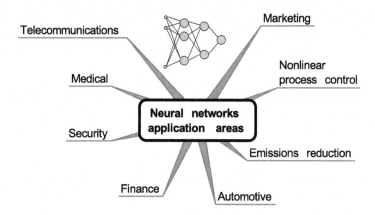

Fig. 4.32 Key application areas of neural networks

customers in large databases, developing models of aggregate behavior that can suggest future consumption trends, making real-time price adjustments as incentives to customers with discovered patterns in purchase behavior, etc.

For example, Williams-Sonoma Inc., retailer of fine home furnishings for every room in the home, segments customers into groups and details the potential profitability of each segment. The results help marketing executives decide which customers should receive a new catalog. Using an active database of more than 33 million households, the Williams-Sonoma modeling team begins each new model build by selecting data from the previous year's mailing results. By using several modeling techniques, including neural networks, the team determines the relative importance of each variable and allows the analysts to segment similar customers into groups that include 30,000 to 50,000 households. The model predicts profitability for each group, based primarily on the previous year's total average purchases per household.[14]

- *Nonlinear Process Control* – With its unique capability to capture nonlinearity from available process data, neural networks are at the core of nonlinear process control systems. The two key vendors of these systems – Aspen Technology and Pavilion Technology – have implemented hundreds of nonlinear control systems in companies like Fonterra Cooperative, Eastman Chemical, Sterling Chemicals, BP Marl, etc. with total savings of hundreds of millions dollars. The sources of savings are: consistently operating units in the optimal range through cost-effective debottlenecking and identification of optimal operating conditions, transition strategies and new grades, minimizing inventory through accurate demand forecasting, optimum planning and scheduling of grades and agile manufacturing.

- *Emissions Reduction* – Environmental limits are one of the significant constraints in optimal control. Unfortunately, most of the hardware emissions sensors are very expensive (in the range of several hundred thousand dollars). An alternative software sensor, based on neural networks, is a much cheaper solution and has been applied in companies like The Dow Chemical Company, Eastman Chemical, Elsta, Michigan Ethanol Company, etc. The economic impact of the continuous NOx, CO, or CO_2 emissions estimation is a significant reduction of compliant cost, increasing production by 2–8%, reduced energy cost per ton by up to 10%, and reduced waste and raw materials by reducing process variability by 50–90%.[15]

- *Automotive* – Neural networks are beginning their invasion in our cars. In the Aston Martin DB9, for example, Ford has used a neural network for detecting misfires in the car's V12 engine. To discover the misfire patterns, the development team drove a vehicle under every imaginable condition. They then forced

---

[14]http://www.sas.com/success/williamssonoma_em.html

[15]http://www.pavtech.com

the vehicle into every conceivable type of misfire. The data collected during the test drives is returned to the lab and fed into a training program. By examining the patterns, the neural network is trained in a simulation to distinguish the normal patterns from the misfires. The final result determines a critical threshold number of misfires into the system. Once the neural network detects the critical number, the misfiring cylinder is shut down to avoid damage such as a melted catalytic converter. Neural networks were the only cost-effective way Ford could make the DB9 meets the demanding Californian emissions requirements.

- *Finance* – Neural networks became the foster child of empirical modeling in almost any activity in the financial industry. The list of users includes almost any big banks, such as Chase, Bank One, Bank of New York, PNC Bank, etc. and insurance giants like State Farm Insurance, Merrill Lynch, CIGNA, etc. The age of the Internet allows accumulation of huge volumes of consumer and trading data that paves the way for neural network models to forecast almost everything in the financial world, from interest rates to the movement of individual stocks.

  One of the most popular neural network applications is fraud detection. For example, the London-based HSBC, one of the world's largest financial organizations, uses neural network solutions (the software package Fair Isaac's Fraud Predictor) to protect more than 450 million active payment card accounts worldwide. The empirical model calculates a fraud score based on the transaction data at the cardholder level as well as the transaction and fraud history at the merchant level. The combined model significantly improves the percentage of fraud detection.

- *Security* – The increased security requirements after September 11, 2001 gave a new boost to biometrics, which uses physiological traits or patterns of individuals to recognize and confirm identity. With their unique pattern recognition capabilities, neural networks are the basis of almost any biometrics device, such as fingerprint detectors, retinal scanners, voice prints identifiers, etc. Security companies like Johnson Controls and TechGuard Security are actively implementing neural networks in their products.

- *Medical* – Hospitals and health-management organizations use neural networks to monitor the effectiveness of treatment regimens, including physician performance and drug efficacy. Another application area is identifying high-risk health plan members and providing them with appropriate interventions at the right time to facilitate behavior change. For example, Nashville-based Healthways, a health provider to more than 2 million customers in all 50 states of the USA, builds predictive models that assess patient risk for certain outcomes and establishes starting points for providing services.

- *Telecommunications* – Most of the leading telecommunications companies use neural networks in many related applications. Typical examples are: optimal traffic routing by self-organization, scheduling problems of radio networks, dynamic routing with effective bandwidth estimation, traffic trend analysis, etc.

4.6.2 Typical Applications of Support Vector Machines

Unfortunately, the technical advantages of SVMs have not been adequately explored for industrial applications. Most of the applications are in the classification area, such as handwritten letter classification, text classification, face recognition, etc. Recently bioinformatics has appeared as one of the most perspective application areas of SVMs. Examples are gene expression-based tissue classification, protein function and structure classification, protein subcellular localization, etc.

One of the few pioneers in using SVMs in medical diagnosis is Health Discovery Corporation. SVMs are at the basis of several products for blood cell and tissue analysis to detect leukemia and prostate and colon cancer as well as of image analysis of mammograms and tumor cell imaging.[16]

In regression applications, SVMs have been implemented in inferential sensor applications in The Dow Chemical Company. Examples are given in Chap. 11.

4.7 Machine Learning Marketing

The key message in machine learning marketing is the unique ability of these systems to automatically discover patterns and dependencies by learning from data. The specific message for each technology, however, is different. Examples of neural networks and SVMs marketing are given below.

4.7.1 Neural Networks Marketing

Neural networks require relatively low marketing efforts due to the easy-to-explain method and the impressive industrial record in many application areas. The proposed marketing slide with the suggested format is shown on Fig. 4.33. The key slogan of applied neural networks: "Automatic modeling by mimicking the brain activity" captures the essence of value creation.

The left section of the marketing slide represents a very simplified view of the nature of neural networks - transferring data into black-box models by learning. The middle section of the marketing slide shows the key advantages of neural networks, such as the capability for adaptive learning from samples, potential for self-organizing the data when no knowledge is available, fitting data to any nonlinear dependence at any desired degree of accuracy, etc. The key application areas of neural networks are shown in the right section. Only the most popular application areas, such as marketing, finance, emission reduction, and security with the

---

[16]http://www.healthdiscoverycorp.com/products.php

corresponding industrial leaders, such as Procter & Gamble, Chase, Dow Chemical, and Johnson Controls, are selected.

The proposed elevator pitch for communicating neural networks to managers is shown in Fig. 4.34.

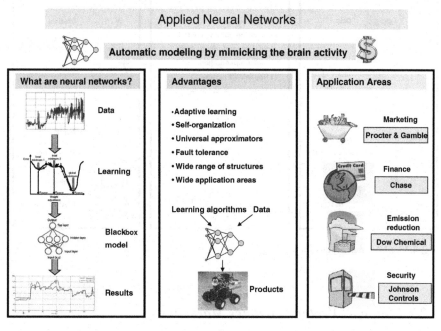

Fig. 4.33 Neural networks marketing slide

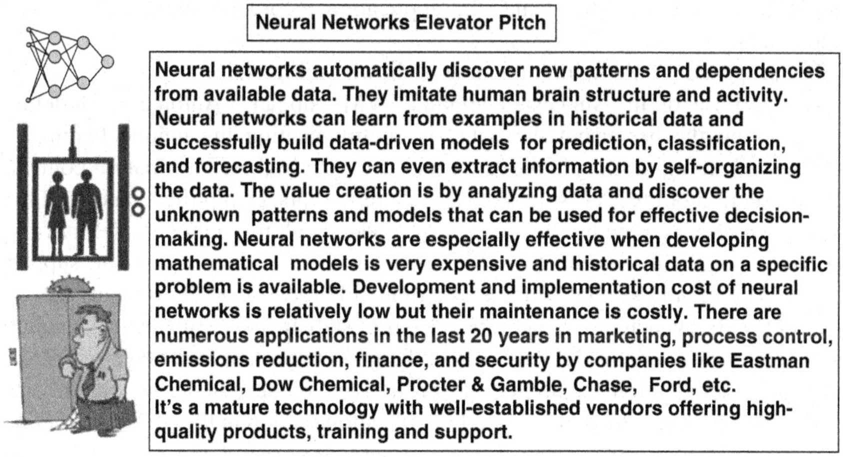

Neural Networks Elevator Pitch

Neural networks automatically discover new patterns and dependencies from available data. They imitate human brain structure and activity. Neural networks can learn from examples in historical data and successfully build data-driven models for prediction, classification, and forecasting. They can even extract information by self-organizing the data. The value creation is by analyzing data and discover the unknown patterns and models that can be used for effective decision-making. Neural networks are especially effective when developing mathematical models is very expensive and historical data on a specific problem is available. Development and implementation cost of neural networks is relatively low but their maintenance is costly. There are numerous applications in the last 20 years in marketing, process control, emissions reduction, finance, and security by companies like Eastman Chemical, Dow Chemical, Procter & Gamble, Chase, Ford, etc.
It's a mature technology with well-established vendors offering high-quality products, training and support.

Fig. 4.34 Neural networks elevator pitch

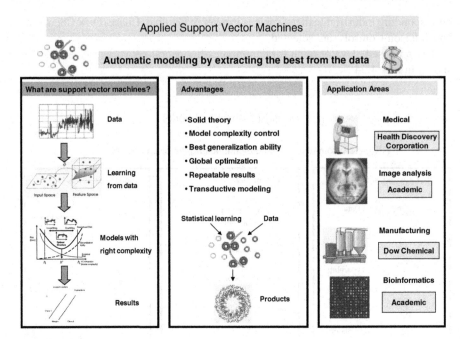

Fig. 4.35 SVM marketing slide

4.7.2   Support Vector Machines Marketing

Marketing SVMs is a real challenge. On the one hand, explaining the approach in plain English without the mathematical vocabulary of statistical learning theory is very difficult. On the other hand, the application record is very short and still unconvincing.

The proposed SVMs marketing slide is shown on Fig. 4.35.

The purpose of the suggested applied SVMs slogan, "Automatic modeling by extracting the best from the data" is communicating the unique feature of the approach to build models based on data with the highest information content. The left section of the SVMs marketing slide represents a simplified version of the nature of SVMs as a method that develops models with the right complexity from the data by optimal learning from the most informative data points, the support vectors. The middle section of the marketing slide focuses on the obvious advantages of SVMs, such as: solid theoretical basis of statistical learning theory, explicit model complexity control which leads to the most important advantage of SVMs, the best generalization capability of the derived models under unknown process conditions. Unfortunately, the right application section in the SVMs marketing slide is not as impressive as in the neural networks case. Only two application areas (medical and manufacturing) are supported by industrial applications at

Support Vector Machines Elevator Pitch
Support vector machines automatically learn new patterns and dependencies from finite data. The name is too academic and there is a lot of math behind the scenes. As a result, the derived models have optimal complexity and improved generalization ability in unknown operating conditions, which reduces maintenance cost. The key idea is simple: we don't need all the data to build a model but only the most informative data points, called support vectors. The value creation is by extracting the "golden" data into unknown patterns and models used for effective decision-making. Support vector machines are especially effective when developing mathematical models is very expensive and only finite historical data on a specific problem is available.   **There are a few applications in the area of bioinformatics, medical analysis by Health Discovery Corporation, and process monitoring by Dow Chemical.**   **It's an emerging technology still in the research domain but with great potential to penetrate in industry in the next decade.**

Fig. 4.36 SVMs elevator pitch

respectively Health Discovery Corporation and Dow Chemical. Most of the applications are still in academia or the research labs.

The SVMs elevator pitch is shown in Fig. 4.36 and requires a lot of effort to grab the attention of any manager.

4.8 Available Resources for Machine Learning

4.8.1 Key Websites

Neural networks:
http://www.dmoz.org/Computers/Artificial_Intelligence/Neural_Networks/

Machine learning tutorials (including neural networks, statistical learning theory, and SVMs):
http://www.patternrecognition.co.za/tutorials.html

SVMs:
http://www.support-vector-machines.org/
http://www.kernel-machines.org/

4.8.2 Key Software

The most popular options for neural networks software are:

The neural network toolbox for MATLAB:
http://www.mathworks.com/products/neuralnet/

The neural network toolbox for Mathematica:
http://www.wolfram.com/products/applications/neuralnetworks/

Neurosolutions package of Neurodynamics (block-based development and interface with Excel):
http://www.nd.com/

Stuttgart Neural Networks Simulator (SNNS), developed at the University of
Stuttgart (free for non-commercial use)
http://www.ra.cs.uni-tuebingen.de/SNNS/

The software options for SVMs are limited. Some examples are given below:
Least-Squares SVM MATLAB toolbox, developed at the University of Leuven
(free for noncommercial use)
http://www.esat.kuleuven.ac.be/sista/lssvmlab/

SVM MATLAB toolbox, developed by Steve Gunn (free for non-commercial use)
http://www.isis.ecs.soton.ac.uk/resources/svminfo/

4.9 Summary

Key messages:

*Defining the right learning capacity of a model is the key issue in machine
learning.*

Artificial neural networks are based on human brain structure and activities.

*Neural networks learn from examples from historical data or self-organize the
data and automatically build empirical models for prediction, classification and
forecasting.*

*Neural networks interpolate well but extrapolate badly outside the known
operating conditions.*

*Neural networks have been successfully applied in marketing, process control,
emission reduction, finance, security, etc. by companies like Procter & Gamble,
Chase, Dow Chemical, Ford, etc.*

*Support vector machines are based on statistical learning theory, which is a
generalization of classical statistics in the case of finite data.*

*Support vector machine models capture the most informative data in a given
data set, called support vectors.*

*Support vector machine models have optimal complexity and improved general-
ization ability in new operating conditions, which reduces maintenance cost.*

Support vector machines are still in the research domain with a few business applications in medical analysis and process monitoring.

The Bottom Line

Applied machine learning systems have the capability to create value through automatic discovery of new patterns and dependencies by learning the behavior of a dynamic complex environment.

Suggested Reading

A good survey of all machine learning methods is the book by T. Mitchell, *Machine Learning*, McGraw-Hill, 1997.

The undisputed bible of neural networks is the book:
S. Haykin, *Neural Networks: A Comprehensive Foundation*, 2nd edition, Prentice Hall, 1999.

A survey with a good explanation of neural network learning methods can be found in the book:
M. Negnevitsky, *Artificial Intelligence: A Guide to Intelligent Systems*, Addison-Wesley, 2002.

The undisputed bible of statistical learning theory is the book (requires mathematical background):
V. Vapnik, *Statistical Learning Theory*, Wiley, 1998.

One of the best sources for statistical learning theory and SVM is the book:
V. Cherkassky and F. Mulier, *Learning from Data*: Concepts, *Theory, and Methods*, 2nd edition, Wiley, 2007.

Chapter 5
Evolutionary Computation: The Profitable Gene

It is not the strongest of the species that survive, nor the most intelligent, but the one most responsive to change.

Charles Darwin

Another key source of enhancing human intelligence is inspired by evolution in nature. Biological evolution has been particularly successful in the design and creation of amazingly complex organisms driven by several simple mechanisms. According to Darwin, the driving force behind natural evolution is the capability of a population of individuals to reproduce and deliver new populations of individuals, which are fitter for their environment. The fundamental evolutionary step is survival of the fittest, which implies some sort of competition, combined with recombination acting on the chromosomes, rather than on the living organisms themselves. Evolutionary computation uses an analogy with natural evolution to perform a search by evolving solutions (equations, electronic schemes, mechanical part, etc.) to problems in the virtual environment of computers. One of the important features of evolutionary computation is that instead of working with one solution at a time in the search-space, a large collection or population of solutions is considered at once. The better solutions are allowed to "have children" and the worse solutions are quickly eliminated. The "child solutions" inherit their "parents' characteristics" with some random variation, and then the better of these solutions are allowed to "have children" themselves, while the worse ones "die", and so on. This simple procedure causes simulated evolution. After a number of generations the computer will contain solutions which are substantially better than their long-dead ancestors at the start.

The value creation of evolutionary computation is based on its unique ability to generate novelty automatically. Novelty recognition is not automatic, however, and is done by humans either by analyzing the final results after simulated evolution termination or in interactive mode at each evolutionary step.

Evolutionary computation is one of the computational intelligence approaches with the fastest penetration rates in industry. The universal appeal of the evolutionary concept, combined with the unique technical capabilities of the different evolutionary

A.K. Kordon, *Applying Computational Intelligence*,
DOI 10.1007/978-3-540-69913-2_5, © Springer-Verlag Berlin Heidelberg 2010

computation algorithms and increased computing power have opened many doors in industrial optimization, design, and model development. We observe a clear trend of transforming the "Selfish Gene"[1] into the "Profitable Gene" and it is the main theme in this chapter.

5.1 Evolutionary Computation in a Nutshell

In the last 20 years the pace of development of evolutionary computation has been almost exponential. It is a challenge even for members of the academic communities to keep track of all of the new ideas. Therefore, from the point of view of understanding the big implementation potential of evolutionary computation, we recommend the reader focuses on the following key topics, shown in the mind-map in Fig. 5.1 and discussed below.

5.1.1 Evolutionary Algorithms

Evolutionary computation is a clear result of Darwin's dangerous idea that *evolution* can be understood and represented in an abstract and common terminology as an *algorithmic process*. It can be lifted out of its home base in biology.[2] In fact, the generic evolutionary algorithm borrows the same major idea from natural evolution. A population of species reproduces with inheritance and is the subject of

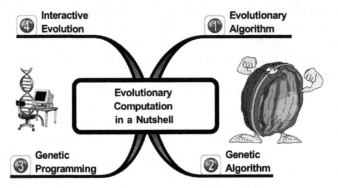

Fig. 5.1 Key topics related to evolutionary computation

---

[1]The phrase became popular from the bestseller of R. Dawkins, *The Selfish Gene*, Oxford University Press, 1976.

[2]D. Dennett, *Darwin's Dangerous Idea: Evolution and the Meaning of Life*, Simon & Schuster, 1995.

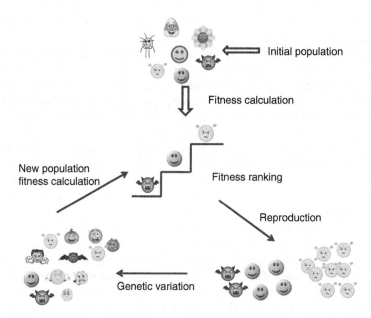

Fig. 5.2 Visualization of the key evolutionary algorithm principle

variation and natural selection. The key evolutionary algorithm principle is illustrated in Fig. 5.2.

The first step in simulated evolution is creating an initial population of individuals (solutions) either randomly or through some other means, reflecting *a priori* knowledge on specific problems. Examples of solutions in the business world are mechanical parts, analog circuits, production schedules, portfolio design, etc. The size of the population is one of the critical parameters in evolutionary algorithms. A too large population size can significantly slow down the simulated evolution while a too small population can lead to nonoptimal solutions.

After initialization, the simulated evolution loop is started until some prespecified criteria, such as defined accuracy of prediction, cost, or number of generations, triggers the termination. The simulated evolution loop includes the three well-known steps from natural evolution, shown in Fig. 5.2: (1) fitness evaluation (natural selection); (2) reproduction; and (3) genetic variation. Fitness evaluation requires defining a fitness function which directs the artificial evolution towards better solutions. In natural evolution, the continuously changing environmental conditions define the fitness to biological species and direct their own evolution. In simulated evolution, humans must specify in advance a quantitative fitness function, such as travel cost, rate of return on investment, profit, error between derived models and real data, etc. If the fitness function is unavailable, the selection of the winning solutions is done interactively. As a result of fitness calculation, evolutionary algorithms try to optimize the fitness scores of individuals, by allowing the fitter solutions to have more children than less fit solutions. For example, the top-ranked species in Fig. 5.2 have more offspring than the lower ranked individuals. The least fit solutions are ruthlessly killed.

Reproduction is the second step in the simulated evolution loop. It is assumed that on the one hand the generated new individuals (solutions) inherit characteristics from their parents but at the same time there is sufficient variability to ensure that children are different from their parents. The success of simulated evolution depends on the right trade-off between the preservation of useful characteristics of parent solutions in child solutions and the diversity of the individuals. Typically, both recombination[3] operators (requiring two or more parents) as well as mutation operators (requiring a single parent only) are used. These operations are performed during the third step in simulated evolution – genetic variation. Different mechanisms of mutation and crossover will be described in the cases of genetic algorithms and genetic programming.

This simple algorithm works very well for many practical applications. There are five main representations of evolutionary algorithm in use today, shown in the mind-map in Fig. 5.3. Historically, three of these were independently developed more than 40 years ago. These algorithms are: evolutionary programming (EP) created by Lawrence Fogel in the early 1960s, evolutionary strategies (ES), invented by Ingo Rechenberg and Hans-Paul Schwefel in the early 1960s, and the genetic algorithm (GA) created by John Holland in the early 1970s. The other two major evolutionary algorithms were created in the 1990s. Genetic programming (GP) was developed by John Koza in the early 1990s and differential evolution (DE) was defined by Rainer Storn and Kenneth Price in 1995.

Navigating through the details of operation and differences between the evolutionary computation algorithms is beyond the scope of this book.[4] We'll focus our

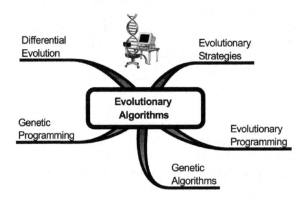

Fig. 5.3 Key representations of evolutionary algorithm

---

[3]The appearance in offspring of new combinations of allelic genes not present in either parent, produced from the mixing of genetic material, as by crossing-over.

[4]A good review of all methods is given in the book: A. Eiben and J. Smith, *Introduction to Evolutionary Computing*, Springer, 2003.

attention on genetic algorithms and genetic programming since these two approaches deliver most of the value from evolutionary computation applications.

5.1.2 Genetic Algorithms

A key feature of genetic algorithms (GA) is their use of two separate spaces: the search space and the solution space. The search space is a space of coded solutions to the specific problem and the solution space is the space of the actual solutions. Coded solutions, or genotypes, must be mapped onto actual solutions, or phenotypes, before the quality or fitness of each solution can be evaluated. An example of GA mapping is shown in Fig. 5.4 where the phenotypes are different chairs. The genotypes consist of a sequence of 1 and 0 that code the parameters to generate or describe an individual phenotype. The level of detail of this description is related to the optimization problem that needs to be solved. For example, if the objective of the GA search is to find a chair with minimal wooden material used and maximal strength, the genotype coding may include the chair geometry parameters, type of material in each part, and material strength.

One of the requirements for a successful GA implementation is the effective definition of the genotype. Fortunately, these sorts of encodings are very common in all kinds of computational or engineering problems. Although the original simple GA does insist on binary encodings i.e. strings of ones and zeros, many other types of encoding are also possible (e.g. vectors of real numbers or more complicated structures like lists, trees, etc.) The algorithm works equally well with more complicated encodings.

In the language of genetic algorithms, the search for a good solution to a problem is equivalent to the search for a particular binary string or chromosome. The universe of all admissible chromosomes can be considered as a fitness landscape. An example of a fitness landscape with multiple optima is shown in Fig. 5.5. Conventional optimization techniques for exploring such kinds of multimodal landscapes will invariably get stuck in a local optimum. In real-world applications,

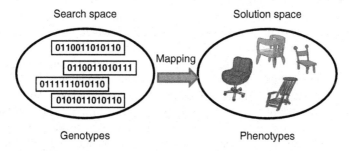

Fig. 5.4 Mapping genotypes in search space with phenotypes in solution space

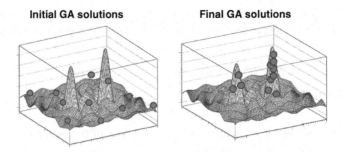

Fig. 5.5 A multimodal fitness landscape with GA solutions at the start and at the end of the run

Fig. 5.6 Crossover operator between two GA chromosomes

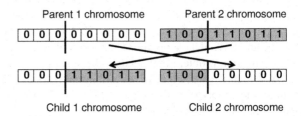

such types of search spaces are usually very large and a challenge to any kind of exhaustive search techniques. Genetic algorithms approach this problem by using a multitude of potential solutions (chromosomes) that samples the search space in many places simultaneously. An example of the initial random distribution of the potential solutions in such a landscape is shown in the left part of Fig. 5.5 where each chromosome is represented by a dark dot. At each step of the simulated evolution, the genetic algorithm directs the exploration of the search space into areas with higher elevation than the current position.

This remarkable ability of genetic algorithms to dedicate most resources to the most promising areas is a direct consequence of the recombination of chromosomes containing partial solutions. In successive generations, all strings in the population are first evaluated to determine their fitness. Second, the strings with a higher ranking are allowed to reproduce by crossover, i.e. portions of two parent strings are exchanged to produce two offspring. An example of a crossover operator between two GA chromosomes is shown in Fig. 5.6 where the selected sections of the parents' chromosomes have been swapped in the offspring.

These offspring are then used to replace the lower ranking strings in the population, maintaining a constant size of the overall population.

Third, mutations modify a small fraction of the chromosomes. An example of the mutation operator on a GA bit string is shown in Fig. 5.7. In a simple GA, mutation is not an important operator but is mainly used to keep diversity and safeguard the algorithm from running into a uniform population incapable of further evolution.

Fig. 5.7 Mutation operator
of a GA chromosome

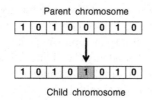

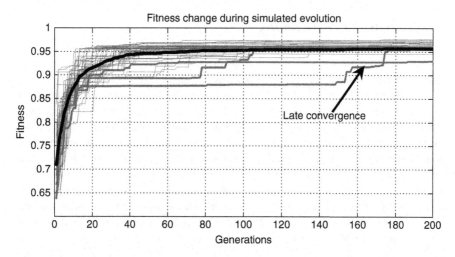

Fig. 5.8 Fitness change during a simulated evolution of 200 generations

As a result of the key genetic operations – crossover and mutation – the average fitness of the whole population and above all, the fitness of the best species, gradually increases during simulated evolution. A typical curve of fitness change is shown on Fig. 5.8. The thick line represents the average fitness from 50 independent runs of simulated evolution which continues 200 generations. As shown in Fig. 5.8, the average fitness converges (i.e. optimal solutions have been found) within 40–60 generations. However, some runs converge late in generation 170.

In the final phases, the GA concentrates the population of potential solutions around the most significant optima (see the right part of Fig. 5.5). However, there is no guarantee that the global optimum will be found. The good news is that according to the *Schema Theorem*,[5] under general conditions, in the presence of

---

[5]D. Goldberg, *Genetic Algorithms in Search, Optimization, and Machine Learning*, Addison-Wesley, 1989.

differential selection, crossover and mutation, any bit string solution that provides above-average fitness will grow exponentially. Or in other words, good bit string solutions (called schemata) will appear more frequently in the population as the genetic search progresses. The "good guys" grab more and more attention and gain influence with the progression of simulated evolution. As a result, the chances of finding optimal GA solutions increase with time.

5.1.3   Genetic Programming

Genetic programming (GP) is an evolutionary computing approach to generate soft structures, such as computer programs, algebraic equations, and electronic circuit schemes. There are three key differences between genetic programming and the classical genetic algorithm. The first difference is in the solution representation. While GA uses strings to represent solutions, the forms evolved by GP are tree structures. The second difference relates to the length of the representation. While standard GA is based on a fixed-length representation, GP trees can vary in length and size. The third difference between standard GA and GP is based on the type of alphabets they use. While standard GA uses a binary coding to form a bit string, GP uses coding of varying size and content depending on the solution domain. An example of GP coding is representing mathematical expressions as hierarchical trees, similar to the taxonomy of the programming language Lisp.

The major distinction between the two evolutionary computation approaches is that the objective of genetic programming is to evolve computer programs (i.e. perform automatic programming) rather than evolving chromosomes of a fixed length). This puts genetic programming at a much higher conceptual level, because suddenly we are no longer working with a fixed structure but with a dynamic structure equivalent to a computer program. Genetic programming essentially follows a similar procedure as for genetic algorithms: there is a population of individual programs that is initialized, evaluated and allowed to reproduce as a function of fitness. Again, crossover and mutation operators are used to generate offspring that replace some or all of the parents.

Another difference between GA and GP is in the mechanism of the genetic operation of crossover. When we select a hierarchical tree structure as a representation, crossover can be used to interchange randomly chosen branches of the parent trees. This operation can occur without disrupting the syntax for the child trees. We'll illustrate the crossover between two algebraic expressions in Fig. 5.9.

The first expression (or Parent 1) has a tree structure in Fig. 5.9, which is equivalent to the following simple mathematical expression:

$$\frac{\sqrt{a^2 + (a - b)} - \log(a)}{ab}$$

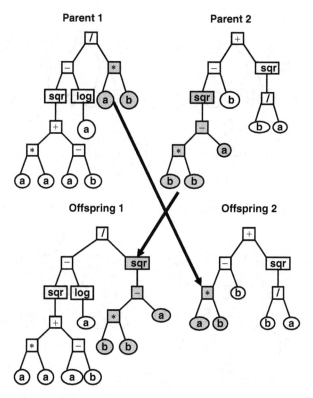

Fig. 5.9 Crossover operation in genetic programming

The tree structure for Parent 2 is equivalent to another mathematical expression:

$$(\sqrt{b^2} - a - b) + \sqrt{\frac{a}{b}}$$

In GP, any point in the tree structure can be chosen as a crossover point. In this particular case, the function (*) is selected to be the crossover point for Parent 1. The selection for the crossover point for Parent 2 is the function *sqr* (square root). The crossover sections in both parents are shown in grey color in Fig. 5.9. On the bottom part of Fig. 5.9 are shown the two offspring, generated by the crossover operation. Replacing the crossover section of the first parent with the crossover material from the second parent creates Offspring 1. In the same way, the second offspring is created by inserting the crossover section from the first parent in place of the crossover section of the second parent.

The mutation operator can randomly change any function, terminal, or subtree by a new one. An example of a mutation operator on the two parents' expressions is shown in Fig. 5.10. The randomly changed function (for Parent 1) and terminal (for Parent 2) are shown with arrows.

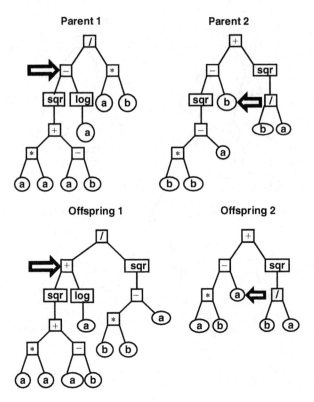

Fig. 5.10 Mutation operation in genetic programming

Because the structures are dynamic, one of the problems that can occur is an excessive growth of the size of the trees over a number of generations. Very often, these structures possess large portions of redundant or junk code (also called introns).

The primitives of GP are functions and terminals. Terminals provide a value to the structure while functions process a value already in the structure. Together, functions and terminals are referred to as nodes. The terminal set includes all inputs and constants used by the GP algorithm and a terminal is at the end of every branch in a tree structure. The function set includes the available functions, operators, and statements that are available as building blocks for the GP algorithm. The function set is problem-specific and can include problem domain-specific relationships like Fourier transforms (for time series problems, for example), or the Arrhenius law (for chemical kinetics problems). An example of a generic functional set, including the basic algebraic and transcendental functions, is given in:

$$ F = \{+, \ -, \ *, \ \div, \ln, \ \exp, \ \text{sqr}, \ \text{power}, \ \cos, \ \sin\}. $$

The basic GP evolution algorithm is very similar to a GA and has the following key steps:

Step 1: GP parameter selection.

Includes definitions of the terminal and functional sets, the fitness function, and selection of population size, maximum individual size, crossover probability, selection method, and maximum number of generations.

Step 2: Creating the initial population, which is made by random functions.

Step 3: Evolution of the population.

This includes: fitness evaluation of the individual structures, selection of the winners, performing genetic operations (copying, crossover, mutation, etc.), replacing the losers with the winners, and so on.

Step 4: Selection of the "best and the brightest" structure (functions) as the output from the algorithm.

Usually the GP algorithm is very computationally intensive, especially in cases with large search spaces and without surpassing the growth of very complex solutions. The lion's share of the computational burden is taken by the fitness evaluation of each individual member of the population.

Usually, the average fitness of the population increases from generation to generation. Often, the final high-fitness models are inappropriate for practical applications since they are very complex, difficult to interpret, and crash on even minor changes in operating conditions. In practice, simplicity of the applied solution is as important as the accuracy of model predictions. Unfortunately, in GP, manual selection of models with the best trade-off between lower complexity and acceptable accuracy requires time consuming screening through a large number of models.

Recently, a special version of GP, called Pareto-front GP, offers a solution by using multiobjective optimization to direct the simulated evolution toward simple models with sufficient accuracy.[6] An example of Pareto-front GP results is given in Fig. 5.11.

In Pareto-front GP the simulated evolution is based on two criteria – prediction error (for example, based on $1-R^2$) and complexity (for example, based on the number of nodes in the equations). In Fig. 5.11 each point corresponds to a certain model with the x-coordinate referring to model complexity and the y-coordinate is the model error. The points marked with circles form the Pareto front of the given set of models. Models at the Pareto front are nondominated by any other model in both criteria simultaneously. Compared to other models, elements of the Pareto front are dominant in terms of at least one criterion; they have either lower complexity or lower model error. All models on the Pareto front are chosen as best members of a population. The Pareto front itself is divided into three areas (see Fig. 5.11). The first area contains the simple under-fit models that occupy the upper

---

[6]G. Smits and M. Kotanchek, Pareto-front exploitation in symbolic regression, In *Genetic Programming Theory and Practice,* U.-M. O'Reilly, T. Yu, R. Riolo and B. Worzel (Eds), Springer, pp. 283–300, 2004.

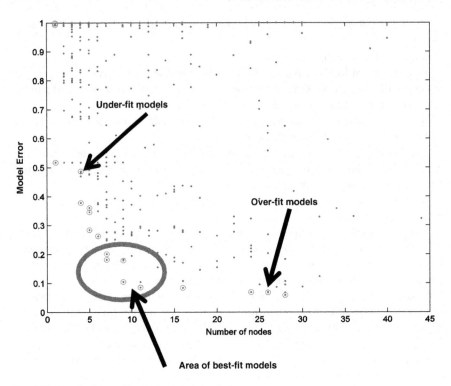

Fig. 5.11 Results from a Pareto-front GP simulation

left part of the Pareto front. The second area of complex over-fit models lie on the bottom right section of the Pareto front. The third and most interesting area of best-fit models is around the tipping point of the Pareto front where the biggest gain in model accuracy for the corresponding model complexity is.

In the Pareto-front GP, an algorithm is used to construct an archive of elite models obtained so far. The archive plays an important role in the simulated evolution and this is the key feature of the Pareto-front GP approach. At the first step, after the initial generation is created randomly, the archive contains all models from the Pareto front of this generation. The archive is updated at each next step of evolution. All models from the current archive are the "elite" and have the right to breed by means of mutation, cloning or crossover with any model of the current generation.

After a new generation is created, its Pareto front is constructed and added to the old archive. The Pareto front of the resulting set of models determines a new updated archive of the new generation. In the Pareto-front GP all the elite models are stored in the archive and all can propagate through generations without losses. At the end of the simulated evolution, we have an automatic selection of the "best and the brightest" models in terms of trade-off between accuracy and complexity. The "sweet zone" of models with the biggest gain in accuracy for the smallest possible complexity (encircled in Fig. 5.11) contains a limited number of solutions. In the case of Fig. 5.11 it is clear that there is no significant gain in accuracy with models of complexity higher than 15 nodes and the number of interesting solutions is between four and five. As a result, model selection is fast and effective and model development efforts are low.

5.1.4 *Interactive Evolutionary Computation*

In some cases in complex design, visual inspection, or creating evolutionary art, the selection of "good" solutions cannot be done automatically by the computer and the human becomes part of the simulated evolution. There are many kind of interaction between the user and the evolutionary computing system. The most popular and obvious way is when the user selects directly the individuals to be reproduced by purely subjective and aesthetic criteria. An example of this type of interaction is shown in Fig. 5.12.

The example is based on the popular artificial art website,[7] called Mondrian evolver, where nine pictures in the style of the famous Dutch painter Piet Mondrian are generated with initially random patterns (see the left hand side of Fig. 5.12). At each step the user selects three parents for breeding. The chosen solutions are transformed by recombination and mutation into nine new individuals for the next generation. The interactive process continues until the user selects the version that fully satisfies her/his own criteria for successful design.

Interactive evolution has several advantages, such as: handling situations with no clear objective functions, no rewriting of the fitness function when preferences change, redirecting simulated evolution if the search gets stuck, and increasing the diversity of selected solutions. There are obvious disadvantages as well. The interactive mode significantly slows down the simulated evolution. On top of that,

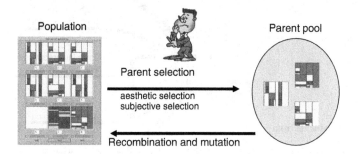

Fig. 5.12 Interactive evolution by direct parent selection from the user

---

[7]http://www.cs.vu.nl/ci/Mondriaan/

the user can get bored and tired easily and her/his selection capability changes unpredictably with the evolutionary process.

Interactive evolution has been applied in evolutionary design, evolutionary art, hot rolled steel surface inspection, image retrieval, hearing aid fitting, and interactive design of food for astronauts.

5.2 Benefits of Evolutionary Computation

The key benefits of evolutionary computation are represented in the mind-map in Fig. 5.13 and discussed below.

- *Reduced Assumptions for Model Development* – Evolutionary computation model development can be done with fewer assumptions than some of the other known methods. For example, there are no assumptions based on the physics of the process, as is in case of first-principles modeling. Some statistical models based on the popular least-squares method require assumptions, such as variable independence, multivariate normal distribution and independent errors with zero mean and constant variance. However, these assumptions are not required for models generated by GP. This "assumption liberation" establishes a technical superiority of generating models from the data with minimal effort from experts.[8] The cost savings are in the experts' reduced time for defining and especially for validating the model assumptions. In case of mechanistic models for chemical processes, which may require defining and validating the assumption space of hundreds of parameters by several experts, the savings could be large.

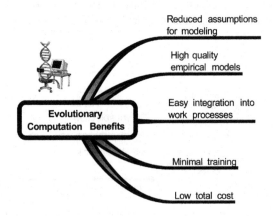

Fig. 5.13 Key benefits from applied evolutionary computation

---

[8]However, all data preparation procedures, such as data cleaning, dealing with missing data, and outlier removal are still valid.

- *High-Quality Empirical Models* – The key evolutionary approach for empirical model building is symbolic regression (nonlinear algebraic equations), generated by genetic programming (GP). A well-known issue of the conventional GP algorithm, however, is the complexity of the generated expressions. Fortunately, Pareto-front GP allows the simulated evolution and model selection to be directed toward structures based on an optimal balance between accuracy and expression complexity. A survey of industrial applications in The Dow Chemical Company shows that the selected models, generated by Pareto-front GP, are simple.[9] The derived models are more robust during process changes than conventional GP or neural network-based models.
- *Easy Integration into Existing Work Processes* – In order to improve efficiency and reduce implementation cost, the procedures for development, implementation, and maintenance in industry are standardized by work processes and methodologies. From that perspective, evolutionary computation in general, and symbolic regression in particular, has a definite competitive advantage. The technology could be integrated under Six Sigma with minimal efforts as an extension of the existing statistical methods with the additional modeling capabilities of symbolic regression. Another advantage of this type of solution is that there is no need for a specialized software environment for their run-time implementation (as is the case of mechanistic and neural network models). This feature allows for a relatively easy software integration of this specific evolutionary computation technology into most of the existing model deployment software environments.
- *Minimal Training of the Final User* – The explicit mathematical expressions, generated by GP, are universally acceptable by any user with mathematical background at high school level. This is not the case, either with the first-principles models (where specific physical knowledge is required) or with the black-box models (where some advanced knowledge on neural networks is a must). In addition, a very important factor in favor of symbolic regression is that process engineers prefer mathematical expressions and very often can find an appropriate physical interpretation. They usually don't hide their distaste toward black boxes.
- *Low Total Cost of Development, Deployment and Maintenance* – Evolutionary computation has a clear advantage in marketing the technology to potential users. The scientific principles are easy to explain to almost any audience. We also find that process engineers are much more open to implement symbolic regression models in manufacturing plants. Most of the alternative approaches are expensive, especially in real-time process monitoring and control systems. As was discussed earlier, symbolic regression models do not require special run-time versions of the software and can be directly implemented in any existing

---

[9]A. Kordon, F. Castillo, G. Smits, and M. Kotanchek, Application issues of genetic programming in industry, In *Genetic Programming Theory and Practice III*, T. Yu, R. Riolo and B. Worzel (eds): Springer, Chap. 16, pp. 241–258, 2005.

process monitoring and control system, i.e. the deployment cost is minimal. In addition, the simple symbolic regression models require minimal maintenance. Model redesign is very rare and most of the models perform with acceptable quality even 20% outside the range of originally development data.

5.3 Evolutionary Computation Issues

The next topic in presenting evolutionary computation is analyzing its limitations. First, we'll focus on the key issue in genetic algorithms – defining the chromosome. Unfortunately, in GA, coding the problem as a bit string may change the nature of the problem. In other words, there is a danger that the coded representation becomes a problem that is different from the one we wanted to solve. Sometimes the mapping between the genotype and the phenotype is not so obvious.

Another main issue of evolutionary computation algorithms is premature convergence and local minima. In principle, GA and GP have less chance of being trapped in a local minimum due to the effects of mutation and crossover. However, if the GA or GP explore a search area extensively, that region of the search space may be almost completely dominated by a set of identical or very similar solutions from that region. It is said that the simulated evolution has converged prematurely and begins to reproduce clones of the "locally best and brightest". In some cases escaping from such a local minimum is slow and additional measures have to be taken, such as increasing the mutation rate.

Exactly the opposite problem happens if the mutation rate is too high. In this case even if good solutions are found, they tend to be rapidly destroyed by frequent and aggressive mutation. There is a danger that due to the unstable population the simulated evolution will never converge. Finding a mutation rate that delivers the right balance between exploration and exploitation is a "tricky business" and often requires experimentation.

Of special importance is the main drawback of all evolutionary computation algorithms, especially GP – the slow speed of model generation due to the inherent high computational requirements of this method. For industrial-scale applications the calculation time maybe on the order of hours and days, even with current high-end workstations.

5.4 How to Apply Evolutionary Computation

Evolutionary computation can be applied in two different modes: automatic simulated evolution and interactive evolution. In the first mode, the user defines the objectives of the artificial evolutionary process and runs the virtual competition several times (it is recommended at least 20 repeats to eliminate the effect of random initialization and to achieve statistically consistent results). The final solutions are selected by the user from the fittest winners. The selection process could be significantly reduced if the algorithm of simulated evolution is of Pareto-front type.

In the second interactive mode, the user is continuously involved in the loop of simulated evolution by making the critical decisions of parents' selection.

5.4.1 *When Do We Need Evolutionary Computation?*

The dominant theme of applied evolutionary computation is novelty generation and recognition. The most profitable ways of novelty generation are shown in Fig. 5.14.

On the top of the list is the unique capability of most of the evolutionary algorithms, especially GA, to optimize complex surfaces, where most of the derivative-based optimizers fail.

Another unique capability for novelty generation is deriving empirical models through symbolic regression. Symbolic regression (i.e. nonlinear function identification) involves finding a mathematical expression that fits a given sample of data. Symbolic regression differs from conventional linear regression with its capability to automatically discover both the functional form and the numeric coefficients for the model. GP-generated symbolic regression is a result of the evolution of a large number of potential mathematical expressions. The final result is a list of the best analytical forms according to the selected objective function. These mathematical expressions can be used as empirical models, which represent the available data. Another advantage is the potential of these analytical functions to be used as proto-forms for models based on first-principles, as will be shown in Chap. 14.

Deriving novel mathematical expressions is one of the possible ways to generate novelty by GP. In principle, any problem that could be represented in structural form can benefit from this capability. The broad application area is called evolving hardware and includes examples such as evolved antennas, electronic circuits, and optical systems.

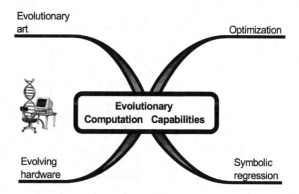

Fig. 5.14 Key capabilities of evolutionary computation

Evolutionary computation can generate novelty even in the aesthetic world of art by interactive evolution. In this case, however, it is assumed that a human evaluator will take control of the selection process and the quality of simulated evolution will depend on her/his aesthetic values.

5.4.2 Applying Evolutionary Computation Systems

The application sequence of an evolutionary computation systems (in this case, genetic programming) is shown in Fig. 5.15. Most of the key blocks are similar to the machine learning approach but the tuning parameters and the nature of model generation (learning versus simulated evolution) is different.

As any data-driven approach, GP depends very strongly on the quality and consistency of the data. A necessary precondition for applying GP to real problems is that the data have been successfully preprocessed in advance. Of special importance is removing the outliers and increasing the information contents of the data by removing insignificant variables and duplicate records. That is why it is necessary to use genetic programming in collaboration with other methods, such as neural networks, support vector machines, and principal component analysis. The data are usually divided by training, used in the simulated evolution, and test data for model validation.

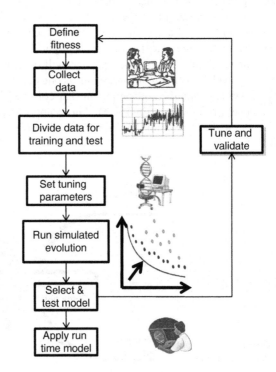

Fig. 5.15 Key steps in applying genetic programming

The preparatory step before running the GP algorithm includes determination of the following parameters:

- Terminal set, usually the inputs used for model generation.
- Functional set (selected functions used for genetic operations): the generic set includes the standard arithmetic operations +, −, *, and /, and the following mathematical functions: sqr, ln, exp, and power.
- Genetic operator tuning parameters:
    Probability for random vs. guided crossover;
    Probability for mutation of terminals;
    Probability for mutation of functions;
- Simulated evolution control parameters:
    Number of generations;
    Number of simulated evolutions;
    Population size.

Usually the genetic operators' parameters are fixed for all practical applications. They are derived after many simulations and represent a good balance between the two key genetic operators: crossover and mutation. In order to address the stochastic nature of GP, it is suggested to repeat the simulated evolution several times (usually 20 runs are recommended). The other two parameters that control the simulated evolution – population size and the number of generations – are problem-size dependent. The advantages of using bigger populations are that they increase genetic diversity, explore more areas of the search space, and improve convergence. However, they increase significantly the calculation time. According to Koza, a population size between 50 and 10,000 can be used for model development of almost any complexity.[10] In the case of Pareto-front GP, the recommended population size, based on optimal design of experiments on typical industrial data sets,[11] is 300. The number of generations depends on the convergence of the simulated evolution. Usually it is obtained experimentally with a starting range of 30–100 generations and is gradually increased until a consistent convergence is achieved.

5.4.3 Applying Evolutionary Computation Systems: An Example

The specific steps for applying GP will be illustrated with generating models from the data set for emissions estimation, described in Chap. 4. Since the objective is to develop a parsimonious inferential sensor, a Pareto-front GP with two objectives,

---

[10]J. Koza, Genetic Programming: On the Programming of Computers by Natural Selection, MIT Press, 1992.

[11]F. Castillo, A. Kordon and G. Smits, Robust Pareto front genetic programming parameter selection based on design of experiments and industrial data, In *Genetic Programming Theory and Practice IV*, R. Riolo, T. Soule and B. Worzel (Eds), Springer, pp. 149–166, 2007.

minimal modeling error $(1-R^2)$ and model complexity, based on the total number of nodes in the evaluated expression, are used. The results after 20 runs of 900 generations and a population size of 300 mathematical expressions are as follows.

The first key result from the simulated evolution is the sensitivity of the individual inputs (process measurements like temperatures, pressures, and flows) relative to the emissions variable, shown in Fig. 5.16.

This parameter reflects how frequently a specific input is selected by any function during the evolutionary process. If the input is weakly related to the output (the emissions), it will be gradually eliminated in the fierce evolutionary struggle for survival. This measure could be used further for variable selection and reduction of the search space with the strongly related inputs only. For example, only inputs two, five, six, and eight are substantial for developing a good predictive model for emissions, according to Fig. 5.16. It has to be taken into account that there is no correspondence between the linear statistical significance and the GP-based sensitivity. Inputs which are statistically insignificant may have very high sensitivity and vice versa.[12]

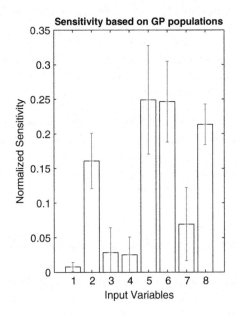

Fig. 5.16 Sensitivity of the eight inputs related to emissions

---

[12]G. Smits, A. Kordon, K. Vladislavleva, E. Jordaan and M. Kotanchek, Variable selection in industrial data sets using Pareto genetic programming, In *Genetic Programming Theory and Practice III,* T. Yu, R. Riolo and B. Worzel (Eds), Springer, pp. 79–92, 2006.

The results from the simulated evolution are shown in Fig. 5.17 where each model is represented by a dot in the accuracy-complexity plane.

The interesting models on the Pareto-front are encircled. An analysis of the Pareto-front shape shows that the accuracy limit of symbolic regression models is around $(1-R^2)$ of 0.12, i.e., the highest possible R^2 is 0.88. After exploring the performance of five potential solutions on the Pareto front around the interesting area with $(1-R^2)$ of 0.13–0.15 and expression complexity of 20, the following model was selected (Model 89 in the population, shown with an arrow on Fig. 5.17).

$$Emissions = 1.1 + 4.3 \frac{x_2 x_6 x_8^2}{x_5^4}$$

The model includes only the four most sensitive process inputs: x2, x5, x6, and x8. The expression is very compact and easy to implement in any software environment.

The performance of the selected model (with number 89 in this specific population) on training data set is shown on Fig. 5.18.

The more important performance of the symbolic regression model on the test data is shown in Fig. 5.19(a) and compared with the best SVM (5.19b) and neural network (5.19c) models. While not as accurate as the SVM model, it has

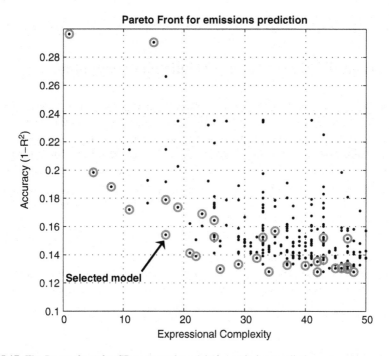

Fig. 5.17 The Pareto front for GP-generated models for emission prediction

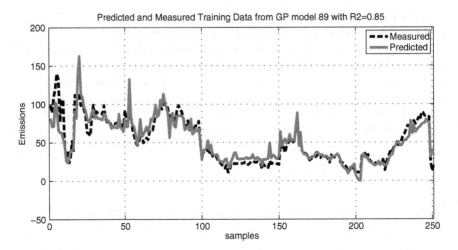

Fig. 5.18 Performance of the selected symbolic regression model for emissions estimation on training data

predicted very well outside the range of training data in the interesting high emissions area shown by the arrow around sample 60. In addition, implementing a simple mathematical function is much easier than SVM based on 74 support vectors.

5.5   Typical Applications of Evolutionary Computation

The business potential of evolutionary computing in the area of engineering design was rapidly recognized in the early 1990s by companies like General Electric, Rolls Royce, and BAE Systems plc. In a few years, many industries, like aerospace, power, chemical, etc., transferred their interest in evolutionary computing into practical solutions. Evolutionary computing entered industry in a revolutionary rather than evolutionary way!

The key selected application areas of evolutionary computation are shown in the mind-map in Fig. 5.20 and discussed below.

- *Optimal Design* – It is one of the first and most popular implementation areas of evolutionary computation. A typical example is the application of genetic algorithms to the preliminary design of gas turbine blade cooling hole geometries at Rolls Royce.[13] The objective was to minimize fuel consumption of the gas turbine.

---

[13]I. Parmee, Evolutionary and Adaptive Computing in Engineering Design, Springer, 2001.

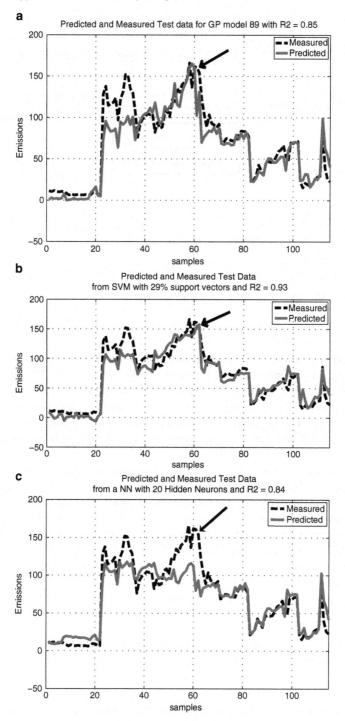

Fig. 5.19 Performance of GP model for emissions estimation (**a**) on test data with respect to the support vector machines model (**b**) and the best neural network model (**c**)

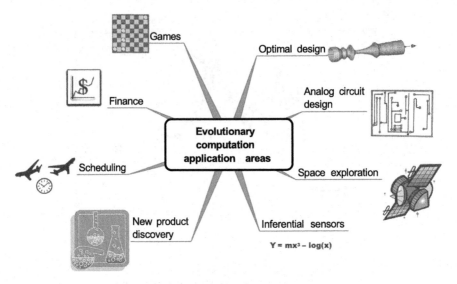

Fig. 5.20 Key evolutionary computation application areas

Another example of a real-life application of a genetic algorithm is a built-in optimizer in the program called ColorPro at Dow Chemical. This program is used to optimize mixtures of polymer and colorants to match a specific color. The optimizer allows a simultaneous optimization of multiple objectives (like the color-match for a specific light source, transmittance, mixing and cost). The program has been in use since the mid-1990s, creating significant value in the order of several million dollars.

- *Analog Circuit Design* – This is an application area explored very intensively by John Koza.[14] The analog circuits that have been successfully synthesized by GP include both passive components (capacitors, wires, resistors, etc.) and active components (transistors). The function set contains three basic types: (1) connection-modifying functions; (2) component creating functions; and (3) automatically defined functions. At each moment, there are several writing heads in the circuit which point to those components and wires that will be changed. Component-modifying functions can flip a component upside down or duplicate a component. Component-creating functions insert a new component at the location of the writing head. For simulation of the electronic circuit, the SPICE package from University of California at Berkeley is used. One of the biggest accomplishments is that GP duplicated the functionality of six current patents in the area of analog circuit design.

---

[14]J. Koza, *et al., Genetic Programming IV: Routine Human-Competitive Machine Intelligence,* Kluwer, 2003.

- *Evolutionary Antenna Design for NASA* – An evolved X-band designed antenna was deployed on NASA's Space Technology 5 (ST5) spacecraft.[15] The ST5 antenna was evolved to meet a challenging set of mission requirements, such as the combination of wide beamwidth for a circularly polarized wave and wide bandwidth. Two evolutionary algorithms were used: (1) GA with representation that did not allow branching in the antenna arms; (2) GP tree representation that allowed branching in the antenna arms. Interestingly, the highest performance antennas from both algorithms yielded very similar performance. Both antennas were comparable in performance to a hand designed antenna produced by the antenna contractor for the mission. However, it was estimated that the evolutionary antenna took three person-months to design and fabricate the first prototype as compared with five person-months for the conventionally designed antenna.

- *Inferential Sensors* – One area with tremendous potential for evolutionary computation and especially for symbolic regression, generated by GP, is inferential or soft sensors. The current solutions on the market, based on neural networks, require frequent retraining and specialized run-time software.

  An example of an inferential sensor for propylene prediction based on an ensemble of four different models is given in Jordaan *et al.*[16] The models were developed from an initial large manufacturing data set of 23 potential input variables and 6900 data points. The size of the data set was reduced by variable selection to seven significant inputs and the models were generated by 20 independent GP runs. As a result of the model selection, a list of 12 models on the Pareto front was proposed for further evaluation by process engineers. All twelve models have high performance (R^2 of 0.97–0.98) and low complexity. After evaluating their extrapolation capabilities with "What-If" scenarios, a diversity of model inputs, and by physical considerations, an ensemble of four models was selected for on-line implementation. Two of the models are shown below:

$$GP\_Model = A + B \left(\frac{Tray64\_T^4 Vapor^3}{Rflx\_flow^2} \right)$$

$$GP\_Model2 = C + D \left(\frac{Feed^3 \sqrt{Tray46\_T - Tray56\_T}}{Vapor^2 Rflx\_flow^4} \right)$$

[15]L. Jason, G. Hornby, and L. Derek, Evolutionary antenna design for a NASA spacecraft, In *Genetic Programming Theory and Practice II,* In U.-M. O'Reilly, T. Yu, R. Riolo and B. Worzel (Eds), Springer, pp. 301–315, 2004.

[16]E. Jordaan, A. Kordon, G. Smits and L. Chiang, Robust inferential sensors based on ensemble of predictors generated by genetic programming, *In Proceedings of PPSN 2004*, pp. 522–531, Springer, 2004.

where $A, B, C,$ and D are fitting parameters, and all model inputs in the equations are continuous process measurements.

The models are simple and interpretable by process engineers. The difference in model inputs increases the robustness of the estimation scheme in case of possible input sensor failure. The inferential sensor has been in operation in Dow Chemical since May 2004.

- *New Product Development* – GP generated models can reduce the cost of new product development by shortening fundamental model development time. For example, symbolic regression proto-models can significantly reduce the hypothesis search space for potential physical/chemical mechanisms. As a result, new product development effort could be considerably reduced by eliminating unimportant variables, enabling rapid testing of new physical mechanisms, and reducing the number of experiments for model validation. The large potential of this type of application was demonstrated in a case study for structure-property relationships, which is discussed in detail in Chap. 14. The generated symbolic solution was similar to the fundamental model and was delivered with significantly less human effort (10 hours vs. 3 months).

- *Scheduling* – From the many applications in this area the most impressive is optimal production and scheduling of liquid industrial gases at Air Liquide.[17] A GA was used to schedule production at 40 plants producing liquid gas and to interact with an ant colony optimizer (described in the next chapter). A top-level optimizer asks the GA and the ant colony optimizer to generate production and distribution schedules. It then evaluates the combination of both schedules and feedback is sent based on their joint result. The ant colony optimizer and the GA adapt in conjunction with each other to derive integrated schedules, even though neither system is explicitly aware of the operations of the other.

- *Finance* – Evolutionary computation, and especially GA and symbolic regression are widely used in various financial activities. For example, State Street Global Advisors is using GP for derivation of a stock selection models for low active risk investment style. Historical simulation results indicate that portfolios based on GP models outperform the benchmark and portfolios based on traditional models. In addition, GP models are more robust in accommodating various market regimes and have more consistent performance than the traditional models.[18]

- *Games* – One of the best examples of using evolutionary computation in games is the famous game[19] Blondie24. It is based on evolving neural networks. The game begins with 15 parents with neural network weights randomly initialized.

---

[17]C. Harper and L. Davis, *Evolutionary Computation at American Air Liquide*, SIGEVO newsletter, *1*, 1, 2006.

[18]Y. Becker, P. Fei, A. Lester, Stock selection: An innovative application of genetic programming methodology, In R. Riolo, T. Soule, B. Worzel (eds), *Genetic Programming Theory and Practice IV*, pp. 315–335, Springer, 2007.

[19]www.digenetics.com

During the simulated evolution, each parent generates one offspring and all 30 players compete with five randomly selected players from the population. 15 players with the greatest total points are selected as parents for the next generation, and so on for 100 generations. Recently, a new type of neural network, called an object neural network, is used in simulated evolution for chess.

5.6 Evolutionary Computation Marketing

The generic concept of evolutionary computation is easy to explain. We suggest defining different marketing strategies for the specific evolutionary computation techniques. An example of a marketing slide for using GP for robust empirical modeling is given in Fig. 5.21.

The key slogan of symbolic regression: "Transfer data into profitable equations" captures the essence of the value creation basis of the approach. The left section of the marketing slide represents a very simplified view of the nature of genetic programming, the method that generates symbolic regression, which is a simulated evolution of competing mathematical functions. The presentation also emphasizes that the selected models have optimal accuracy and complexity, which makes them robust in changing operating conditions.

The middle section of the marketing slide represents the key advantages of GP-generated symbolic regression, such as automatically derived robust simple models

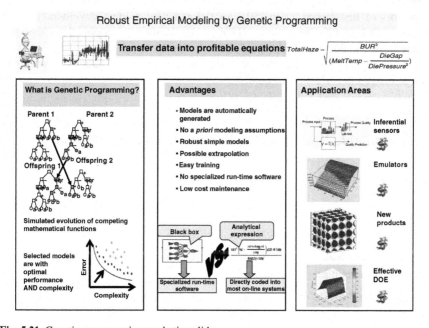

Fig. 5.21 Genetic programming marketing slide

Evolutionary Computation Elevator Pitch
Evolutionary computation automatically generates novelty from available data and knowledge. Solutions to many practical problems can be found by simulating natural evolution in a computer. However, instead of biological species, artificial entities like mathematical expressions, electronic circuits, or antenna schemes are created. The novelty is generated by the competition for high fitness of the artificial entities during the simulated evolution. The value creation is by transferring data into newly discovered empirical models, structures, schemes, designs, you name it. Evolutionary computation is especially effective when developing mathematical models is very expensive and there is some historical data and knowledge on a specific problem. Development and implementation cost of evolutionary computation is relatively low. There are numerous applications in optimal design, inferential sensors, new product invention, finance, scheduling, games, etc. by companies like Dow Chemical, GE, Boeing, Air Liquide, SSgA, Rolls Royce, etc.

Fig. 5.22 Evolutionary computation elevator pitch

with potential for extrapolation from no *a priori* assumptions, easy training, no need for specialized run-time software, and low maintenance cost. A direct comparison with neural network-based models is shown with emphasis on the model representation (black-box vs. mathematical expression), the need of run-time software, and the level of extrapolation. These clear advantages of symbolic regression are visually demonstrated. The key application areas of GP are shown in the right section of the marketing slide on Fig. 5.21. The slide includes the most valuable symbolic regression application areas in Dow Chemical, such as inferential sensors, emulators of first-principles models, accelerated new product development, and effective Design Of Experiments (DOE). The number of applications and the created value is given in the internal slide.

The proposed generic elevator pitch for communicating evolutionary computation to managers is shown in Fig. 5.22.

5.7   Available Resources for Evolutionary Computation

5.7.1   Key Websites

Illinois Genetic Algorithms Laboratory:
http://www.illigal.uiuc.edu/web/

John Kozas's website:
http://www.genetic-programming.org/

Special Interest Group on Evolutionary Computation (SIGEVO) within the Association for Computer Machinery (ACM):
http://www.sigevo.org/

5.7.2 Key Software

Genetic Algorithm and Direct Search Toolbox for MATLAB:
http://www.mathworks.com/products/gads/index.html

GP professional package Discipulus by RML Technologies:
http://www.rmltech.com/

Single and Multiobjective Genetic Algorithm Toolbox for MATLAB in C++, developed at the University of Illinois (free for non-commercial use):
http://www.illigal.uiuc.edu/web/source-code/2007/06/05/single-and-multiobjective-genetic-algorithm-toolbox-for-matlab-in-c/

Java-based TinyGP, developed at the University of Essex (free for non-commercial use):
http://cswww.essex.ac.uk/staff/rpoli/TinyGP/

GP Studio by BridgerTech (free for non-commercial use):
http://bridgertech.com/gp_studio.htm

DataModeler by Evolved Analytics (a Mathematica-based package with the most advanced features for development of symbolic regression models)
http://www.evolved-analytics.com

5.8 Summary

Key *messages*:

Evolutionary computation mimics natural evolution in the virtual world of computers.

Evolutionary algorithms handle a population of artificial species which reproduce with inheritance and are a subject of variation and selection.

Genetic algorithms can find optimal solutions based on intensive exploration and exploitation of a complex search space by a population of potential solutions.

Genetic programming can generate novel structures which fit a defined objective.

Evolutionary computation has demonstrated industrial success in optimal design, inferential sensors, scheduling, and new product discovery.

The Bottom Line

Applied evolutionary computation systems have the capability to create value through automatically generating novelty from simulated evolution.

Suggested Reading

W. Banzhaf, P. Nordin, R. Keller, F. Francone, *Genetic Programming*, Morgan Kaufmann, 1998.

L. Davis, *Handbook of Genetic Algorithm*, Van Nostrand Reinhold, New York, 1991.

A. Eiben and J. Smith, *Introduction to Evolutionary Computing*, Springer, 2003.

D. Fogel, *Evolutionary Computation: Toward a New Philosophy of Machine Intelligence*, 3rd edition, 2005.

D. Goldberg, *Genetic Algorithm in Search, Optimization, and Machine Learning*, Addison-Wesley, 1989.

J. Koza, *Genetic Programming: On the Programming of Computers by Natural Selection*, MIT Press, 1992.

M. Negnevitsky, *Artificial Intelligence: A Guide to Intelligent Systems*, Addison-Wesley, 2002.

I. Parmee, *Evolutionary and Adaptive Computing in Engineering Design*, Springer, 2001.

R. Poli, W. Langdon, and N. McPhee, *A Field Guide to Genetic Programming*, free electronic download from http://www.lulu.com, 2008.

Chapter 6
Swarm Intelligence: The Benefits of Swarms

> *Dumb parts, properly connected into a swarm, yield smart results.*
>
> Kevin Kelly

It is a well-known fact that an individual ant is not very bright and almost blind, but ants in a colony, operating as a team, do remarkable things like effective foraging, optimal brood sorting, or impressive cemetery formation. Many other biological species, such as insects, fish, birds, can present similar intelligent collective behavior although they are composed of simple individuals. At the basis of the increased "intelligence" is the shared "information" discovered individually and communicated to the swarm by different mechanisms of social interaction. In this way, intelligent solutions to problems naturally emerge from the self-organization and communication of these simple individuals. It is really amazing that the seamless coordination of all individual activities does not seem to require *ANY SUPERVISOR!*[1]

Swarm intelligence is the emergent collective intelligence of groups of simple individuals, called agents. The individual agents do not know they are solving a problem, but the "invisible hand" of their collective interaction leads to the problem solution. The biological advantages of swarm intelligence for survival of the species in their natural evolution are obvious. Recently, some of the energy savings as a result of the collective behavior of biological swarms have been quantified by proper measurements. For example, a study of great white pelicans has found that birds flying in formation use up to a fifth less energy than those flying solo.[2]

The objective of this chapter is to identify the key benefits of using artificial swarms. The value creation capabilities of swarm intelligence are based on exploring the emerging phenomena driven by social interaction among the individual

---

[1]Even the famous queen ant has reproductive rather than power-related functions.

[2]H. Weimerskirch, *et al.*, Energy saving in flight formation, *Nature*, 413, (18 October 2001), pp. 697–698, 2001.

A.K. Kordon, *Applying Computational Intelligence*,
DOI 10.1007/978-3-540-69913-2_6, © Springer-Verlag Berlin Heidelberg 2010

agents. These emerging phenomena can derive unique routes, schedules, and optimal trajectories, applicable in areas like supply chains, vehicle routing, and process optimization.

Meanwhile, the "dark side" of swarm intelligence is currently a hot topic in science fiction. The famous Michael Crichton novel *Prey* about a swarm of microscopic machines (self-replicating nanoparticles) destroying humans has captured the attention of millions of readers.[3] Unfortunately, the popular negative artistic image of swarm intelligence as a threat to humanity can raise concerns and alienate potential users. One of the objectives of this chapter is to describe the nature of smarm intelligence and to demonstrate the groundlessness of the fears about this emerging technology as the next scientific Frankenstein.

6.1   Swarm Intelligence in a Nutshell

Swarm intelligence is a computational intelligence technique based around the study of collective behavior in decentralized, self-organized systems. The expression "swarm intelligence" was introduced by Beni and Wang in the late 1980s in the context of cellular robotic systems, where many simple robots are self-organized through nearest-neighbor interactions.[4] The research field has grown tremendously since 2000, especially after publishing of the key books, related to the two main development areas, Ant Colony Optimization[5] (ACO) and Particle Swarm Optimization[6] (PSO).

Swarm intelligence systems are typically made up of a population of simple agents interacting locally with one another and with their environment. This interaction often leads to the emergence of global behavior, which is not coded in the actions of the simple agents. Analyzing the mechanisms of collective intelligence that drives the appearance of new complexity out of interactive simplicity requires knowledge of several research areas like biology, physics, computer science, and mathematics. From the point of view of understanding the big implementation potential of swarm intelligence, we recommend the following key topics, shown in the mind-map in Fig. 6.1.

---

[3]M Crichton, *Prey*, HarperCollins, 2002.

[4]G. Beni and J. Wang, Swarm Intelligence, In *Proceedings 7th Annual Meeting of the Robotic Society of Japan*, pp. 425–428, RSJ Press, Tokyo, 1989.

[5]E. Bonabeau, M. Dorigo, and G. Theraulaz, *Swarm Intelligence: From Natural Evolution to Artificial Systems*, Oxford University Press, 1999.

[6]J. Kennedy and R. Eberhart, *Swarm Intelligence*, Morgan Kaufmann, 2001.

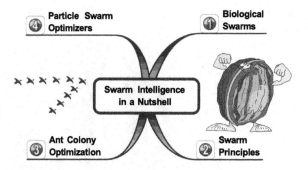

Fig. 6.1 Key topics related to swarm intelligence

6.1.1 Biological Swarms

Swarming as a type of collective interaction is a popular behavior in many biological species. The list includes, but is not limited to: ants, bees, termites, wasps, fish, sheep, and birds. Of special interest are the enormous swarming capabilities of insects. It is a well-known fact that 2% of insects are social. Some swarming insects, like bees, directly create value (or to be more specific, honey) and have been used by humans since ancient time.

The high productivity of bees is based on a unique blend between colony cooperation and effective specialization by division of labor. As a result, food sources are exploited according to quality and distance from the hive, not to mention that the regulation of hive temperature can compete with the most sophisticated digital controllers . . .

The other insects similar to bees – wasps – demonstrate amazing capabilities for "intelligent design" of complex nests. The structure consists of horizontal columns, a protective covering, and a central entrance hole and is built by a flying escadrille of pulp foragers, water foragers and builders.

The building champions among insect swarms, however, are the termites. The building process has two major phases:

- Random walk (uncoordinated phase);
- Coordinated phase.

As a result of this effective self-organization, unique structures, like the termite "cathedral" mound, shown in Fig. 6.2, are built with tremendous speed. The interior "design" is also spectacular with cone-shaped outer walls and ventilation ducts, brood chambers in the central hive, spiral cooling vents, and support pillars.

The famous social insects are ants, which represent about 50% of these biological species. We'll focus on their behavior in more detail in Sect. 6.1.3 but here are some highlights. First, we share a little known fact that the total weight of all ants added together is equal to the total weight of humans (the average weight of an ant is between 1 and 5 mg). However, ants began their evolutionary battle for survival 100 million years ago, much earlier than our ancestors.

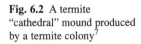

Fig. 6.2 A termite "cathedral" mound produced by a termite colony[7]

The ants' efficiency through social interaction continues to surprise researchers.[8] Examples of such are: capabilities like organizing "highways" to and from their foraging sites by leaving pheromone[9] trails, forming chains from their own bodies to create "bridges" to pull and hold leaves together with silk, and the almost perfect division of labor between major and minor ants. Some ant colonies have networks of nests several hundreds of meters in span. The most advanced army allocation pattern, shown in Fig. 6.3, belongs to the tropical ant *Eciton burchelli*. It includes as many as 200,000 blind workers and its structure consists of a 15 m-wide swarm front, a dense ant phalange 1 meter behind, and a complex trail that converges to a single straight line "highway" to the bivouac. The Art of War of this army of ants doesn't need generals.

Fish schooling is another known form of swarm intelligence (see Fig. 6.4). Schools are composed of many fish of the same species moving in more or less harmonious patterns throughout the water. A very prevalent behavior, schooling is exhibited by almost 80% of the more than 20,000 known fish species during some phase of their life cycle.

Why do fish school? One of the key reasons is that some species of fish secrete a slime that helps to reduce the friction of water over their bodies. In addition, the fish swim in precise, staggered patterns when traveling in schools and the motion

---

[7]http://www.scholarpedia.org/article/Swarm_intelligence

[8]A very interesting book on this topic is D. Gordon, *Ants at Work: How an Insect Society is Organized*, W. Norton, NY, 1999.

[9]A pheromone is a chemical used by animals to communicate.

Fig. 6.3 Forging patterns of an army of Eciton burchelli[10]

Fig. 6.4 Fish school

of their tails produces tiny currents called vortices.[11] Each individual can use the tiny whirlpool of its neighbor to assist in reducing the water's friction on its own body.

---

[10]http://www.projects.ex.ac.uk/bugclub/raiders.html

[11]Swirling motions similar to little whirlpools.

Another reason for schooling is the increased safety against predators. A potential predator looking for a meal might become confused by the closely spaced school, which can give the impression of one vast fish.

We'll finish this section by answering the generic question: Why do animals swarm? The four key reasons, according to biologists, are: (1) defense against predators by enhanced predator detection and minimizing the chances of being captured; (2) improved foraging success rate; (3) better chances to find a mate; and (4) decrease of energy consumption.

6.1.2   Principles of Biological Swarms

The next topic of interest is defining the generic principles behind the different forms of swarm behavior. It will lead to the design of artificial swarms. Firstly, let's define the key characteristics of a swarm as:

- Distributed: no central data source;
- No explicit model of the environment;
- Perception of the environment (sensing capability);
- Ability to change the environment.

Secondly, we'll focus on the key issue of self-organization of biological swarms. It is the complexity and sophistication of self-organization that allows functioning with no clear leader. The essence of self-organization is the appearance of a structure without explicit external pressure or involvement.

The obvious result from self-organization is the creation of different structures, such as social organization based on division of labor, foraging trails, and all of these remarkable designs of nests.

Thirdly, the mechanism of the unique indirect way of communication of biological swarms, called stigmergy, will be discussed. Stigmergy is defined as the indirect interaction of two individuals when one of them modifies the environment and the other responds to the new environment at a later time. Since no direct communication takes place between individuals, information is communicated through the state or changes in the local environment. In some sense, environmental modification serves as external memory and the work can be continued by any other individual. Stigmergy is the basis of coordination by indirect interaction, which in many cases for biological swarms is more appealing than direct communication.

The final topic in this section is the basic principles of swarm intelligence, as defined by Mark Millonas from Santa Fe Institute:[12]

---

[12]M. Millonas. Swarms, phase transitions, and collective intelligence. In C.G. Langton (Ed.), *Artificial Life III*, pp. 417-445, Santa Fe Institute Studies in the Sciences of the Complexity, Vol. XVII, Addison-Wesley, 1994.

- *Proximity Principle*: individuals should be able to interact so as to form social links.
- *Quality Principle*: individuals should be able to evaluate their interactions with the environment and one another.
- *Diverse Response Principle*: the population should not commit its activities along excessively narrow channels.
- *Stability Principle*: the population should not change its mode of behavior every time the environment changes.
- *Adaptability Principle*: the population must be able to change behavior mode when necessary.

From that perspective, a swarm system is composed of a set of individuals which interact with one another and the environment. Swarm intelligence is defined as an emerging property of the swarm system as a result of its principles of proximity, quality, diversity, stability, and adaptability.[13]

There are two key directions in research and applied swarm intelligence: (1) Ant Colony Optimization (ACO), based on insect swarm intelligence; and (2) Particle Swarm Optimizers (PSO), based on social interaction in bird flocking. Both approaches will be discussed in the next two sections.

6.1.3 Ant Colony Optimization

Individual ants are simple insects with limited memory and capable of performing simple actions. However, an ant colony generates a complex collective behavior providing intelligent solutions to problems such as: carrying large items, forming bridges, finding the shortest routes from the nest to a food source, prioritizing food sources based on their distance and ease of access, sorting corpses. Moreover, in a colony each ant has its prescribed task, but the ants can switch tasks if the collective needs it. For example, if part of the nest is damaged, more ants do nest maintenance work to repair it

One of the fundamental questions is: How do ants know which task to perform? When ants meet, they touch with their antennas. It is a well-known fact that these are organs of chemical perception and the ant can perceive the colony-specific odor from all members of the nest. In addition to this odor, ants have an odor specific to their task, because of the temperature and humidity conditions in which it works, so that an ant can evaluate its rate of encounter with ants of a certain task. In addition, the pattern of ant influences the probability of performing a specific task.

How can ants manage to find the shortest path? The answer from biology is simple – by applying the stigmetry mechanism of indirect communication based on pheromone deposition over the path they follow. The scenario is as follows:

---

[13]L. de Castro, *Fundamentals of Natural Computing*, Chapman & Hall, 2006.

– An isolated ant moves at random, but when it finds a pheromone trail, there is a high probability that this ant will decide to follow the trail.

– An ant foraging for food deposits pheromone over its route. When it finds a food source, it returns to the nest reinforcing its trail.

– Other ants have greater probability to start following this trail and laying more pheromone on it.

– This process works as a positive feedback loop system because the higher the intensity of the pheromone over a trail, the higher the probability of an ant to start traveling through it.

The short-path ant algorithm is demonstrated in Fig. 6.5 in the case of two competing routes, one of which is significantly shorter. Let's assume that in the initial search phase, an equal number of ants moves to both routes. However, the ants on the short path will complete the travel more times and thereby lay more pheromone over it. The pheromone concentration on the short trail will increase at a higher rate than on the long trail, and in the advanced search phase the ants on the long route will choose to follow the short route (the amount of pheromone is proportional to the thickness of the routes in Fig. 6.5). Since most ants will no longer travel on the long route, and since the pheromone is volatile, the long trail will start evaporating. In the final search phase only the shortest route will remain.

Surprisingly, this simple algorithm is at the basis of a method for finding optimal solutions in real problems, called Ant Colony Optimization (ACO). Each artificial

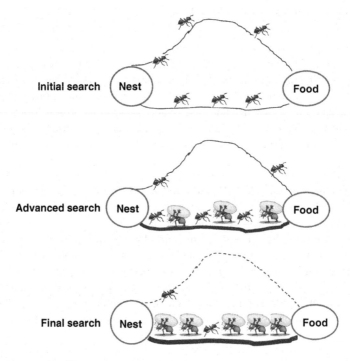

Fig. 6.5 Ant route handling

ant is a probabilistic mechanism that constructs a solution to the problem, using artificial pheromone deposition, heuristic information about pheromone trails and a memory for already visited places.

In Ant Colony Optimization, a colony of artificial ants gradually constructs solutions for a defined problem, using artificial pheromone trails, which are modified accordingly during the algorithm. In the solution construction phase, each ant builds a problem-specific solution – for example, selection of the next route for the supply chain.

The choice of solution fragment by an artificial ant at each step of the construction stage is proportional to the amount of artificial pheromone deposited on each of the possible solutions. In the next step of the algorithms, after all ants have found a solution, the pheromone deposits on each solution fragment are updated. The high-quality solution fragments are supported by stronger pheromone reinforcement. After several iterations, better solution fragments are more frequently used by the artificial ant colony and the opposite, less successful, solutions gradually disappear. The pheromone trails also evaporate during the update in order to forget the least used solutions.

6.1.4 Particle Swarm Optimizers

In contrast to the insect-driven ant colony optimization, the second key direction in swarm intelligence, invented by Jim Kennedy and Russ Eberhart in the mid-1990s,[14] is inspired mostly by the social behavior of bird flocking and fish schooling.

One of the key questions in analyzing flock behavior is: How does a large number of birds produce a seamless, graceful flocking dance, while often, but suddenly changing direction, scattering and regrouping? The impression is that even though the individual birds change the shape and the direction of their motion, they appear to move as a single coherent organism. The analyses of flock behavior of various types of birds have led to defining the main flock principles, as shown in the mind-map on Fig. 6.6 and summarized below.

Flock principles:[15]

1. *Velocity Matching*: attempt to match velocity with nearby flock mates.
2. *Flock Centering*: attempt to stay close to nearby flock mates.
3. *Collision Avoidance*: avoid colliding with nearby flock mates.
4. *Locality*: its nearest flock mates only influence the motion of each bird, i.e., vision is the most important sense for flock organization.

---

[14]The original paper is: J. Kennedy and R. Eberhart, Particle swarm optimization, *Proc. of the IEEE Int. Conf. on Neural Networks*, Perth, Australia, pp. 1942–1948, 1995.

[15]S. Das, A. Abraham, and A. Konar, Swarm intelligence algorithms in bioinformatics, In *Computational Intelligence in Bioinformatics*, A. Kelemen, *et al.* (Eds), Springer, 2007.

Fig. 6.6 Key flock principles

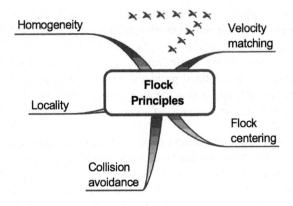

5. *Homogeneity*: each bird in the flock has the same behavior. The flock moves without a leader, even in cases when temporary leaders appear.

The defined flock principles are at the core of the Particle Swarm Optimization (PSO) algorithm. The analogy is very direct: in PSO, each single solution is like a 'bird' in the search space, which is called a "particle". A selected number of solutions (particles) form a flock (swarm) which flies in a D-dimensional search space trying to uncover better solutions. For the user the situation recalls the simulation of bird flocking in a two-dimensional plane. Each particle is represented by its position on the XY plane as well as by its velocity (V_x as the velocity component on the X-axis and V_y as the velocity component on the Y axis). All particles in the swarm have fitness values which are evaluated by a defined fitness function to be optimized, and have velocities which direct the flying of the particles. (The particles fly through the problem space by following the particles with the best solutions so far.)

Each particle also has a memory of the best location in the search space that it has found (*pbest*) and knows through social interaction the best location found to date by all the particles in the flock (*gbest* or *lbest*). The way the best location found is obtained depends on the swarm topology. There are different neighborhood topologies used to identify which particles from the swarm can influence the individuals. The most common ones are known as the *gbest* or fully connected topology and *lbest* or ring topology and are illustrated in Fig. 6.7.

In the *gbest* swarm topology, shown in Fig. 6.7a, the trajectory of each particle is influenced by the best individual found in the entire swarm (shown as a big bird). It is assumed that *gbest* swarms converge fast, as all the particles are attracted simultaneously to the best part of the search space. However, if the global optimum is not close to the best particle, it may be impossible for the swarm to explore other areas and, consequently, the swarm can be trapped in local optima.

In the *lbest* swarm topology, shown in Fig. 6.7b, each individual is influenced by a smaller number of its neighbors (which are seen as adjacent members of the swarm ring). Typically, *lbest* neighborhoods comprise two neighbors: one on the

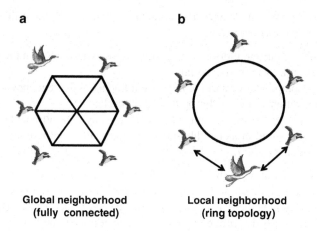

a   **b**

Global neighborhood   **Local neighborhood**
(fully connected)   **(ring topology)**

Fig. 6.7 Graphical representation of *gbest* (*a*) and *lbest* (*b*) swarm topologies

right side and one on the left side (a ring lattice). This type of swarm will converge slower but can locate the global optimum with a greater chance.

PSO is initialized with a group of random particles (solutions) and then searches for optima by updating each generation. At each generation, each particle is updated by the following two "best" values – of the particle itself and of the neighborhood. The first one is the best previous location (the position giving the best fitness value) a particle has achieved so far. This value is called *pbest*. At each iteration the *P* vector of the particle with the best fitness in the neighborhood, designated *lbest* or *gbest*, and the *P* vector of the current particle, are combined to adjust the velocity along each dimension, and that velocity is then used to compute a new position for the particle. The two equations, for particle velocity and position update, that drive PSO are given below:

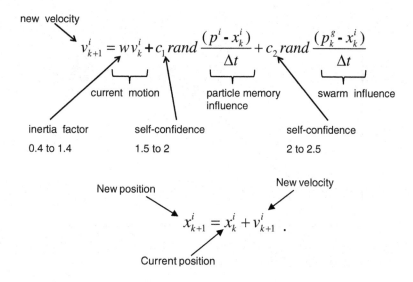

new velocity

$$v_{k+1}^i = w v_k^i + c_1 \, rand \, \frac{(p^i - x_k^i)}{\Delta t} + c_2 \, rand \, \frac{(p_k^g - x_k^i)}{\Delta t}$$

current motion   particle memory   swarm influence
                                        influence

inertia factor   self-confidence   self-confidence

0.4 to 1.4   1.5 to 2   2 to 2.5

New position   New velocity

$$x_{k+1}^i = x_k^i + v_{k+1}^i \ .$$

Current position

New velocity (which denotes the amount of change) of the i-th particle is determined by three components:

(1) momentum or current motion – the current velocity term to push the particle in the direction it has traveled so far;
(2) cognitive component or particle memory influence – the tendency to return to the best position visited so far by the i-th particle;
(3) social component or swarm influence – the tendency to be attracted towards the best position found in its neighborhood either by the ring topology *lbest* or by the star topology *gbest*.

A visual interpretation of the PSO algorithm is given in Fig. 6.8.

Let's assume that a particle (visualized by a bird) has a current position $X(k)$ at time k and its current velocity is represented by the vector $V(k)$. The next position of the bird at time $k+1$ is determined by its current position $X(k)$ and the next velocity $V(k+1)$. The next velocity, according to the new velocity PSO equation, is a vector blending of the current velocity $V(k)$ with the acceleration component towards the swarm-best (represented by the big bird) with its velocity of V_{gbest} and the other acceleration component towards the particle best with its velocity of V_{pbest}. As a result of these velocity adjustments, the next position $X(k+1)$ is closer to the global optimum.

The other parameters in the PSO algorithm have the following interpretation. The inertia or momentum factor w controls the impact of the particle's previous velocity on its current velocity and plays a balancing role between encouraging a more intensive local search of already discovered perspective regions (low w values) and exploring new diverse areas (high w values).

The purpose of the two random numbers *rand* in the PSO new velocity equation is to ensure that the algorithm is stochastic and neither the cognitive nor the social components are dominant. The direct control of the influence of the social and

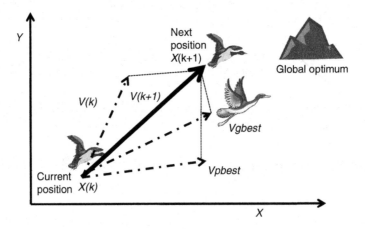

Fig. 6.8 Visualization of particle position update diagram

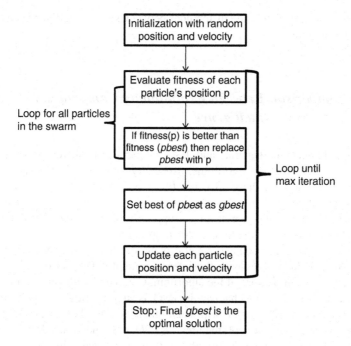

Fig. 6.9 A flowchart of the PSO algorithm

cognitive components is done by the c_1 and c_2 weight coefficients (called also self-confidence). Low values of these coefficients allow each particle to explore far away from already uncovered high-fitness points, high values of these parameters push towards more intensive search of high fitness regions.

Another important feature of the PSO algorithm is that particle velocities are clamped to the range $[-V_{max}, V_{max}]$ which serves as a constraint to control the global exploration ability of the particle swarm. Thus, the likelihood of particles leaving the search space is reduced. Note that this is not to restrict the values of Xi within the range $[-V_{max}, V_{max}]$; it only limits the maximum distance that a particle will move during one iteration.

The flow chart of the PSO algorithm, which is self-explanatory, is given in Fig. 6.9.

6.2 Benefits of Swarm Intelligence

The value creation capabilities of swarm intelligence based on social interactions may generate tremendous benefits, especially in the area of nontrivial optimization of complex problems. The specific advantages of using swarm intelligence by both ACO and PSO methods are discussed in this section. Firstly, we'll begin by clarifying some similarities and differences between swarm intelligence and evolutionary computation.

6.2.1 Comparison Between Evolutionary Computation and Swarm Intelligence

Swarm intelligence, in general, and PSO in particular, shares many common features with evolutionary algorithms, especially with genetic algorithms (GA). For example, both algorithms (GA and PSO) start with a group of a randomly generated population. Both have fitness values to evaluate the population. Both update the population and search for the optimum with random techniques. Both systems do not guarantee finding the global optimum.

However, PSO does not have genetic operators like crossover and mutation. Particles update themselves with the internal velocity. They also have memory, which is an important feature of the algorithm. Compared with genetic algorithms, the information sharing mechanism in PSO is significantly different. In GAs, chromosomes share information with each other. So the whole population moves like a one group towards an optimal area. In PSO, only the swarm leader (*gbest* or *lbest*) spreads the information to others. It is a one-way information sharing mechanism. Compared with GA, in most cases all the particles tend to converge to the best solution quickly even in the local version.

Other differences between swarm intelligence (PSO) and evolutionary algorithms can be defined as:

- A particle has a position (contents of a candidate solution) and a velocity whilst an individual in evolutionary algorithms typically has just the contents of a candidate solution.
- In evolutionary algorithms individuals compete for survival and most "die" while the population of the swarm is constant.
- The key difference is based on the driving forces of novelty generation. In the case of evolutionary algorithms the new solutions emerge from the strong struggle for high fitness. In the case of swarm intelligence, the novelty is generated by social interaction between the individuals.

6.2.2 Benefits of Swarm Intelligence Optimization

The key benefits of applying both ACO and PSO are captured in the mind-map, shown in Fig. 6.10 and discussed below.

- *Derivative-Free Optimization* – The search for an optimal solution in swarm intelligence is not based on functional derivatives but on different mechanisms

Fig. 6.10 Key benefits from
swarm intelligence

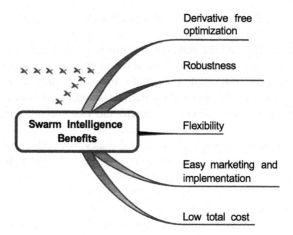

of social interaction between artificial individuals. In this way the chances of
being entrapped in local minima are significantly reduced (but not eliminated!).

- *Robustness* – The population-based ACO and PSO algorithms are more protective towards individual failure. The poor performance of even several members
of the swarm is not a danger for the overall performance. The collective behavior
compensates the laggards and the optimum solution is found independently of
the variations in individual performance.

- *Flexibility* – Probably the biggest benefit from swarm intelligence is its capability
to operate in a dynamic environment. The swarm can continuously track even for
fast-changing optima. In principle, there is no significant difference in functioning of the algorithm in steady-state or in dynamic mode. In the case of classical
methods, different algorithms and models are required for these two modes.

- *Easy Marketing and Implementation* – The principles of biology-inspired swarm
intelligence are easy to communicate to a broad audience of potential users and
there is no need for a heavy math or statistical background. In addition, the
implementation of both ACO and especially PSO on any software environment
is trivial. The tuning parameters are few and easy to understand and adjust. In
some cases, implementing PSO can even be transparent for the final user and be
a part of the optimization options of a larger project.

- *Low Total Cost* – In summary, the low marketing and implementation cost, as
well as potentially low maintenance cost due to the built-in adaptability in
changing operating conditions, result in low total-cost-of-ownership.

6.3 Swarm Intelligence Issues

PSO have two major algorithmic drawbacks. The first drawback is that PSO usually
suffers from premature convergence when problems with multiple optima are being
optimized. The original PSO is not a local optimizer and there is no guarantee that

the solution found is a local optimum. At the basis of this problem is that, for the *gbest* PSO, particles converge to a single point, which is on the line between the global best and the personal best positions.

The second drawback of PSO is that its performance is very sensitive to parameter settings. For example, increasing the value of the inertia weight, w, will increase the speed of the particles resulting in more exploration (global search) and less exploitation (local search) and vice versa. Tuning the proper inertia is not an easy task and is problem-dependent.

Beyond these specific technical issues, both ACO and PSO lack a solid mathematical foundation for analysis, especially for realistic algorithm convergence conditions and a generic methodology for parameter tuning. An addition, there are some questions about the "dark side" of swarm intelligence. On the technical front, questions are asked about some serious issues, such as the nature of predictability in distributed bottom-up approaches, the efficiency of the emergent behavior, and the dissipative nature of self-organization. On the social front, there is a growing resistance towards two potential application areas of swarm intelligence – military/law enforcement and medical. In the efforts of fighting the war on terror, the first application area has recently been explored very actively. As a result, designing and using flocks of flying and interacting smart micro-robots with miniature cameras for surveillance and in military actions is not science fiction anymore. Very soon it may change the nature of war and intelligence. However, the perspective of a new technological Big Brother as continuously tracking and spying smart swarms is chilling.

Even scarier looks the other big potential application area – using miniature nanoswarms for fighting diseases, especially cancer. The initial inspiring idea of designing a swarm of nanoparticles carrying specific medicine and moving it to a target area with cancer cells was diverted in the negative direction by the novel *Prey*. The potential for internal destruction of the human body by a nanoswarm killer, so vividly described in fiction, creates an attitude to prevent this happening in nonfiction.

6.4 How to Apply Swarm Intelligence

Due to the simple algorithms, implementing swarm intelligence on any software environment is trivial. However, applying ACO and PSO require different tuning parameters and problem formulation and will be discussed separately.

6.4.1 When Do We Need Swarm Intelligence?

The key capabilities of swarm intelligence that may create value are shown in the mind-map in Fig. 6.11. No doubt the most valuable feature of swarm intelligence is its potential in optimization. However, both swarm intelligence methods have

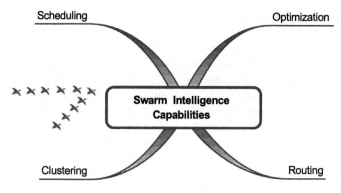

Fig. 6.11 Key capabilities of swarm intelligence

different optimization markets. ACO algorithms are in general more suitable to combinatorial optimization. The goal of combinatorial optimization is to find values for discrete variables (structures) that optimize the value of an objective function. Let's not forget that at the basis of ACO are artificial ants walking on an artificial graph. Graphs are a typical example of discrete structures.

On the other hand, PSO algorithms are in general more suitable to functional optimization where the goal is to find the optimal value of a certain function of real-valued variables. PSO shines especially in solving optimization problems which are unsuitable or infeasible for analytical or exact approaches. Examples are nasty functions with multiple optima, hard to model nonstationary environments, distributed systems with limited measurements, and problems with many variables and sources of uncertainty. Of special importance is the capability of PSO for dynamic optimization, an area where it has a competitive advantage versus the other methods, especially GA.

Many real-world applications of swarm intelligence are driven by the unique capability of ACO to use artificial ants for optimal routing. Most of the successful routing applications are based on different versions of the ant-foraging models and are specific for each implementation area, such as routing in telephone networks, data communication networks, and vehicles. However, beyond some level of complexity, there is a limitation since routing algorithms are generally difficult to analyze either mathematically or visually. Unfortunately, convergence to the optimal solution is not guaranteed, the speed of adaptation to fast changes could be unacceptable, and oscillatory behavior of the algorithm cannot be excluded.

Both ACO and PSO are capable of performing sophisticated clustering. An example is the PSO clustering, which overcomes some of the limitations of the popular clustering algorithms like K-means.[16] For example, K-means has no "global view" of the clustering solution: each centroid (cluster center) moves to the centre of its assigned examples, regardless of other centroids. In addition,

---

[16]K-means is an algorithm for data partitioning into K clusters so that the within-cluster distance is minimized.

different initial centroids (generated at random) can lead to different clustering results; so it is recommended to run K-means many times with a different set of initial conditions at each run and to check if (almost) all runs lead to similar results.

Fortunately, PSO handles these limitations and improves clustering efficiency significantly. PSO moves the centroids[17] according to its global search procedure, i.e. each particle has a "global view" of its entire clustering solution and the fitness function takes into account the positions of all centroids in the particle. The PSO population also contains a number of different sets of centroids. This is similar to multiple runs of the K-means algorithm but, instead of multiple independent runs, during the PSO search the particles "communicate" with each other, allowing them to share information about areas of the search space with high fitness.

Using artificial ants for scheduling and task allocation is another competitive capability of swarm intelligence. Examples are scheduling paint booths in a truck factory and optimal ordering of pickers at a large distribution center of a major retail chain.[18]

6.4.2   Applying Ant Colony Optimization

The generic application sequence for ACO is shown in Fig. 6.12. It begins with one of the most time-consuming steps of defining an appropriate representation of the problem. It includes specifying the components that an ant will use to incrementally construct a candidate solution and especially paying attention to enforcing the construction of valid solutions. For example, in the case of finding optimal routes, candidate solutions could be the distances to specific locations.

In the same way as data collection is critical for applying fuzzy, machine learning, and evolutionary computation systems, selecting a representative test case is decisive for ACO implementation. On the one hand, the selected case must drive the optimization algorithm development and adjustment. On the other hand, it has to cover the broadest possible conditions to validate the ACO performance.

The next application step is defining a problem-dependent heuristic function (η) that measures the quality of each component that can be added to a partial candidate solution. It has to specify how to update the amount of pheromone (τ) associated with each component in a path followed by an ant. Usually pheromone increases in proportion to the quality of the path (solution). As a result, a probabilistic transition rule based on the value of the heuristic function η and the current amount of pheromone τ associated with each candidate solution component is defined.

---

[17] A particle is defined as a set of centroids.

[18] E. Bonabeau and C. Meyer, Swarm intelligence: A whole new way to think about business, *Harvard Business Review*, May 2001.

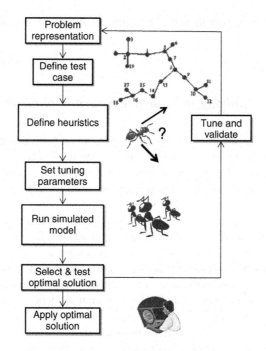

Fig. 6.12 Key steps in applying ant colony optimization

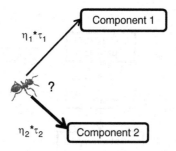

Fig. 6.13 Each ant selects the next component based on the product of pheromone τ and a problem-specific heuristic function η

This rule is used to decide which solution component is chosen to be added to the current partial solution. Typically, the probability of choosing a component i is proportional to the product $\eta_i \times \tau_i$. An illustration of such a rule is shown in Fig. 6.13.

The next step in the ACO application sequence is setting the tuning parameters. Some of these parameters, such as the number of ants, the stopping criteria, based either on accuracy or prescribed number of iterations, or the number of repetitive runs, are generic. The rest of the tuning parameters, such as pheromone ranges and rate of evaporation are algorithm-specific.

Usually selection of an ACO algorithm requires several simulation runs on the test case with refined adjustment of the tuning parameters until an acceptable performance is achieved. Then the developed algorithm can be used on other similar applications.

6.4.3 Applying the Particle Swarm Optimizer

The PSO application sequence is shown in Fig. 6.14. It begins with the definition of a PSO particle. In the same way as defining the GA chromosome within the context of the solved problem is critical for application success, implementing PSO successfully depends on mapping the problem solution into the PSO particle. For example, if the objective is to optimize some process variable, such as to maximize ethylene production in a cracking furnace, which depends on key variables like steam-to-naphtha ratio, the outlet furnace temperature, and the outlet furnace pressure, the PSO particle **p** is defined as a vector with three components (the three related process variables) and a value of the produced ethylene. It is very important to have estimates of the ranges of each component.

The second step of the PSO application sequence includes the preparation of a representative test case that will help to develop, tune, and validate the derived solution. Ideally, it is preferable to use data and knowledge for the full range of expected operation of the solution.

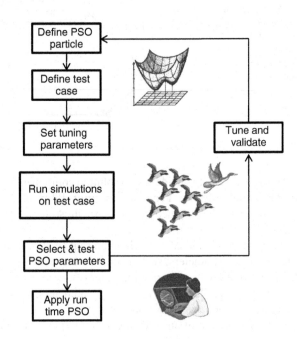

Fig. 6.14 Key steps in applying particle swarm optimization

A typical PSO setting includes the following parameters:

- Neighborhood structure: the global version is faster but might converge to a local optimum for some problems. The local version is a little bit slower but with lower probability to be trapped into a local optimum. One can use the global version to get a quick result and use the local version to refine the search.
- Number of particles in the population: the typical range is from 20 up to 50. Actually for most of the problems 10 particles are sufficient to get good results. For some difficult or special problems, one can try 100 or 200 particles as well.
- Dimension of particles: this is determined by the problem to be optimized.
- Range of particles: this is also determined by the problem to be optimized; you can specify different ranges for different dimensions of particles.
- V_{max}: this determines the maximum change one particle can take during one iteration. Usually the range of V_{max} is related to the upper limit of X_{max}.
- Learning factors: c_1 and c_2 are usually equal to 2. However, other settings were also used in different references. But usually c_1 equals c_2 and is in the range $[0, 4]$.
- Inertia weight w: usually is in the range 0.4 to 1.4.
- Maximum number of iterations: problem-dependent, the typical range is 200 up to 2000 iterations.

Due to the stochastic nature of PSO, it is recommended that the algorithm is run for at least 20 simulations for a given set of parameters. The final solution is selected after an exhaustive tuning and validation in all possible conditions. It could be applied as a run-time optimizer for similar problems.

6.4.4 Applying the Particle Swarm Optimizer: An Example

The PSO application sequence is illustrated with a simple example of optimizing a function with multiple optima:

$$y = \sin(n * x_1) + \cos(n * x_1 * x_2) + n * (x_1 + x_2)$$

where the number of optima n can be a setting parameter.

The particle is defined as the maximum value of y with dimensionality of two $[x_1, x_2]$. The ranges of the both dimensions are between -1 and $+1$. The PSO tuning parameters are as follows: population size $= 50$, $V_{max} = +1$, $c_1 = 3$, $c_2 = 1$, $w = 0.73$, maximum number of iterations $= 250$.

The objective of the study is to explore the PSO performance when the search space becomes complex due to the high number of multiple optima. Of special interest are the cases if PSO can distinguish between two geographically close optima with small differences relative to the global optimum. In order to validate the reproducibility of the results the PSO was run 20 times.

The PSO reliably identified the global optimum up to the case with 32 multiple optima. A 2D grid of the fitness landscape with 16 optima is shown in Fig. 6.15.

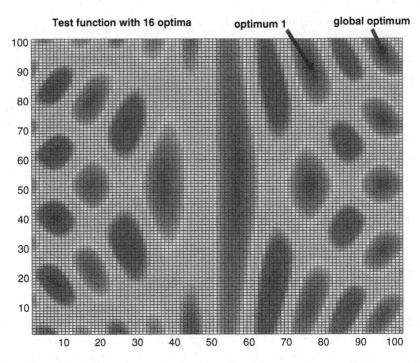

Fig. 6.15 A 2D grid of a function with 16 optima. The differences between optimum 1 and the global optimum (shown with arrows) is very small

Optimum 1 is geographically close to the global optimum and the difference in their fitness is small. A typical distribution of the global solutions *gbest*, generated during the PSO run of a function with 16 optima is shown in Fig. 6.16. The distribution reliably represents the fitness landscape, shown in Fig. 6.15, and identified the global optimum in 100% of the cases.

However, the results for optimization of a function with 32 optima are not so impressive. The 2D grid of the fitness landscape with 32 optima is shown in Fig. 6.17. In this case optimum 1 is geographically closer to the global optimum and the difference in their fitness is almost negligible. In 60% of the runs PSO cannot identify correctly the global optimum and converges to optimum 1 (see Fig. 6.18). Only after tuning the parameters by increasing the population size to 100 and the number of maximal iteration to 500, ca PSO identify the global optimum with 90% success rate (see Fig. 6.19).

6.5 Typical Swarm Intelligence Applications

The application record of swarm intelligence is not as impressive as that of the more-established computational intelligence methods. However, the speed of adoption in real-world applications, especially of PSO, is growing. The key selected

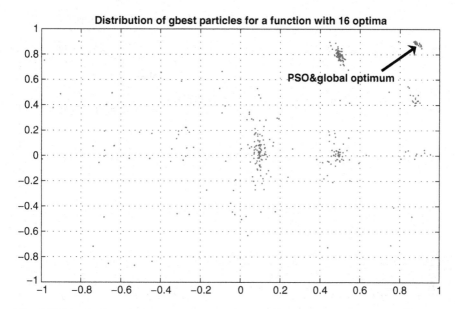

Fig. 6.16 Distribution of *gbest* particles after the final iteration in case of a function with 16 optima

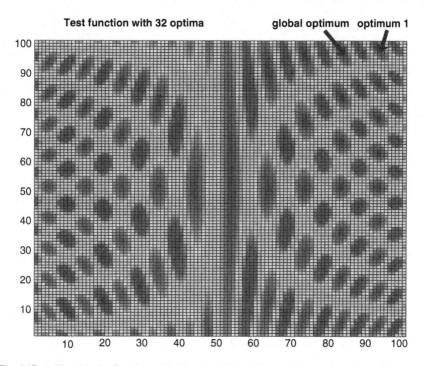

Fig. 6.17 A 2D grid of a function with 32 optima. The differences between optimum 1 and the global optimum (shown with arrows) is negligible

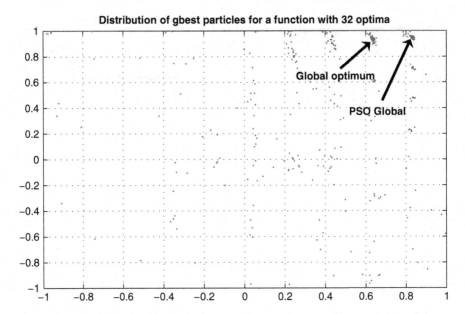

Fig. 6.18 Distribution of *gbest* particles after the final iteration in case of a function with 32 optima and different global optimum and PSO global solution

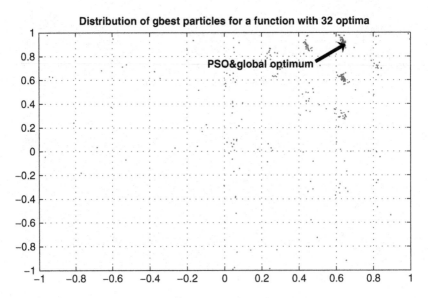

Fig. 6.19 Distribution of *gbest* particles after the final iteration in the case of a function with 32 optima and convergence of the PSO global solution to the global optimum

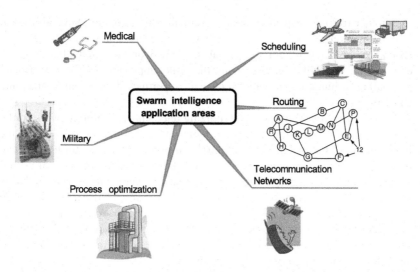

Fig. 6.20 Key swarm intelligence application areas

application areas of swarm intelligence are shown in the mind-map in Fig. 6.20 and discussed below.

- *Scheduling* – The unique features of ACO for task allocation and job scheduling have been successfully used in several industrial applications. In the previous chapter we already discussed the impressive performance of a joint ACO and GA algorithm for optimal scheduled deliveries of liquid gas to 8000 customers at American Air Liquide. The cost savings and operational efficiencies of this application for one of their 40 plants are more than $6 million dollars per year.

  Another application of swarm intelligence-based scheduling of truck painting at General Motors claims at least $3 million per annum. An interesting industrial scheduling problem in an Alcan aluminum casting center was successfully resolved by using ACO which generates 60 optimal schedules in less than 40 seconds.[19]

- *Routing* – One of the first successful ACO application was finding optimal routs for Southwest Airlines cargo operations. The derived solutions looked strange, since it was suggested to leave cargo on a plane headed initially in the wrong direction. However, implementing the algorithm resulted in significant cutback on cargo storage facilities and reduced wage costs. The estimated annual gain is more than $10 million.[20]

---

[19]M. Gravel, W. Price, and C. Cagne, Scheduling continuous casting of aluminum using a multiple objective ant colony optimization metaheuristic, *European Journal of Operating Research, 143*, pp. 218–229, 2002.

[20]E. Bonabeau and C. Meyer, Swarm Intelligence: A whole new way to think about business, *Harvard Business Review*, May 2001.

Another fruitful application area is optimal vehicle routing. Several impressive applications have been implemented by the Swiss company AntOptima.[21] Examples are: DyvOil, for the management and optimization of heating oil distribution; OptiMilk, for improving the milk supply process, and AntRoute, for routing of hundreds of vehicles of main supermarket chains in Switzerland and Italy.

- *Telecommunication Networks* – Optimizing telecommunication network traffic is a special case of routing with tremendous value creation potential due to the large volume. This is a difficult optimization problem because traffic load and network topology vary with time in unpredictable ways and the lack of central coordination. All of these features suggest that ACO could be a proper solution for this type of problem. A special routing algorithm, called AntNet, has been developed and tested on different networks under traffic patterns. It proved to be very robust and in most cases better than the competitive solutions.[22] ACO has been used by leading telecommunication companies like France Telecom, British Telecom, and the former MCI WorldCom.

- *Process Optimization* – Recently PSOs have been applied in several process optimization problems. Some applications in The Dow Chemical Company include using PSO for optimal color matching, foam acoustic optimal parameter estimation, crystallization kinetics optimal parameter estimation, and optimal neural network structure selection for day-ahead forecasting of electricity prices.[23] Examples of other interesting applications in this area include numerically controlled milling optimization, reactive power and voltage control, battery pack state-of-charge estimation, and cracking furnace optimization.

  A very broad potential application area is using PSO for optimizing data derived from statistical design of experiments. A PSO application for ingredient mix optimization in a major pharmaceutical corporation demonstrated that the fitness of the PSO-derived optimal solution is over twice the fitness found by the statistical design of experiments.[24]

- *Military* – The most well-known military application of swarm intelligence is developing a "swarm" of small unmanned aerial vehicles (UAV) with the capabilities to carry out key reconnaissance and other missions at low cost. For example, a swarm of surveillance UAVs could keep watch over a convoy, taking turns to land on one of the trucks for refueling. Working together as a team, they would ensure complete surveillance of the area around the convoy. Other applications include indoor surveillance. In recent tests up to five radio-controlled

---

[21] www.antoptima.com

[22] M. Dorigo, M. Birattari, and T. Stützle, Ant colony optimization: artificial ants as a computational intelligence technique, *IEEE Computational Intelligence Magazine*, *1*, pp. 28–39, 2006.

[23] A. Kalos, Automated neural network structure determination via discrete particle swarm optimization (for nonlinear time series models), *Proc. 5th WSEAS International Conference on Simulation, Modeling and Optimization*, Corfu, Greece, 2005.

[24] J. Kennedy and R. Eberhart, *Swarm Intelligence*, Morgan Kaufmann, 2001.

helicopters are being used to collaboratively track small ground vehicles and land on the back of small moving platforms.

A different approach is the "cooperative hunters" concept, where a swarm of UAVs are searching after one or more "smart targets", moving in a predefined area while trying to avoid detection. By arranging themselves into an efficient flight configuration, the UAVs optimize their combined sensing and are thus capable of searching larger territories than a group of uncooperative UAVs. Swarm control algorithms can optimize flying patterns over familiar terrain and introduce fault tolerance to improve coverage of unfamiliar and difficult terrain.[25]

- *Medical* – One of the first PSO medical applications is for successful classification of human tremors, related to Parkinson's disease (see pp. 382–389 in the reference in Footnote 24). A hybrid clustering approach based on self-organizing maps and PSO was applied in different cases for gene clustering of microarrays. Recently, the idea of using a swarm of nanoparticles to fight cancer cells has come close to reality. A research team led by scientists at The University of Texas M.D. Anderson Cancer Center and Rice University has shown in preclinical experiments that cancer cells treated with carbon nanotubes can be destroyed by noninvasive radio waves that heat up the nanotubes while sparing untreated tissue. The technique completely destroyed liver cancer tumors in rabbits without side effects.[26]

6.6   Swarm Intelligence Marketing

As in the case of evolutionary computation, the generic concept of swarm intelligence is easy to explain. However, one potential source of confusion could be the existence of two different approaches – ACO and PSO. It is recommended to emphasize and demonstrate the common basis of both methods – social interaction. An example of a swarm intelligence marketing slide, organized in this philosophy, is given in Fig. 6.21.

The motto of swarm intelligence: "Transfer social interactions into value" captures the essence of the value creation basis of the approach. The left section of the marketing slide represents the generic view of the approach and presents the two key methods (ACO and PSO), inspired by ants and bird flocks. It is focused on the three key phases of swarm intelligence: (1) analyses of social interactions in biology (represented by ants and birds); (2) derived algorithms for optimization; and (3) the results of finding the optimal solution.

---

[25] Smart Weapons for UAVs, Defense Update, January 2007.

[26] C. Gannon, *et al.*, Carbon nanotube-enhanced thermal destruction of cancer cells in a noninvasive radiofrequency field, *Cancer, 110*, pp. 2654–2665, 2007.

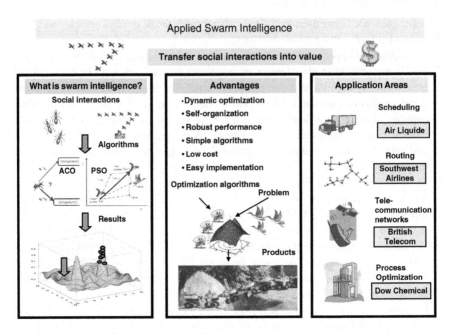

Fig. 6.21 Swarm intelligence marketing slide

Swarm Intelligence Elevator Pitch

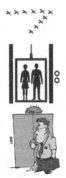

Surprisingly, we can improve the way we schedule, optimize, or classify by learning from social behavior of ants, termites, birds, and fish. Swarm intelligence captures the emerging wisdom from the social interactions of these biological species and translates it into powerful algorithms. For example, Ant Colony Optimization uses a colony of virtual ants to construct optimal solutions by digital pheromone deposition. Another powerful algorithm, Particle Swarm Optimization, mimics a flock of interacting particles to find optimal solutions in difficult conditions of a rugged fitness landscape with multiple optima. Swarm intelligence is especially effective when operating in a dynamic environment. The total-cost-of-ownership of swarm intelligence algorithms is relatively low due to their simplicity. It is a new technology but with growing interest in industry. Swarm intelligence has been successfully applied in scheduling, telecommunication and data network routing, and process optimization by companies like Air Liquide, Dow Chemical, Southwest Airlines, France Telecom, and British Telecom.

Fig. 6.22 Swarm intelligence elevator pitch

The middle section of the marketing slide represents the key advantages of swarm intelligence, such as derivative-free optimization, self-organization, robust performance, simple algorithms, low cost, and easy implementation. The visualization section represents the problem as a fitness landscape with optima, both optimization algorithms as ants and birds targeting the optimum, and a line of swarming mini-robots, as products.

The key application areas of swarm intelligence are shown in the right section of the marketing slide on Fig. 6.21. The slide includes the most valuable swarm intelligence application areas in scheduling, routing, telecommunication networks, and process optimization. Examples of leading companies in the specific application areas are given.

The proposed elevator pitch for inspiring managers about the great capabilities of swarm intelligence is shown in Fig. 6.22.

6.7 Available Resources for Swarm Intelligence

6.7.1 Key Websites

The key PSO site:
http://www.swarmintelligence.org/

Another PSO site with information and free code:
http://www.particleswarm.info/

The key ACO site:
http://www.aco-metaheuristic.org/

6.7.2 Selected Software

PSO Matlab toolbox (free for noncommercial use)
http://www.mathworks.com/matlabcentral/fileexchange/loadFile.do?objectId=7506

Several ACO packages, free for noncommercial use are available on the site:
http://www.aco-metaheuristic.org/aco-code/public-software.html

6.8 Summary

Key messages:

Swarm intelligence is coherence without choreography and is based on the emerging collective intelligence of simple artificial individuals.

Swarm intelligence is inspired by the social behavior of ants, bees, termites, wasps, birds, fish, sheep, even humans.

Ant Colony Optimization (ACO) uses a colony of artificial ants to construct optimal solutions for a defined problem by digital pheromone deposition and heuristics.

Particle Swarm Optimization (PSO) uses a flock of communicating artificial particles searching for optimal solutions of a defined problem.

Swarm intelligence-based systems have been successfully applied in scheduling, telecommunication and network routing, process optimization, and different military and medical applications.

The Bottom Line

Applied swarm intelligence has the capability to transfer the algorithms derived from social interaction of artificial individuals into value.

Suggested Reading

The following books give detailed technical descriptions of the different swarm intelligence techniques:
E. Bonabeau, M. Dorigo, and G. Theraulaz, *Swarm IntelligenceSwarm Intelligence: From Natural Evolution to Artificial Systems*, Oxford University Press, 1999.
L. de Castro, *Fundamentals of Natural Computing*, Chapman & Hall, 2006.
M. Dorigo and T. Stutzle, *Ant Colony OptimizationOptimization*, MIT Press, 2004.
A. Engelbrecht, *Fundamentals of Computational Swarm IntelligenceSwarm Intelligence*, Wiley, 2005.
J. Kennedy and R. Eberhart, *Swarm Intelligence*, Morgan Kaufmann, 2001.

Chapter 7
Intelligent Agents: The Computer Intelligence Agency (CIA)

We made our agents dumber and dumber and dumber... until finally they made money.

Oren Etzioni

One of the key challenges facing the global economy is the management of a transition from systems comprising many relatively isolated, small-scale elements to large-scale, physically distributed systems. Some obvious examples are: the Internet, dynamic inventory systems based on Radio-Frequency Identification (RFID) tagging, wireless sensors networks, and marketing new products into various countries with different cultures. Unfortunately, the traditional engineering approaches for decomposing a problem into modular, hierarchical, and ultimately centralized command-and-control regimes do not scale well in this new environment. A possible alternative is the new growing area of complex systems analysis and one of its leading methods – intelligent agents.

The need for complex systems in practical applications is illustrated by their unique place in the certainty-agreement diagram, shown in Fig. 7.1. The systems with high degrees of certainty and agreement occupy the upper-right corner of the diagram and represent most of the existing technical systems, which we can plan, monitor and control. On the opposite side are the chaotic systems with low certainty and agreement, which we try to avoid. The vast range between ordered and chaotic systems belongs to the zone of complexity. In this region, there is not enough stability to have repetition or prediction, but not enough instability to create anarchy or to disperse the system.

Most of the complex systems that include business components, such as supply-chain, enterprise planning, scheduling and control, have relatively high degrees of certainty and agreement and are a natural extension to the ordered technical systems. On the other extreme are the complex systems that include social components, such as financial markets, different social organizations, terrorist networks, etc. Their behavior has much lower degrees of certainty and agreement, and in some cases, like the looting in Baghdad in 2003 or the economic collapse of Albania in 1997, the social systems were on the edge of anarchy.

A.K. Kordon, *Applying Computational Intelligence,*
DOI 10.1007/978-3-540-69913-2_7, © Springer-Verlag Berlin Heidelberg 2010

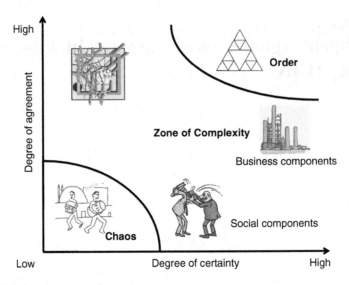

Fig. 7.1 Certainty-agreement diagram

Intelligent agents represent complex systems by the interactions between different simple entities with defined structure, called agents. Intelligent agents have the
following features: the relationships among elements are more important than the
elements themselves, complex outcomes can emerge from a few simple rules, small
changes in the behavior can have big effects on the system, and patterns of order can
emerge without control.

The objective of this chapter is to identify the potential for value creation by
using intelligent agents for complex systems analysis. Most of these capabilities are
related to handling very complex enterprises in real time and exploring the consequences of social responses to specific actions. It is our opinion that the value
creation potential of intelligent agents will grow with the increasing role of human
behavior-based business modeling.

7.1   Intelligent Agents in a Nutshell

Intelligent agents[1] is a computational intelligence technique of bottom-up modeling
that represents the behavior of a complex system by the interactions of its simple
components, defined as agents. The field is very broad and related to other areas,
like economics, sociology, object-oriented programming, to name a few. For many

---

[1]Another popular name of this technology is agent-based modeling.

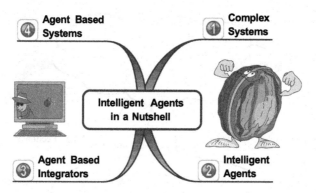

Fig. 7.2 Key topics related to intelligent agents

researchers, intelligent agents are the new face of the good old AI.[2] The notion of an intelligent agent is so different for the various communities that the only limitation we may propose is to exclude the professional CIA and KGB agents.

For the purposes of applying intelligent agents in real-world problems, we recommend focusing on the following topics, captured in the mind-map in Fig. 7.2 and discussed below.

7.1.1 Complex Systems

A complex system is a system with a large number of elements, building blocks, or agents, capable of exchanging stimuli with one another and with their environment.[3] The interaction between elements may occur with immediate neighbors or with distant ones; the agents can be all identical or different; they may move in space or occupy fixed positions, and can be in one state or multiple states. The common characteristic of all complex systems is that they display organization without any external coordination principle being applied. Adaptability and robustness are often a byproduct to such an extent that even when part of the system is altered or damaged, the system may still be able to function. The key feature of a complex system, however, is its emergent behavior.

Emergence is the arising of new, unexpected structures, patterns, or processes in a self-organized mode. These emergent phenomena can be understood by analyzing the higher-level components that appear from self-organization. Emergent phenomena seem to have a life of their own with their rules, laws, and behavior unlike the low-level components. A generic structure of a complex adaptive system with emerging macropatterns from agents' microinteractions is shown in Fig. 7.3. The adaptive feature is based on the feedback to the agents.

---

[2]S. Russell and P. Norvig, *Artificial Intelligence: A Modern Approach*, 2nd edition, Prentice Hall, 2002.

[3]J. Ottino, Complex systems, *AIChE Journal*, *49*, pp. 292–299, 2003.

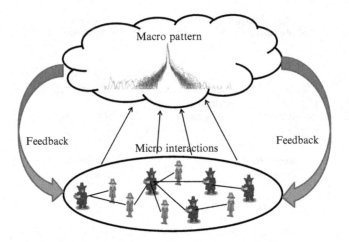

Fig. 7.3 A generic structure of a complex system

A typical example of emerging phenomena is materials synthesis by self-assembly where the desired structure of the material forms spontaneously, moving to a minimum free-energy state by building an ensemble of possible configurations of molecules. This new research area has already demonstrated applicability in bio-materials, porous materials for catalysis and separation, electronic, and photonic devices.

One of the leading gurus in complex systems, John Holland, has identified three key properties and four mechanisms of complex adaptive systems:[4]

1. *Nonlinearity Property*: Components or agents exchange resources or informa-tion in ways that are not simply additive. For example, a driver suddenly changes lanes.
2. *Diversity Property*: Agents or groups of agents differentiate from one another over time. For example, new kinds of cars or trucks appear over time on a highway.
3. *Aggregation Property*: A group of agents is treated as a single agent at a higher level. For example, trucks form a convoy.

The four key mechanisms of complex adaptive systems are as follows:

1. *Flows Mechanism*: resources or information are exchanged between agents such that the resources can be repeatedly forwarded from agent to agent. For example, speeders flash their lights to warn about police.
2. *Tagging Mechanism*: The presence of identifiable flags that let agents identify the traits of other agents. For example, a sports car suggests, but doesn't guarantee, aggressive driving.

---

[4]J. Holland, *Hidden Order,* Addison-Wesley, 1995.

3. *Internal Models Mechanism*: Formal, informal, or implicit representations of the world that are embedded within agents. For example, a driver's idea of traffic conditions.
4. *Building Blocks Mechanism*: Used when an agent participates in more than one kind of interaction where each interaction is a building block for larger activities. For example, a police car can pull over other cars and produce traffic jams, but can also get into accidents just like everyone else.

There are many available methods for complex systems analysis. The mathematical techniques include nonlinear dynamics, chaos theory, cellular automata, graph and network theory, game theory, etc. However, three approaches have dominated the field: (1) nonlinear dynamics, (2) evolving networks, and (3) intelligent agents. The first approach, based on nonlinear dynamics and chaos theory, is widely used in deterministic complex systems analysis. There are many successful applications in different areas of chemical engineering, such as mixing, the dynamics of reactors, fluidized beds, pulsed combustors, and bubble columns. However, this approach is limited to an analytical description of the problem, which is not always possible, especially in analysis of nontechnical complex systems, which include humans.

Recently, the second approach, based on studies of network topologies, dynamics, and evolution is a fast-growing research and application area.[5] Some complex systems, such as the Internet or complex reaction networks, can be analyzed by this approach. However, it has limited capabilities to represent human-centric systems and emerging phenomena outside network dynamics.

The third approach, intelligent agents, is based on the assumption that some phenomena can and should be modeled directly in terms of computer objects, called agents, rather than in terms of equations. The central idea is to have agents that interact with one another according to prescribed rules. This new type of modeling has started to compete with and, in many cases, replace equation-based approaches in complex systems, such as ecology, traffic optimization, supply networks, and behavior-based economics. It is a universal approach, open to technical and human-related systems. Most of the applications of complex systems are implemented by intelligent agents.[6] Based on these obvious advantages, we will focus in the forthcoming sections on intelligent agents technology as the most generic approach for complex systems analysis.

7.1.2   *Intelligent Agents*

Probably the definitions of intelligent agents are as many as the researchers in this field. There is a consensus, however, that autonomy, i.e. the ability to act without the intervention of humans or other systems, is the most important feature of an

---

[5]A. Barabasi, *Linked: The New Science of Networks*, Perseus Publishing, Cambridge, 2003.
[6]M. Luck, P. McBurney, O. Shehory, and S. Willmott, *Agent Technology Roadmap*, AgentLink III, 2005.

agent. Beyond that, different attributes take on different importance based on the domain of the agent. Some of the key generic attributes are summarized below:

- *Autonomy* – An agent is autonomous if its behavior is determined by its own experience (with the ability to learn and adapt). It is assumed that an agent has its own initiative and can act independently. As a result, from the viewpoint of other agents its behavior is neither fully predictable, nor fully controllable.
- *Sociality* – An agent is characterized by its relationships with other agents, and not only by its own properties. Relationships among agents are complex and not reducible. Conflicts are not easily resolvable and cooperation among agents cannot be taken for granted.
- *Identity* – Agents can be abstract or physical and may be created and terminated. Their boundaries are open and changeable.
- *Rational Self-interest* – An agent aims to meet its goals. However, self-interest is in a context of social relations. Rationality is also limited to agent's perceptions. It is assumed that the right action is the one that will cause the agent to be most successful.

Figure 7.4 illustrates a high-level view of an agent interacting with the environment. An agent receives input as events from its environment and, through a set of actions, reacts to it in order to modify it.

Agents can be defined as the decision-making components in complex adaptive systems. Their decision power is separated into two levels, shown in Fig. 7.4. The first decision level includes the rule base (rule 1, rule 2,. . ., rule *n*) which specifies the response to a specific event. An example of such a rule for a bidding agent is:

If current profit < last profit then
    Increase our price
Else
    Do nothing

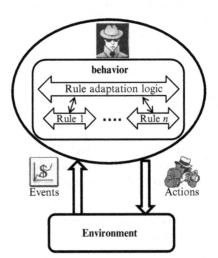

Fig. 7.4 A generic intelligent
agent structure

The second-level rules, or the rules to change the rules, provide the adaptive capability of the agent. Different methods and strategies can be used, but we need to be careful not to introduce too much complexity. One of the key issues in intelligent agents design is the complexity of the rules. It is not a surprise that defining simple rules is strongly recommended. Above all, simple agent rules allow easy decoupling between the agent's behavior and the interaction mechanisms of the multi-agent system. As a result, models are quickly developed, the validation time is shortened, and the chance for model use is increased. Let's not forget the quote at the beginning of this chapter, defined by Oren Etzioni, sharing his painful experience from practical applications. According to this experience, only simple-minded agents behave efficiently and make the applied agent-based system profitable. And we also know that Agent 007 is not an intellectual...

7.1.3   Agent-Based Integrators

Individual agents are carriers of a specific microbehavior. The macrobehavior is represented by a multi-agent system. A multi-agent system is one that consists of a number of agents, which *interact* with one-another.[7] In the most general case, agents in the multi-agent system will be acting on behalf of users with different goals and motivations. To successfully interact, they will require the ability to *cooperate*, *coordinate*, and *negotiate* with each other, in a similar way as people do.

From the applications perspective, we define two types of multi-agent systems, dependent on the nature of the agents and their interactions. The first type, called agent-based integrators, includes business components, such as technical systems and their related organizations. An important feature of agent-based integrators is the minimal presence of social interactions, i.e. the degree of uncertainty is relatively low. The objective of these multi-agent systems is to manage business processes through effective coordination. The second type, called agent-based systems, includes social components, such as the different types of social roles in an organization. It is the classical multi-agent system, where the emergent patterns appear as a result of social interactions. Agent-based systems will be discussed in the next section.

One of the key potential application areas of agent-based integrators is modern flexible and reconfigurable manufacturing. Of special importance is the emergence of so-called holonic[8] manufacturing systems, which are highly decentralized. Such a system integrates the entire range of manufacturing activities from order booking through design, production and marketing. Generally speaking, intelligent agents may represent any way of implementing decentralized decision policies. Of special

---

[7]M. Wooldridge, *An Introduction to Multi-Agent Systems*, Wiley, 2002.

[8]A holon is an autonomous and cooperative building block of a manufacturing system for transforming, transporting, storing, and validating information on physical objects.

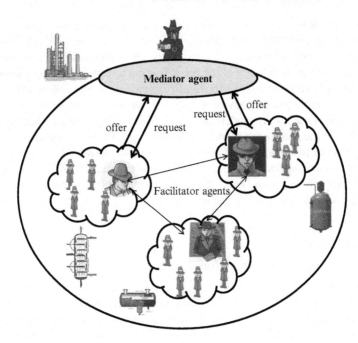

Fig. 7.5 A typical structure of an agent-based integrator

importance are three application areas: product design, process operation at the planning and scheduling level, and process operation at the lower level of real-time control.[9]

A typical architecture of an agent-based integrator for holonic manufacturing is given on Fig. 7.5 where the physical equivalent is integrated control of a process unit that includes three components – a distillation tower, a heat-exchanger, and a reactor.

The equivalents of the agents could be the process operators, responsible for the specific components and corresponding equipment on them, such as sensors, pumps, valves, and pipes. Agents are organized in small communities or clusters, depending on their functions. Other agents play the role of coordinators of the activities by facilitating and mediating. Agents are classified into clusters according to some shared property. The facilitator agent allows direct communication only of agents belonging to the same defined cluster, and communications among different clusters are implemented by exchanging messages through their facilitators. The mediator agent distributes requests to the appropriate set of providers directly or through lower level mediators, collects the offers from alternative providers, and selects the best one on the basis of a performance index.

At the basis of agent-based integrators is the combination of autonomy, robustness, reactiveness, and proactiveness of the intelligent agents. An agent can be

---

[9]M. Paolucci and R. Sacile, *Agent-Based Manufacturing and Control Systems*, CRC Press, 2005.

relied upon to persist in achieving its goals and trying alternatives that are appropriate to the changing environment. When an agent takes on a goal, the responsibility for achieving that goal rests on the agent. Continuous checking and supervision is not needed. Such an agent can be viewed as an employee that has a sense of responsibility and shows initiative.

7.1.4 Agent-Based Systems

The key difference between agent-based integrators and agent-based systems is the decisive role of the social component in the latter. As such, agent-based systems are one of the key modeling techniques to explore social-based complex systems. In some sense, intelligent agents try to play the role of the "invisible hand" that directs social relationships.

Agent-based systems use sets of agents and frameworks for simulating the agent's decisions and interactions. These systems can show how a system could evolve through time, using only a description of the behaviors of the individual agents. Above all, agent-based systems can be used to study how macro-level patterns emerge from rules at the micro-level. There are two value creation opportunities from the emergent macropatterns. The first opportunity is from the solution of the direct problem, i.e. by understanding, analysis and using the unknown patterns, generated by the agent-based system simulation. The second opportunity, with higher potential for return, is from the solution of the inverse problem. In this case the performance of the explored complex system can be improved by redesign of agents' structure and microbehavior. Often, the final result could be introducing adequate operating procedures or organizational changes in the explored business structure.

Agent-based systems are classic multi-agent systems, i.e. a collection of agents and their relationships. Usually the collection of agents reflects the real-world structure of the specific problem under exploration. For example, if the objective of the agent-based system is to develop a supply-chain model, the following agent types can be defined: factory, wholesaler, distributor, retailer, and customer. They represent the real five-stage process of supply-chain that includes a sequence of factories, distributors, wholesalers, and retailers who respond to customers' demand.[10] The mapping of the real-world supply-chain into the agent-based system is illustrated in Fig. 7.6.

Each agent type is defined by its attributes and methods or rules. For example, the factory agent is represented by the following attributes: the agent's name; inventory level; desired inventory level; amount in pipeline; desired amount in pipeline; the amounts received, shipped, ordered, and demanded; various decision

---

[10]M. North and C. Macal, *Managing Business Complexity: Discovering Strategic Solutions with Agent-Based Modeling and Simulation*, Oxford University Press, 2007.

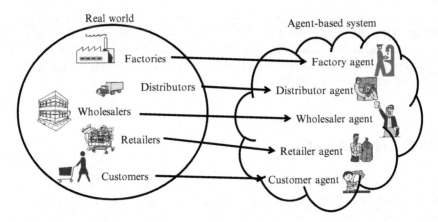

Fig. 7.6 An example of mapping a real-world supply-chain system into an agent-based system

parameters; and the costs incurred of holding inventory or backorders. The values of these variables at any point in time constitute the agent state. Examples of agent rules in the case of the factory agent type include a rule for determining how much to order, and from whom at any point in time and a rule for forecasting demand.

In addition to the collection of agents, the multi-agent system consists of agent interactions. Each agent relation involves two agents. For example, the factory-distributor relation includes the attributes of the number of items in-transit from factory to distributor and the order in-transit from distributor to factory. Agent relations also have methods that operate on them just as human agents have. For example, get shipments, get orders, get upstream agent, and get downstream agent are useful methods for agent interactions.

The key issues in defining multi-agent systems are how the agents communicate and cooperate. For agent communication, several standard formats for the exchange of messages, called agent communication languages, have been developed. The best-known agent communication language is KQML, which is comprised of two parts: (1) the Knowledge Query and Manipulation Language (KQML) and (2) the Knowledge Interchange Format (KIF). KQML is the type of so called "outer" language, which defines various acceptable communicative verbs, or *performatives*. Here are some examples of *performatives*:

- ask-if ("is it true that . . .");
- perform ("please perform the following action . . .");
- reply ("the answer is . . .");

The knowledge interchange format KIF is a language for expressing message content and includes the following:

- properties of things in a domain (e.g. "Kroger is retailer");
- relationships between things in a domain (e.g. "Procter & Gamble is Wal-Mart's supplier");

- general properties of a domain (e.g. "All customer orders are registered by at least one retailer").

An example of a KQML/KIF dialogue between two agents A and B discussing if the size of part 1 is bigger than the size of part 2 with confirmation that the size of part 1 is 20 and of part 2 is 18 is given below:

A to B: (ask-if (> (size part1) (size part2)))
B to A: (reply true)
B to A: (inform (= (size part1) 20))
B to A: (inform (= (size part2) 18))

More recently, the Foundation for Intelligent Physical Agents (FIPA) developed a new agent communication language with 20 performatives, such as: accept-proposal, agree, cancel, confirm, inform, request, etc.[11]

In order to be able to communicate, agents must have agreed on a common set of terms about the objects, concepts, and relationships in the domain of interest, known as ontology. The objective of ontology is to represent domain knowledge in a generic way. It provides the necessary vocabulary for communication about a domain. Defining effective ontologies is one of the key research areas in intelligent agents.

The second key issue in agent-based system design is the way the agents cooperate. As is the case with cooperation among humans, the issue is not trivial. At the basis of agents cooperation is the famous Prisoner's Dilemma,[12] which can be defined as follows:

- Two men are collectively charged with a crime and held in separate cells, with no way of meeting or communicating. They are told that:
  - if one confesses (or defects to the authorities) and the other does not, the defector will be freed, and the other will be jailed for three years;
  - if both defect to the authorities, then each will be jailed for two years.

- Both prisoners know that if they remain loyal to each other, i.e., they don't defect to the authorities, then each of them will be jailed for one year only.

The options are illustrated in Fig. 7.7.

The driving force in the dilemma is the temptation to defect to the authorities and testify against the other prisoner. But if both prisoners defect, they will both get longer sentences than if neither collaborates (two years vs. one year in jail). At the same time, there is the hope for a shorter period in jail if each prisoner refuses to testify (defect). However, by instinct, the individual rational action is to defect and testify against the other and hope for the chance to be free. Unfortunately, the winning strategy of cooperation between both prisoners assumes a high risk-taking culture from both of them. It is not the obvious option for a society of self-interested

---

[11]http://www.fipa.org/

[12]R. Axelrod, *The Evolution of Cooperation*, Basic Books, New York, NY, 1984.

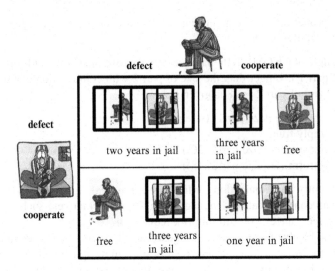

Fig. 7.7 The options in the Prisoner's Dilemma

individuals. The solution is in the iterative Prisoner's Dilemma where the individuals have the chance to learn from their mistakes and to build relationships of
mutual trust. Cooperation in social systems is not given by default and often needs
time to evolve. The same is true in the cooperation attempts of intelligent agents.

7.2 Benefits of Intelligent Agents

Before discussing the specific benefits from applying intelligent agents we'll reduce
the potential sources of confusion and clarify the differences of this approach with
some similar methods, such as swarm intelligence, object-oriented programming,
and expert systems. There are a lot of similarities between agent-based systems and
swarm intelligence. Of special importance is the same source of value creation –
emerging patterns. However, swarm intelligence is limited to specific types of
agents (ants or birds, defined as particles in PSO) with their fixed interactions,
such as pheromone tracing or position and velocity update.

The key differences between agents and objects as well as expert systems are
identified in the next two sections.

7.2.1 Comparison Between Agents and Objects

- Agents are autonomous: agents embody a stronger notion of autonomy than
  objects, and, in particular, they decide for themselves whether or not to perform
  an action on request from another agent.

- Agents are "smart": they are capable of flexible (reactive, pro-active, and social) behavior. In contrast, the standard object model doesn't have this capability.
- Agents are active: in the object-oriented case, the source of decision-making lies with the object that invokes the method. In the agent case, the decision lies with the agent that receives the request.
- When they act, objects do it for free. Agents do it because they want to and very often are doing it for money.

7.2.2 Comparison Between Agents and Expert Systems

- Expert systems typically represent "expertise" about some (abstract) domain of knowledge (e.g. flammable materials). They do not interact directly with any environment but through a user acting as a middle man.
- Expert systems are not generally capable of reactive or proactive behavior.
- Classical AI ignores the social aspects of agency. Expert systems are not generally equipped with social ability in the sense of cooperation, coordination, and negotiation.

7.2.3 Benefits of Agent-Based Modeling

The key benefits from intelligent agents are captured in the mind-map in Fig. 7.8 and discussed below.

- *Emergent Phenomena Generation* – The driving force behind most of the applications is the assumption that macropatterns will emerge from a given set of agents with specified microbehavior. The important factor here is that the emergent phenomena properties are independent of the properties of the agents. As a result, the features of macropatterns could be difficult to understand and predict. It is also possible that nothing new emerges from the agent interactions.

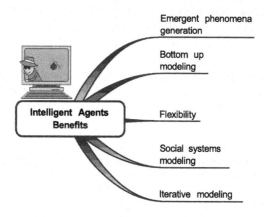

Fig. 7.8 Key benefits from intelligent agents

Fortunately, the potential for emerging phenomena could be assessed in advance if the following features in the agent's microbehavior[13] exist:

- *Individual behavior is nonlinear and can be characterized by thresholds, if-then rules, or nonlinear coupling.*
- *Individual behavior exhibits memory, path-dependence, and hysteresis, or temporal correlations, including learning and adaptation.*
- *Agent interactions are heterogeneous and can generate network effects.*
- *Averages will not work. Aggregate differential equations tend to smooth out fluctuations. In contrast, agent-based systems amplify fluctuations under certain conditions: the system is linearly stable but unstable to larger perturbations.*

- *Bottom-up Modeling* – Agent-based systems begin at the level of the decision-making individual or organization and build upwards from there. It is relatively easy to put a person into the role of an agent and to define microbehavior. Obviously this bottom-up modeling is quite different from the existing equation-based methods. It is not always appropriate, however, and some criteria to use bottom-up modeling are needed. The most important requirements for bottom-up modeling are given below (see previous reference):

  - *Individual behavior is complex. Everything can be done with equations, in principle, but the* complexity *of differential equations increases exponentially as the complexity of the behavior increases. At some point describing complex individual behavior with equations becomes intractable.*
  - *Activities are a* more *natural way of describing the system than processes.*
  - *Validation and calibration of the model through expert judgment is crucial. Agent-based systems are often the most appropriate way of describing what is actually happening in the real world, and the experts can easily understand the model and have a feeling of ownership.*

- *Flexibility* – Agent-based systems are very flexible in controlling the level of complexity by changing agent microbehavior and agent interactions. It is also easy to change the dimensionality of the system by adding or removing agents.
- *Social Systems Modeling* - The ability of agent-based systems to mimic human behavior allows modeling and simulation of social systems in a very natural way. This is probably the approach with the best potential to represent the diverse complexity of the social world. It gives the social analysts a proper medium to explore with details the emerging social phenomena of interest. An agent-based approach may become the economic *E. coli* for social systems analysis.[14] The idea is taken from biology where the ability to experiment with animal models like *E. coli* has led to great advances in understanding the nature

---

[13]The criteria are defined in: E. Bonabeau, Agent-based modeling: methods and techniques for simulating human systems, *Proc.Natl.Acad.Sci.*, *99*, 3, pp. 7280–7287, 2002.

[14]J. Miller and S. Page, *Complex Adaptive Systems: An Introduction to Computational Models of Social Life*, Princeton University Press, 2007.

of biological phenomena. In the same way, artificial adaptive agents could be the new way to develop and validate new social theories.

Using social components in modeling business-related problems gives the unique capability of agent-based systems to deliver more realistic solutions, especially related to planning and strategic decision-making. The existing analytical or statistical methods do not reflect potential social response and are limited to technical components only.

- *Iterative Business Modeling* – Agent-based systems are naturally iterative. The model development starts with an initial description of the microbehavior of the defined agents. The next iteration includes computer simulations and initial results analysis, which lead to the next step of agent behavior corrections. The progressive refinement process continues until the results are acceptable. This mode of modeling is very close to business practice and convenient for both model developers and users.

7.3   Intelligent Agents Issues

From all the discussed computational intelligence approaches, agent-based systems are the least prepared technology for the real world. One of the reasons is the confusion about the technology even at a conceptual level. On the one hand, the idea of an agent is extremely intuitive. On the other hand, it encourages developers to believe that they understand the agent-related concepts when they do not. Part of the problem is the diverse notion about the technology even among the different research communities. In addition, the natural complexity of the approach requires knowledge in broad areas, such as object-oriented programming, classical AI, machine learning methods, etc. As a result, the requirements towards agent-based systems developers are much higher relative to the other computational intelligence methods.

This contributes to the next key issue of intelligent agents – high total cost of ownership. The model development cost is high due to the complex and time-consuming process of defining and validating agent behavior and interactions. It requires active participation of the human equivalents of the defined generic agents. An inevitable effect of modeling social behavior is the high probability of disagreement and discussions among the human equivalents about their agent-counterparts' performance and the necessary microbehavior changes. The iterative nature of agent-based system may significantly prolong this process and raise the development cost. The implementation and maintenance cost of agent-based systems is also high due to the lack of standards and user-friendly software. An additional negative factor is the limited capability of the agent-based systems to be integrated within the existing enterprise infrastructure.

The third key issue of intelligent agents' technology is the difficult marketing. One of the reasons is the source of value creation of agent-based system – the expected novelty from the emerging behavior. Potential users are confused with the uncertainty if something valuable will emerge during the simulations. In some cases the nature of agent-based models also creates obstacles for accepting the technology. Some users have difficulties understanding that the key delivery is simulation and not coded equations or statistical regressions, as in the case of classical modeling.

7.4   How to Apply Intelligent Agents

Applying agent-based systems is still a challenging task that requires special skills in broad areas, like object-oriented programming, agent design, effective visualization, and domain expertise. Of special importance is the continuous direct involvement of the final users in all implementation phases. First, we'll assess the needs for agent-based systems.

7.4.1   When Do We Need Intelligent Agents?

The key capabilities of agent-based systems that may create value are captured in the mind-map in Fig. 7.9.

Probably the most unique and important capability of intelligent agents is simulation of the economics. The broad research field is called the artificial economy or Agent-based Computational Economics (ACE).[15] An artificial economy includes three key types of agents: (1) artificial workers and consumers, (2) artificial firms, and (3) artificial markets. Usually, artificial agents with preferences are consumers and their choices for consumption of goods and leisure are constrained by income and wealth. Artificial workers earn wages in firms as workers, migrate between firms, and own shares of firms. Artificial firms make products to

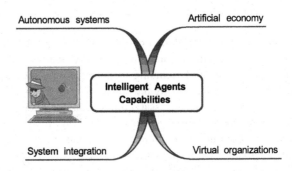

Fig. 7.9 Key capabilities of agent-based systems

---

[15]L. Tesfatsion and K. Judd (Eds), *Handbook of Computational Economics II: Agent-Based Computational Economics*, Elsevier, 2006.

sell to consumers and firms and pay wages to workers. Often, banks are defined as special case of a firm. Agents migrate between firms when it is utility-improving to do so. Sales and profits of the artificial firms are determined by artificial markets, which are the third component of the artificial economy. Different pattern emerge from specific artificial markets. For example, the emerging pattern from the artificial market for consumption and capital goods is product prices. In the case of the artificial market for ownership of firms, the emerging pattern is share prices.

Another unique capability of intelligent agents is the formation of virtual organizations. Virtual organizations are composed of a number of individuals, departments or organizations each of which has a range of capabilities and resources at their disposal. These virtual organizations are formed so that resources may be pooled and services combined with a view of exploiting a perceived market niche. However, in the modern commercial environment it is essential to respond rapidly to changes. Thus, there is a need for robust, agile, flexible systems to support the process of virtual organizations management and this task can be accomplished by agent-based systems.

Using intelligent agents as system integrators is the other unique capability of agent-based models. As was discussed in Sect. 7.1.3, intelligent agents are the backbone of holonic manufacturing where the base element, the "holon", includes both control agents and the physical machine, forming an autonomous technical unit that cooperates with other units. Several low-level holons can cooperate to form a higher-level holon, leading to the emergence of a "holarchy". For example, agent-based integrators may use a society of work-piece agents, machine agents, and transportation agents to manage material flow in a production line. Assignments of work pieces to machines can be done by competitive bidding, while interactions between work pieces and transport agents coordinate their respective goals and manage the flow between machines.

The ultimate utilization of the system integration capabilities of intelligent agents is the design of autonomous systems. These are self-contained systems that can function independently of other components or systems by acquiring, processing and acting on environmental information. The key features of autonomous systems are self-configuration, self-optimization, self-healing, and self-protection. The key application areas are space exploration and industrial units that operate in difficult and unpredictable environments, like the drilling platforms in the ocean.

7.4.2 Applying Intelligent Agents

Applying intelligent agents is significantly different from other computational intelligence techniques. The difference is based on the nontrivial nature of models and software tools. The generic implementation sequence is given in Fig. 7.10 and discussed below. It represents the more complex case of agent-based systems that include social components. The application sequence of agent-based integrators is very close to object-oriented design and will not be discussed.

Fig. 7.10 Key steps in applying agent-based systems

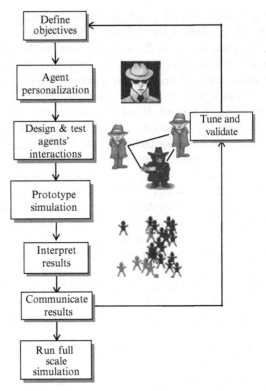

Even the first step in the application sequence, defining the system objectives, is different from that in the other computational intelligence applications. The problem is that the only expected deliverable from an agent-based system is a simulation driven by the defined agents' microbehavior and interactions. It is unknown in advance if the simulation will create value since there are no guarantees if an emergent macrobehavior will appear. That's why it is strongly recommended to define clearly this realistic goal and to avoid anything more specific.

The next step covers agent personalization, which is very similar to object formulation in object-oriented programming. Using a generic object-oriented specification language, such as the Unified Modeling Language (UML), is strongly recommended.[16] Usually agent design includes defining agent properties and methods (rules). The critical factor is the appropriate complexity of the rules. It is always preferable to start with as simple rules as possible in order to decouple agent design

---

[16]There is a special addition to UML, called AgentUML, directly applicable for agent-systems design – see http://www.omg.org/cgi-bin/doc?ec/99-10-03

from the next step, agency design. In addition, simple rules can allow agents to execute rapidly and reduces the time for validation and correction of microbehavior.

Very often defining and validating agents' interactions is the most time consuming step in applying intelligent agents. The challenging part is building agents capable of interacting (cooperating, coordinating, negotiating) with other agents in order to successfully carry out the delegated tasks. It is especially difficult when the other agents cannot share the same interests/goals. Usually, the design of the agents and the agency needs several iterations until a compromised behavior is achieved.

Due to the high complexity of agent-based modeling it is strongly recommend to separate the model building process into two phases – prototype simulation and full-scale simulation. Prototype simulations are based on reduced number of agents with simple rules and interactions that allow getting at the core problem without the details or extras. Their results are also easier to interpret and describe.

Interpretation of agent-based simulation is a very time-consuming process and requires the participation of all key stakeholders. The key objective of this step is capturing and defining the emerging macrobehaviors. Usually it includes the following action items:

- Identify conditions under which the system exhibits various types of behaviors.
- Understand the factors that lead to these behaviors.
- Identify simulation results as stable, unstable, and transient.
- Identify regions in parameter space that exhibit stable and unstable behaviors.
- Identify which particular agent rules lead to various types of system behaviors.

Another challenge is communicating the results from agent-based simulations. The preferable way is organizing a meeting and demonstrating the emerging macropatterns by designing multiple simulation runs and putting them together to illustrate the "big" picture. It is very important to understand that the dynamic patterns of a simulation are often more importance than static end states. That's why in many cases communication of agent-based simulation results in documents is not very effective.

Verification of agent-based models is very difficult and mostly limited to simple behaviors and interactions among agents. Usually it leads to re-defining rules and not changing the agents' structure.

Applying a full-scale simulation requires education of the final user about the realistic capabilities of the model. That includes instructions on how to interpret and use the results from the simulation.

7.5 Typical Applications of Intelligent Agents

Intelligent agents are still knocking at the door of industry. The application record is short and the created value is low relative to the other computational intelligence methods. There is a growing tendency, though, and we believe in the big potential for value creation of this technology in both its directions – as agent-based

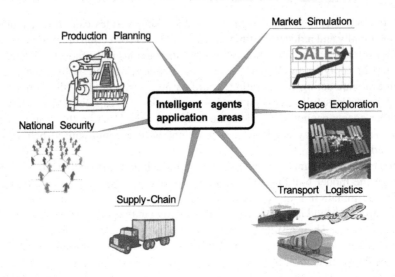

Fig. 7.11 Key intelligent agents application areas

integrators and agent-based systems. The current key application areas are shown in Fig. 7.11 and discussed next.

- *Market Simulation* – Agent-based systems have been used in different types of market simulations. A team at the Ford Motor Company has implemented intelligent agents for studying the potential rise of a hydrogen transportation infrastructure.[17] The objective of the simulation is to investigate the key factors that might lead to the successful emergence of a self-sustaining hydrogen transportation infrastructure. In the model, the agents simulate drivers who choose what types of vehicles to buy (e.g. hydrogen-fueled cars versus gasoline-fueled cars) as well as fueling stations that choose to offer different fuels.

  Another example is the agent-based simulation of electric power markets. An example is the Electricity Market Complex Adaptive Systems (EMCAS) model, developed at Argonne National Laboratory and applied for market stability and efficiency analysis of Illinois state electric power.[18] EMCAS represents the behavior of an electric power system as well as the producers and consumers that operate within it. The agents include generation companies that offer power into the electricity marketplace and demand companies that buy bulk electricity from the marketplace.

- *Space Exploration* – Agent-based systems are used in the simulation of shuttle missions and planning the activities at the International Space Station (ISS). For

---

[17]C. Stephan and J. Sullivan, Growth of a hydrogen transportation infrastructure, *Proceedings of the Agent 2004 Conference*, Argonne National Laboratory, Argonne, IL, USA, 2004.

[18]M. North and C. Macal, *Managing Business Complexity: Discovering Strategic Solutions with Agent-Based Modeling and Simulation*, Oxford University Press, 2007.

that purpose, a specialized multi-agent modeling and simulation language and distributed runtime system, named Brahms, has been developed at NASA Ames Research Center.[19] It can be used to model and run a simulation of the distributed work activities of multiple agents – humans, robots, and software agents - coordinating a mission in one or more locations. A Brahms model also includes geography (mission habitat, territory to be explored), objects (space-suits, generators, vehicles, rocks), and activities inside and outside of the habitat (meals, email, briefing and planning, getting into or out of a spacesuit, exploring a site, preparing a report). A Brahms model and virtual environment for decision support on the ISS has been developed to simulate which tasks could be off-loaded to robot assistants, such as Ames Research Center's Personal Satellite Assistant and Johnson Space Center's Robonaut. The Brahms model is based on data recorded from an actual day of complex operations aboard the ISS, including a space walk. This model may help to reduce waiting time, improve situational awareness, and schedule resources more effectively on the ISS.

- *Transport Logistics* – Handling complex dynamic transport systems of diverse carriers is another appropriate application area for intelligent agents. An example is the Living Systems® Adaptive Transportation Networks (LS/ATN) application for the cost-based optimization system for transport logistics[20] Developed by Whitestein Technologies, originally for DHL, LS/ATN is designed to provide automatic optimization for large-scale transport companies, taking into account the many constraints on their vehicle fleet, cargo, and drivers. Although the agent solution accounts for only 20% of the entire system, agent technology plays a central role in the optimization. Vehicle drivers send information specifying their location and proposed route, and the system determines if that vehicle can collect an additional load, or swap future loads with another vehicle in order to reduce cost. A negotiation is performed automatically by agents, with each agent representing one vehicle, using an auction-like protocol. The vehicle that can provide the cheapest delivery wins the auction, reducing the overall cost of cargo delivery and in most cases, the combined distance traveled for all vehicles. The aim is to find a local optimum, so that only vehicles traveling in close proximity to each other will be involved in negotiations.

- *Supply-Chain* – An important niche for application of intelligent agents is dynamic supply-chain scheduling. An example is the applied system at Tankers International, which operates one of the largest oil tanker pools in the world (see previous reference). The objective is to dynamically schedule the most profitable deployment of ships-to-cargo for its very large crude carrier fleet. An agent-based optimizer, Ocean i-Scheduler, was developed by Magenta Technology for use in real-time planning of cargo assignment to vessels in the fleet. The system can dynamically adapt plans in response to unexpected changes, such as

---

[19]http://www.cict.nasa.gov/infusion

[20]M. Luck, P. McBurney, O. Shehory, and S. Willmott, *Agent Technology Roadmap*, AgentLink III, 2005.

transportation cost fluctuations or changes to vessels, ports or cargo. Agent-based optimization techniques not only provided improved responsiveness, but also reduced the human effort necessary to deal with the vast amounts of information required, thus reducing costly mistakes, and preserving the knowledge developed in the process of scheduling.

- *National Security* – An unexpected application area of agent-based simulations is analyzing the formation and dynamic evolution of terrorist networks. An example is the agent-based system NetBreaker,[21] developed at Argonne National Laboratory. It considers both the social and resource aspects of terrorist networks, providing a view of possible network dynamics. As the simulation progresses an analyst is provided with a visual representation of both the shapes the network could take and its dynamics, an estimate of the threat, and quantified questions illustrating what new information would be most beneficial. NetBreaker's goal is to reduce surprise by providing and quantifying possibilities, not determining which possibility is correct; extrapolation, exploration and estimation, not interpolation. NetBreaker does not try to remove a human analyst from this process, but aids that analyst in seeing all possibilities and acting accordingly.

- *Production Planning* – In many industries, unexpected changes during plant operation invalidate the production planning and schedule within one or two hours, and from that point on it serves as a build list, identifying the jobs that need to be done, while assignment of parts and operations to machines is handled by other mechanisms, collectively called shop-floor control. Modern scheduling systems seek to maintain schedule validity in the face of real-time interruptions, a process known as "reactive scheduling," and often are closely coupled with shop-floor control mechanisms to facilitate the execution of the schedule in real time.

  Agent-based systems can be used to handle this problem. An example is the ProPlantT system for production planning at Tesla-TV a Czech company for production of radio and TV broadcasting stations. It includes four classes of agents: production planning, production management, production, and a meta-agent, which monitors the performance of the agent community. Chrysler has also explored the capabilities of agent-based shop-floor control and scheduling.[22]

7.6 Intelligent Agents Marketing

The concept of intelligent agents is easy to explain and communicate. The deliverables and the value creation mechanisms, however, are confusing to the user and more marketing efforts than the other computational intelligence methods are needed. It is very important to explain to the user upfront that there are no

---

[21]http://www.anl.gov/National_Security/docs/factsheet_Netbreaker.pdf

[22]V. Parunak, A practitioners' review of industrial agent applications, *Autonomous Agents and Multi-Agent Systems*, *3*, pp. 389–407, 2000.

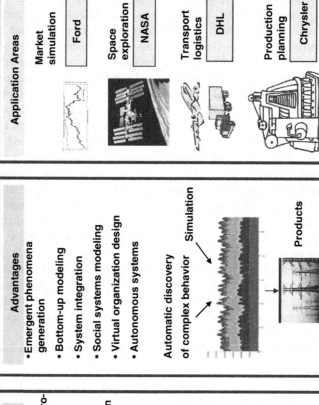
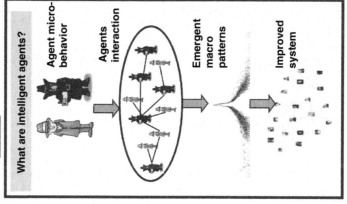

Fig. 7.12 Agent-based systems marketing slide

guarantees of the appearance of emerging patterns, which can lead to more insightful system understanding and eventual improvement. Marketing the value of agent-based integration of complex technical systems is more obvious and we'll focus on selling to potential users the agent-based systems with social components. An example of an intelligent agents marketing slide is given in Fig. 7.12.

The main slogan of intelligent agents, "Transfer emergent patterns from microbehavior into value" communicates the source of the value creation basis of the approach. However, the message may be confusing to a nontechnical user and needs to be clarified with the other sections of the slide. The left section of the marketing slide tries to play this role and represents the generic view of intelligent agents. The key phases of agent-based modeling systems are illustrated: (1) defining agents' microbehavior by rules and actions, (2) defining the multi-agent system by agents interactions, (3) identifying emerging macropatterns from the simulations, and (4) using the discovered macrobehavior for system improvement.

The middle section of the marketing slide represents the key advantages of intelligent systems, such as emergent phenomena generation, bottom-up modeling, system integration, social systems modeling, virtual organization design, and autonomous system design. The visualization section represents a simulation as an illustration of the key value source - automatic discovery of complex behavior (in this case, electric price bidding patterns), and a power grid, as an example of a product where the agent-based system can be applied.

The key application areas of intelligent agents are shown in the right section of the marketing slide in Fig. 7.12. The slide includes the most valuable agent-based application areas in market simulation, space exploration, transport logistics, and production planning. Examples of leading companies in the specific application areas and NASA are given.

The proposed elevator pitch representing intelligent agents is shown in Fig. 7.13.

Intelligent Agents Elevator Pitch

Recently, there is a growing need for estimating social responses to key business decisions. One of the few available technologies that is capable of modeling social behavior is intelligent agents. Agent-based systems is a method for bottom-up modeling where the key social components are defined as artificial agents. Their microbehavior is captured by actions and rules they follow in order to achieve their individual goals. Intelligent agents interact in different ways and can negotiate. The unknown social responses emerge in a computer simulation from the microbehavior of the interacting agents. For example, different price bidding strategies emerge in electricity market simulations of generation, transmission, distribution, and demand companies. Another big application area of intelligent agents is effective coordination of various activities in complex decentralized technical systems like global transport logistics. Agent-based systems have been successfully applied in artificial market simulations, space exploration, supply-chain, planning, etc., by companies like Ford, DHL, Tankers International, etc. and NASA.

Fig. 7.13. Intelligent agents elevator pitch

7.7 Available Resources for Intelligent Agents

7.7.1 Key Websites

A key portal for intelligent agents:
http://www.multiagent.com/

Open Agent-Based Modeling consortium:
http://www.openabm.org/site/

Santa Fe Institute
http://www.santafe.edu/

Complex Adaptive Systems at Argonne National Lab
http://www.dis.anl.gov/exp/cas/

Key site for agent-based computational economics:
http://www.econ.iastate.edu/tesfatsi/acecode.htm

7.7.2 Selected Software

REPAST Symphony, a Java-based tool, developed at Argonne National Lab (free and open source):
http://repast.sourceforge.net/

SWARM, a C-based tool, developed at Santa Fe Institute (free and open source):
http://www.swarm.org/index.php?title=Swarm_main_page

Zeus, a Java-based tool, developed at British Telecom (available under an open source license):
http://labs.bt.com/projects/agents/zeus/

Anylogic, an integrated environment for discrete event, system dynamics, and agent-based modeling developed by a Russian professional vendor – XJ technologies:
http://www.xjtek.com/anylogic/

7.8 Summary

Key messages:

Intelligent agents is a technique for bottom-up modeling that represents the behavior of a complex system by the interactions of its simple components, defined as agents.

An intelligent agent receives inputs from its environment and through a set of actions reacts to modify it in its favor.

Agent-based integrators effectively coordinate various activities in complex decentralized technical systems.

Agent-based systems generate emerging macropatterns from rules at the micro-level of interacting social agents.

Intelligent agents have been successfully applied in artificial market simulations, space exploration, transport logistics, supply chains, national security, and production planning.

The Bottom Line

Applied intelligent agents transfer emerging macropatterns from microbehavior of interacting artificial agents into value.

Suggested Reading

A recommended list of books that cover the key issues in intelligent agents:

J. Miller and S. Page, *Complex Adaptive Systems: An Introduction to Computational Models of Social Life*, Princeton University Press, 2007.

M. North and C. Macal, *Managing Business ComplexityComplexity: Discovering Strategic Solutions with Agent-Based Modeling and Simulation*, Oxford University Press, 2007.

M. Paolucci and R. Sacile, *Agent-Based ManufacturingManufacturing and Control Systems*, CRC Press, Boca Raton, FL, 2005.

S. Russell and P. Norvig, *Artificial Intelligence: A Modern Approach*, 2nd edition, Prentice Hall, 2002.

L. Tesfatsion and K. Judd (Eds), *Handbook of Computational Economics II: Agent-Based Computational Economics*, Elsevier, 2006.

G. Weiss, (Editor), *Multi-Agent Systems: A Modern Approach to Distributed Artificial Intelligence*, MIT Press, 2000.

M. Wooldridge, *An Introduction to Multi-Agent Systems*, Wiley, 2002.

Part II
Computational Intelligence Creates Value

Chapter 8
Why We Need Intelligent Solutions

*The biggest difficulty with mankind today is that our
knowledge has increased so much faster than our wisdom.*
Frank Whitmore

The objective of the second part of the book is to clarify the enormous potential
for value creation from computational intelligence methods. It includes three
chapters. Chapter 8 identifies the top 10 key expected markets for computational
intelligence. Chapter 9 discusses the competitive advantages of computational
intelligence relative to the prevailing modeling approaches in industry, such as
mechanistic modeling, statistics, classical optimization, and heuristics. Chapter 10
summarizes the main issues in applying computational intelligence, which are
critical for creating value.

It is a common perception that human intelligence is the driving force for economic development, productivity gains, and, the end, increasing living standards.
From that perspective, enhancing human intelligence by any means increases the
chances for more effective value creation. Of special importance are the recent
challenges from the data avalanche in the Internet and the dynamic global market
operating in high fluctuations of energy prices and potential political instabilities.
As a result, human intelligence had to learn how to handle an enormous amount of
information and make the right decisions with a high level of complexity in real
time. A significant factor is the growing role of data analysis, pattern recognition,
and learning. Computational intelligence can play a critical role in this process.

The objective of this chapter is to define the potential market for computational
intelligence by identifying the main sources for value creation based on the need for
intelligent solutions. The top ten most important sources are shown in Fig. 8.1 and
are discussed in the following sections.

A.K. Kordon, *Applying Computational Intelligence*,   203
DOI 10.1007/978-3-540-69913-2_8, © Springer-Verlag Berlin Heidelberg 2010

Fig. 8.1 A mind-map of the key sources for value creation from computational intelligence

Fig. 8.2 Key features of a successful competitor in the global economy

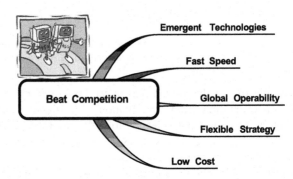

8.1   Beat Competition

The first and critical source for value creation from computational intelligence is the potential for increased competitive advantage by effectively using this technology. The complete analysis of competitive advantages of computational intelligence is presented in the next chapter. At this point we'll focus on the key features which make a global competitor successful in the current business environment. These components are shown in the mind-map in Fig. 8.2 and discussed briefly below. For each component, the potential role of computational intelligence is identified.

8.1.1   Effective Utilization of Emerging Technologies

Without doubt, introducing and exploring the full potential of a new technology for value creation is critical for the competitiveness of many enterprises, especially in the high-tech area. Of special importance is the reliable low-cost transfer of the emerging technology from the research domain to the business domain. Promoting a new technology in the business environment includes resolving technical issues, such as a sound scientific basis, reproducible results on real-world applications, and the potential for scale-up. Of equal importance are the efforts to introduce new infrastructure and work processes in combination with cultural changes and technology acceptance.

Computational intelligence can reduce the cost of the introduction and utilization of new technologies in several ways. Firstly, computational intelligence can reduce the cost through early detection and identification of perspective new technologies through pattern recognition and data analysis (which covers the broad application area of data mining). Secondly, computational intelligence can reduce the cost by minimizing modeling and experimental efforts during the introduction of technology by using a broad variety of robust empirical modeling techniques, such as symbolic regression, neural networks, support vector machines, and fuzzy systems. In contrast to first-principles modeling methods, which are problem-specific, these techniques are universal and depend only on the availability and quality of the data. Thirdly, computational intelligence can reduce the cost by more reliable scale-up of the developed models from laboratory scale to full manufacturing scale. Two computational intelligence techniques – symbolic regression via genetic programming and support vector machines for regression – have unique features to operate with acceptable performance outside the model development range.

8.1.2 Fast Response to Changing Environment

The high speed of response to dynamic changes in the business environment is another important feature of contemporary competitiveness. It requires capabilities to predict future trends, plan efficiently the expected changes, and quickly learn the new operating conditions. Computational intelligence has unique features, which can deliver the required capabilities at low development cost.

Short-term and long-term predictive capabilities relating to the future directions of the specific business environment, for example predicting demands, energy prices, and competitive environment, can be implemented using neural networks and intelligent agents. The abilities to operate continuously in automatic adaptive mode with self-adjustment and tracking the optimal conditions can be delivered by neural networks, support vector machines, and swarm intelligence. An additional factor that has to be taken into account is that, on average, the cost of adaptation is significantly lower than the cost of manual readjustments and the definition of new operating regimes. Quickly learning new operating conditions is the best application area of neural networks and support vector machines.

8.1.3 Effective Operation in the Global Economy

Utilizing the opportunities of globalization is a very important component in the contemporary competitive landscape. In principle, global operation adds an extra level of complexity related to the specifics of local cultures. This by itself requires a higher level of human intelligence. As a result, additional capabilities for analyzing the new local environment, such as exploring the potential for new products, linked to cultural habits, adjusting the global marketing strategy to local preferences, and adopting global work processes in agreement with the national legal systems, are needed. Another effect of outsourcing manufacturing across the globe is the gradually increased cost and complexity of the supply chain.

Fortunately, computational intelligence can satisfy some of the discussed needs for effective global operation. Recently, there is a growing research area of computational economics, which uses computational intelligence methods.[1] One of the key directions is exploring new markets based on simulation of social behavior, including local culture by intelligent agents. However, this is still in the research domain and not yet ready for reliable industrial applications.

At the same time, another important area of global operations – handling the increased complexity and cost of the supply chain – can rely on the advanced optimal scheduling methods of swarm intelligence and evolutionary computation. There is an impressive record of successful industrial applications with significant value creation on this topic discussed in the previous chapters. The benefits of using computational intelligence in a mundane but important function of global operation – machine translation of the necessary documentation into numerous foreign languages – were already demonstrated in the discussed application at the Ford Motor Company in Chap. 1.

8.1.4 Flexible Strategy

Changing and redefining business strategy according to the new economic realities is the next component of contemporary competition where computational intelligence can play an important role. The specific capabilities necessary for flexible strategy development and the corresponding computational intelligence techniques are as follows.

An important capability is defining a business strategy based on solid data and trend analysis. The higher the complexity of the business environment, the more the need for sophisticated methods of high-dimensionality data analysis. An example of a high-dimensional system includes thousands of variables and millions of records. The unique combination of fuzzy systems with neural networks and support vector machines allows us to analyze vague and numerical information and handle the "curse of dimensionality".[2] An additional requirement is blending the diverse

---

[1]L. Tesfatsion and K. Judd (Eds), *Handbook of Computational Economics II: Agent-Based Computational Economics*, Elsevier, 2006.

[2]Model predictive performance degradation with increased number of input dimensions (number of variables).

sources of data from accounting, planning, manufacturing, and the supply chain. Using this type of sophisticated data analysis beyond the capabilities of classical statistics could be of significant competitive advantage in defining a business strategy.

Another important capability necessary for a successful flexible business strategy is predicting new trends by analyzing patents, markets, and demands. The key computational intelligence techniques that can be used for implementing this capability are neural networks, support vector machines, and intelligent agents. The same techniques can be used for constantly adjust the business strategy by analyzing the feedback from customers, capturing new patterns in business environment, and learning new dependencies from data.

8.1.5   Low Cost of Operation

This is the component of contemporary competition where computational intelligence has demonstrated its value creation potential in all key segments of business operation, especially in manufacturing. For example, evolutionary computation, support vector machines, and neural networks have created inferential sensors that reduce out-of spec losses. Fuzzy systems, neural networks and swarm intelligence are useful techniques for implementing advanced nonlinear control, which optimizes manufacturing cost. Evolutionary computation, swarm intelligence and intelligent agents have successfully minimized the cost of complex supply-chain operations.

8.2   Accelerate Innovations

According to a current analysis, only 10% of all publicly traded companies can sustain for more than a few years a growth trajectory that creates above average shareholder returns.[3] As a result, these enterprises enjoy the support of Wall Street and the investors. The key strategy for success in achieving sustainable growth is fast innovation with higher speed than the competition. The three driving forces of fast innovation are: (1) differentiation from the competition, (2) fast time to market, and (3) disruptive innovation. The purpose of differentiation is to provide an offering which the customer believes delivers superior performance per unit of cost. The objective of fast market penetration is to earn high margins, as a temporary monopoly, and quickly create a new innovation to counter the gradual commoditization of the old one. Disruptive innovations is the most substantial driving force since the disruptive technologies redefine the competitive landscape, by making any previous competitive advantages obsolete. Typical examples of

---

[3]M. George, J. Works, and K. Watson-Hemphill, *Fast Innovation*, McGraw-Hill, 2005.

Fig. 8.3 Necessary features
to accelerate innovations in
practice

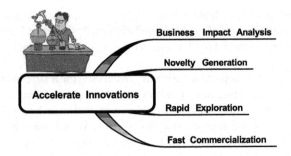

disruptive technologies are personal computers or mp3 musical players like
the iPod.

The necessary features to accelerate innovations across the three driving
forces, related to computational intelligence, are shown in Fig. 8.3 and discussed
below.

8.2.1 Business Impact Analysis of Innovation

An early estimate of the potential business impact of any innovative idea is of
crucial importance. It is expected that adding modeling and analytical capabilities
on this topic will be the key competitive area in the near future.

Computational intelligence offers unique features for business impact analysis
and may become the leading technology in this topic. It includes improved search
techniques based on machine learning for verification novelty on patents and the
available literature. Computational intelligence also allows complex data analysis
based on market surveys, future demand assessment, and the identification of
technology gaps to assess the potential business impact of innovation. Of critical
importance for the success of this task are the capabilities to develop financial and
market penetration models. Although still in its infancy, this new type of modeling
is growing and could be based on evolutionary computation, neural networks,
computing with words, and intelligent agents. It is discussed briefly in the last
chapter of the book.

8.2.2 Automatic Novelty Generation

One of the key strengths of computational intelligence is its capability to derive
unknown solutions within a specific problem domain. The novelty can be generated
automatically by evolutionary computation. For example, the capabilities of genetic
programming to generate novel electronic schemes with given frequency responses

by combining several building blocks (capacitors, resistors, amplifiers, etc.), under known physical laws of electric circuits, have been successfully demonstrated, with reinvention of many patents.[4] Similar results have been achieved with generation of novel optical systems.

8.2.3   Rapid Exploration of New Ideas

Modeling capabilities are critical for exploring the technical features of innovations. Initially fundamental knowledge and human expertise about a new invention are at a very low level. The option of building first-principles models is very time consuming and expensive. On top of that, there are very few data available either on the literature or from laboratory experiments. Often it is also difficult to satisfy the assumptions of classical statistics to develop statistical models with a limited number of observations. Two computational intelligence methods, however, support vector machines and symbolic regression via genetic programming, can generate models with a small number of data points. Though not perfect, these surrogate empirical models can help to identify key factors, mathematical relationships, and to select proper physical mechanisms. In this way the time for hypothesis search in exploring the innovation can be significantly reduced. In a case study for new product development, discussed in Chap. 14, the surrogate symbolic regression empirical models saved three man months of development time.

8.2.4   Fast Commercialization in Practice

Computational intelligence can contribute to accelerating innovation commercialization by reducing the scale-up efforts and by offering a broader range of technologies for process control and optimization. The key decision in scale-up activities is either to build pilot plant facilities or to take the risk and directly jump to full-scale operation. The cost and time savings if the pilot plant phase is avoided are significant. Two capabilities are important in this critical decision – robust modeling and advanced design of experiments to validate the hypothesis of reliable performance after scale-up. As was discussed earlier, support vector machines and symbolic regression models demonstrate impressive performance outside the initial range of model development. Evolutionary computation can also broaden the capabilities of statistical design of experiments by generating nonlinear transforms of the explored factors. This allows us to avoid some of the statistical limitations and build a linearized model without doing additional experiments. The unique feature derived

---

[4]J. Koza *et al.*, *Genetic Programming IV: Routine Human-Competitive Machine Intelligence*, Kluwer, 2003.

by combining evolutionary computation with classical statistics is discussed in
Chap. 11.

The other key contribution of computational intelligence to fast commercialization is with the broader range of process control and optimization technologies that
can handle complex nonlinear systems, and this is the topic of the next section.

8.3   Produce Efficiently

Increasing effectively human intelligence has a direct positive impact in manufacturing by improving the quality and reducing production losses. One of the
advantages of applying computational intelligence in manufacturing is the almost
direct translation of the created value into hard money.

The key areas of value creation from computer-intelligent manufacturing are
shown in Fig. 8.4 and discussed next.

8.3.1   Accurate Production Planning

Planning in manufacturing is an area where even classical AI has created a lot
of value (illustrated in Chap. 1 by the Direct Labor Management System for
assembly line planning at the Ford Motor Company). The computational intelligence technique, which is mostly related to planning activities, is intelligent agents.
They have unique capabilities of integrating planning with continuous adaptation
and learning. It is also possible to integrate the different sources of data and, as
a result, to tightly couple the expected financial objectives with the specific
manufacturing settings.

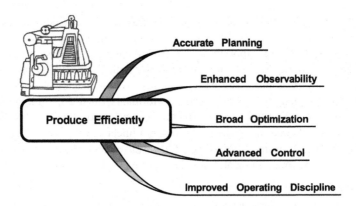

Fig. 8.4 Key areas of value creation from intelligent manufacturing

Another computational intelligence technique, which could be useful for effective manufacturing planning, is swarm intelligence, especially ant colony optimization.

8.3.2 Enhanced Observability of Processes

The observability of a process is defined as a data flow from the process, which gives sufficient information to estimate the current process state and make correct decisions for future actions, including process control. Classical observability was based mostly on hardware sensors and first-principles models. The modern observability, however, includes also empirical model-based estimates of critical process parameters as well as process trend analysis. In this way the decision-making process is much more comprehensive and has a predictive component linked to process trends.

Several computational intelligence approaches contribute to enhanced process observability. The key application is inferential sensors with thousands of applications based on neural networks, and, recently, on symbolic regression via genetic programming. Inferential sensors are discussed in detail in Chaps. 11 and 14. Neural networks and support vector machines are the basic techniques for trend analysis and related fault-detection and process health monitoring systems. These systems create significant value by early detection of potential faults and they reduce the losses from low quality and process shutdowns.

8.3.3 Broad Product and Process Optimization

Optimization is one of the few research approaches that directly translate results into value. Needless to say, many optimization techniques have been tried and used in industry. Most of the classical optimization approaches originated from mathematically derived techniques, such as linear programming, direct search or gradient-based methods. The higher the complexity of a problem, the stronger the limitations of classical optimization methods. In the case of complex high-dimensional landscapes with noisy data, these methods have difficulties in finding a global optimum. Fortunately, computational intelligence offers several approaches that can deliver solutions beyond the limitations of classical optimization.

In new product design, evolutionary computation methods, such as genetic algorithms and evolutionary strategies, have been successfully used by companies like Boeing, BMW, Rolls Royce, etc. An interesting application area is optimal formulation design, where genetic algorithms and recently PSO have been implemented for optimal color matching at the Dow Chemical Company.

In process optimization, genetic algorithms, evolutionary strategies, ant colony optimization, and particle swarm optimizers deliver unique capabilities for finding a steady-state and dynamic optimum.

8.3.4   Advanced Process Control

It is a well-known fact that 90% of the control loops in industry use the classical Proportional-Integral-Derivative (PID) control algorithm. Optimal tuning of these controllers on complex plants is not trivial, though, and requires high-level expertise. The losses due to nonoptimal control are substantial and probably in the order of at least hundreds of millions of dollars. Computational intelligence can reduce these losses through several approaches. For example, genetic algorithms are one of the most appropriate methods for performing automatic tuning of control parameters to their optimal values.

A growing part of the remaining 10% of industrial control loops are based on nonlinear optimal control methods. Three computational intelligence technologies – neural networks, evolutionary computation, and swarm intelligence – have the potential to implement nonlinear control. Neural networks are at the core of advanced control systems. Hundreds of industrial applications, mostly in the petrochemical and chemical industries, have been reported by the key vendors – Pavilion Technologies and Aspen Technologies.

8.3.5   Improved Operating Discipline

Operating discipline integrates all participants in a manufacturing unit into a well-organized work process, which includes managers, engineers, and operators. It can be viewed as the collective knowledge base of a specific manufacturing process. The problem is that it is static and the only link to the dynamic nature of the process is through process operators and engineers in the control room.

Computational intelligence allows us to develop an adaptive operating discipline that responds adequately to the changing process environment in real time. It is based on the available process knowledge with additional capabilities, such as multivariate monitoring, early fault detection combined with intelligent alarm diagnostics, and optimal performance metrics. An example of an improved operating discipline system using neural networks, fuzzy systems, knowledge-based systems, and evolutionary computation will be shown in Chap. 14.

8.4   Distribute Effectively

A company's supply chain comprises geographically dispersed facilities where raw materials, intermediate products, or finished products are acquired, transformed, stored, or sold, and transportation links that connect facilities along with

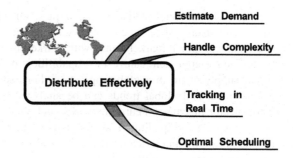

Fig. 8.5 Key components for an effective supply chain

products flow.[5] A supply chain network includes vendors, manufacturing plants, distribution centers and markets. While outsourcing manufacturing in areas with low labor cost reduces the overall production expenses, it may significantly increase distribution operating prices. As a result, the importance of cost reduction in the global supply chain increases significantly. The areas where computational intelligence may contribute to improved efficiency and reduced cost of supply chain operations are shown in Fig. 8.5 and discussed below.

8.4.1 Estimate Demand

Demand forecasting is based on quantitative methods for predicting future demand for products sold by a company. Such predictions are critical for modeling and optimizing product distribution. They are necessary for total cost minimization of product prices, product sourcing locations, and supply chain expenses. The standard approach for demand forecasting is by using statistical time-series analysis, which is limited to linear methods.

Computational intelligence has the capability to broaden the forecast modeling options with nonlinear time-series models. One technique with impressive features to represent nonlinear temporal dependencies is recurrent neural networks. It has been successfully applied in voice recognition, process control, and robotics.

8.4.2 Handle Global Market Complexity

If estimating demand of a local or national market is very difficult, forecasting demand on the global market is an extremely challenging task. The level of complexity added by globalization cannot be appropriately handled by most of

---

[5]J. Shapiro, *Modeling the Supply Chain*, Duxbury, Pacific Grove, CA, 2001.

the existing analytical and statistical techniques. Of special importance is the development of data analysis and knowledge representation methods which can combine local and global market trends with marketing expertise. Some computational intelligence approaches have the capabilities to design such a system for an optimal supply chain in a global operation. For example, the complex high-dimensional local and global trends can be analyzed and automatically learned by support vector machines, neural networks, and evolutionary computation. The marketing expertise and qualitative assessment can be captured by fuzzy systems and integrated into the numerical analysis. The whole global supply chain system could be implemented by intelligent agents.

8.4.3   Real-Time Operation

Recently, an emerging technology, called Radio-Frequency Identification (RFID) has demonstrated the potential to revolutionize the supply chain. RFID is an automatic identification method, relying on storing and remotely retrieving data using devices called RFID tags or transponders. An RFID tag is an object that can be attached to or incorporated into a product, animal, or person for the purpose of identification using radio waves. Many big retail companies, such as Wal-Mart, Target, Tesco, and Metro AG, are in the process of reorganizing their supply chain based on RFID. Since January, 2005, Wal-Mart has required its top 100 suppliers to apply RFID labels to all shipments. One of the key benefits of this technology is the possibility for real-time tracking during the entire distribution process. As a result, the transportation cost and the inventory buffers can be dynamically minimized.

The biggest obstacle, however, is the large dimensionality of the data and the short time horizon required for data analysis and optimization in real-time. Classical statistical methods and optimizers have obvious limitations to solve this type of problem. Fortunately, computational intelligence has some unique capabilities to address the challenges related to high dimensionality of the data and the speed of analysis and optimization. For example, the incremental on-line learning features of neural networks as well as sequential support vector machines can deliver dynamic nonlinear models in real time. Ant colony optimization and particle swarm optimizers can operate fast in a dynamic environment with high dimensionality as well.

8.4.4   Optimal Scheduling

In general, optimal scheduling in industry has two key parts — optimal distribution and production planning. Optimal distribution scheduling includes transportation factors, such as vehicle loading and routing, channel selection, and carrier selection. The inventory factors include safety stock requirements, replenishment quantities, and replenishment time. Optimal production scheduling requires actions like: sequencing of orders on a machine, timing of major and minor changeovers, and management of work-in-progress inventory. The existing optimization models are mostly based on analytical linear programming and mixed integer programming techniques, which have limitations when the problem has high dimensionality and the optimal search landscape is noisy and with multiple optima. Several computational intelligence methods, especially from evolutionary computation and swarm intelligence, are capable of delivering reliable solutions for complex real-world problems. An example is the impressive application of using ant colony optimization and genetic algorithms for optimal scheduling of liquid gas distribution at Air Liquide, discussed in Chaps. 5 and 6.

8.5 Impress Customers

Dealing with local and global customers becomes a critical topic in the global economy. The number of developed methods and offered software products for customer relations management, customer loyalty analysis, and customer behavior pattern detection has significantly grown in the last couple of years. Computational intelligence plays an important role in this process. The key ways to impress customers and win their support are shown in Fig. 8.6 and discussed below with identified contributions from different computational intelligence techniques.

8.5.1 Analyze Customers

The Internet, the enormous data collection capacity of modern warehouses, and the enhanced software capabilities for handling large databases have changed the landscape of customer relation management in the direction of more sophisticated methods for data analysis. Computational intelligence can deliver value in two

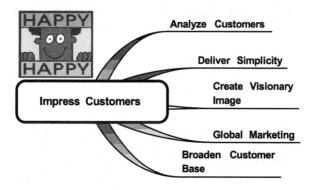

Fig. 8.6 Necessary features to impress customers

principally different ways for customer analysis. The first way, which is predominant, is the passive manner of analyzing all possible information about customers. The basic features include finding classical statistical relationships and customer categorization. Many customer-related decisions must be made quickly even with statistical tools (for example, assessing the likelihood of a customer to switch providers). In other cases, there are potentially thousands of models that could be built (for example, in identifying the best target customer for thousands of products). In this category of complex problems, several computational intelligence methods give unique capabilities for analysis. Of special importance for automatic detection of clusters of new customers are the self-organizing maps, derived by neural networks.

The second method of customer analysis, still in its infancy, is the active analysis of customer behavior by simulations with intelligent agents.

8.5.2   Deliver Simple Solutions

Simplicity is one of the key product features that attracts customers and keeps their loyalty in a natural way for a long time. A typical example is the current iPod-mania, partially inspired by the extremely simple and elegant way of performing complex tasks. It is not a secret that even the most intelligent customers prefer to manipulate products in a simple way without thinking too much. An additional advantage of having simple solutions is their better positioning for global marketing. Usually products with simple use can be manipulated by customers with minimal educational level and from any culture.

The paradox is that a lot of intelligence is needed to transfer the growing technical complexity into solutions with obvious simplicity of use. One of the computational intelligence technologies – multicriteria evolutionary computation – has unique capabilities to automatically generate a variety of solutions with different levels of simplicity dependent on the performance or functionality of the solution. In this case the fitness, i.e. the driving force of simulated evolution, is based on two criteria — performance and simplicity. As a result, the winning solutions are the simplest possible for a given performance. We call this feature emerging simplicity and it is illustrated with real-world applications in the third part of the book.

8.5.3   Create a Visionary Image

Some customers prefer companies and products with high-tech appeal. A typical example is the successful small cleaning robot Roomba with more than 2 million sold.[6] Suddenly, this small device transfers carpet cleaning from a symbol of low

---

[6]http://www.irobot.com/

pride and mundane activity into the first step of introducing robots into our homes. Many customers are buying Roomba not because of its tremendous technical features (a human still cleans much better) but to look more visionary and to gain respect from their neighbors and colleagues.

Computational intelligence is one of the few emerging technologies that naturally fits the visionary image. For example, another successful product in the mass market is the popular neuro-fuzzy rice cooker, manufactured by Panasonic. It is based on two computational intelligence techniques – neural networks and fuzzy logic. According to customers, the neuro-fuzzy-cooked rice is much tastier. It turns out that the visionary image of the product can even improve appetite.[7]

Another interesting opportunity for applied computational intelligence is the hunger of top managers to identify emerging technologies for future growth. Computational intelligence has enough application credibility to look like the real deal, and enormous technical appeal to be viewed as a cutting-edge technology for the 21st century. By embracing computational intelligence, top managers have a very good chance to look visionary with relatively low risk by introducing the technology. Most of the other emerging technologies with high media attention, such as nanotechnology or quantum computing, are too far away from delivering practical solutions in the near future.

8.5.4 Broaden Customer Base

The ultimate goal of customer analysis and modeling is to broaden the existing customer base. One of the possible ways to increase the number of customers is by predicting their future spending habits based on the analysis of their market baskets. Computational intelligence is the core technology in this type of behavior analysis. A typical example is the purchase suggestions from Amazon based on customers' shopping carts and wish lists. Another possible way to broaden the customer base is by geographic expansion. The key issue in this direction is adapting global marketing to local culture. Simulation based on intelligent agents may help in defining the local marketing strategy.

8.6   Enhance Creativity

Creative work is critical in innovation generation and process improvement, which are the key value generators. In principle, creativity is strongly individual and difficult to be directly influenced by better technology or organization. One of the challenges of contemporary competition is to find the fertilized intellectual soil for

---

[7]http://rice-cookers.wowshopper.com/pics-inventory/nszcc-e.gif

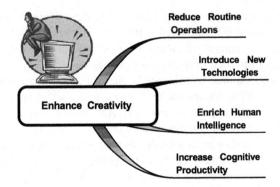

Fig. 8.7 Key components to increase productivity by enhancing creativity

Reduce Routine Operations

Introduce New Technologies

Enhance Creativity

Enrich Human Intelligence

Increase Cognitive Productivity

the blossoming of employees' creativity. Some computational intelligence techniques give opportunities to enhance creativity by reducing routine intellectual work, magnifying imagination, adding intellectual sensors, and increasing cognitive productivity (see the mind-map in Fig. 8.7). All of these options are briefly discussed next.

8.6.1   Reduce Routine Operations Related to Intelligence

Different computational intelligence techniques are at the basis of many routine operations which reduce the time for mundane intellectual labor. The long list includes, but is not limited to, the following activities: pattern analysis, fighting spam, spell and grammar checkers, automatic translation, voice recognition, fault detection, and complex alarm handling. An example is digital mammography which uses automatic scanning of mammograms for potential breast cancer lesion identification with neural networks and support vectors machines. In this way the time for the final decision of the radiologist is significantly reduced. Of special importance is the automatic handling of complex alarms in manufacturing, since there is typically a short response time and the potential for fatal human error is high due to the decision-makers being in a stressful state. Very often process operators have to handle alarm showers with hundreds, even thousands of alarms per minute. This makes the reasoning process of identifying the root cause from multivariate sources in real time extremely difficult. Using automatic intelligent alarm handling with root cause identification can help the operating personnel in finding the solution in a reliable and timely manner and reducing the risk for product losses and potential shutdowns.

8.6.2   Magnify Imagination

For some activities in the area of research, design, and art improving the capabilities to inspire imagination is critical. In science, according to the famous Einstein thought, imagination is more important than knowledge. The key computational intelligence approach that may help in stimulating imagination is evolutionary computation. Its capability for novelty generation can be used automatically or by interactive simulated evolution in all of the mentioned application areas. One area with very impressive results in using evolutionary computation is evolutionary art. The reader can find interesting examples on the Web.[8]

8.6.3   Add Intellectual Sensors

Creativity often depends on factors beyond our biological sensors and usually the phrase "sixth sense" is used to describe the phenomenon. The perception of a "sixth sense" is wildly different and in most cases is based on predicting without sufficient data or knowledge. One of the biggest benefits of computational intelligence is its capability to add new more complicated features based on data. We call them intellectual sensors. Some examples of this type of high-level sensor are pattern recognition, trend detection, and handling high dimensions. Pattern recognition allows automatic definition of unknown patterns from data, which could be very helpful in any intellectual activity. Humans have difficulties in capturing trends, especially in different time-scales. However, it is a task within the reach of several methods, including neural networks. As a result, potential trends of important events, such as market direction or key process variable drifts, can be automatically detected in the very early phase of development.

8.6.4   Increase Cognitive Productivity

The final result of enhanced creativity is increased cognitive productivity. We'll emphasize in three different ways by which computational intelligence improves intellectual labor. The first way is by reducing the time and increasing the productivity of data analysis. The various computational intelligence methods can capture data trends in the very early phase of development, recognize complex patterns, derive empirical models with robust behavior, and handle imprecise information. All of these impressive features are not available using the competitive approaches at that level of high speed and low cost.

The second way of increasing cognitive productivity using CI is by improving the efficiency of exploring new ideas. As was discussed earlier in this chapter, computational intelligence enhances productivity in each phase of the exploration process – from analyzing the business impact and modeling to effective scale-up and commercialization of the new idea.

---

[8]http://www.karlsims.com

The third way of increasing cognitive productivity with CI is by improving the quality of decision-making. The objective nature of computational intelligence solutions, based on data, is a preferable option for automatic cognition in comparison to the alternative of heuristic-based expert systems.

8.7 Attract Investors

The Internet bubble did not deter investors from the attraction of high-tech opportunities. Computational intelligence is one of the rare cases of high-tech which is seen as a catalyst of other technologies. This unique capability broadens significantly the opportunities for attracting investors, and some of them are represented in Fig. 8.8 and discussed below.

8.7.1   Intelligence and Growth

It is a well-known fact that the key attractive point for investors is the potential for growth. Usually this potential is linked to the efficiency of intellectual labor. Since the key role of computational intelligence is enhancing human intelligence, the potential influence to growth is big. In addition, computational intelligence can contribute to various aspects of growth. As was discussed earlier in this chapter, fast innovations are critical for growth, and they could be significantly accelerated using computational intelligence. Another form of growth by geographical expansion is decisive for successful globalization. Through its unique capabilities of market analysis and supply-chain optimization, computational intelligence could significantly reduce the cost of global operation. Computational intelligence also has technical competitive advantage in different areas of manufacturing which can

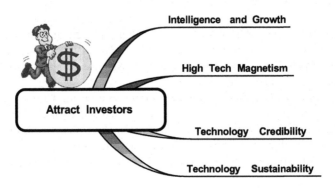

Fig. 8.8 Necessary features to attract investors

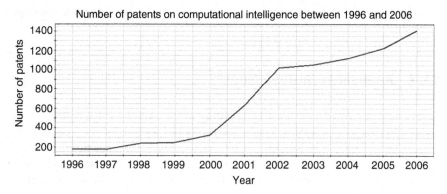

Fig. 8.9 Number of published patents on computational intelligence[9] between 1996 and 2006

deliver productivity growth. Last, but not least, computational intelligence is one of the fastest-growing areas in delivering intellectual capital by filing patents. The growth of computational intelligence-related patents in the period between 1996 and 2006 is clearly seen in Fig. 8.9.

8.7.2 High-Tech Magnetism

The high-tech appeal of computational intelligence is another advantage in its favor. Several companies (Wall Street favorites) have a significant presence in computational intelligence. Examples are: the machine learning group at Google and Microsoft, the data mining group at Amazon and Dow Chemical, and the computational intelligence group at General Electric. The success of computational intelligence in these established businesses makes the task of allocating venture capital for future computational intelligence-related enterprises easier. An additional attractive point to investors is that computational intelligence doesn't require big capital investments relative to other emerging technologies.

8.7.3 Technology Credibility

One of the differences between computational intelligence and the other high-tech alternatives is that it has already demonstrated its potential for value creation in many application areas. In contrast to quantum computing or nanotechnology, computational intelligence is beyond the purely research phase of development and has created mass applications such as Internet search engines, word-processor spell-checkers and the rice cooker. Some events, highly-publicized by the media,

---

[9]The results are based on a search with the key word "computational intelligence" on the website http://www.micropatent.com

like the chess battle between Kasparov and Big Blue, also add to technology credibility.

8.7.4 Technology Sustainability

A very important factor in favor of investing in computational intelligence is its potential for sustainable growth. One of the computational intelligence technologies, evolutionary computation, benefits directly from increasing computational power. It is estimated that the fast-growing processor capacity will reach the brain's potential in the foreseeable future, when the capabilities of evolutionary computation to generate novelty in almost any area of human activities will be practically limitless.

Computational intelligence is also a very fast-growing research area with several new approaches on the horizon, such as computing with words, artificial immune systems, intelligent evolving systems, and co-evolutionary systems. In addition, the internal research development of the field is combined with increased demand for industrial applications. Both of these tendencies are discussed in the last chapter of the book.

8.8 Improve National Defense

It is a well-known secret that both AI and computational intelligence are supported by the defense research agencies in almost any developed country. The military were one of the first users of this technology in many application areas, such as intelligence gathering and exploitation, command and control of military operations, and cyber security. A systemic view of the key components of national defense, influenced by computational intelligence, is shown in Fig. 8.10 and discussed next.

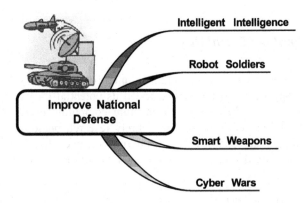

Fig. 8.10 Important components of national defense for the 21st century

8.8.1 Intelligent Intelligence

Intelligence is concerned with recognizing the unusual, which is often in the form of hidden patterns – implicit and explicit connections between events, processes, things and concepts. In addition, these patterns are very weak and are observed over long periods of time. To find, collect, organize and analyze these patterns the intelligence community employs a wide variety of sensors. These key sensors are involved with gathering intelligence from human agents in the field, from intercepting and analyzing communications, and from reconnaissance photographic sources. From these and other sources, intelligence analysts apply a broad spectrum of analytical tools and techniques, most of them based on computational intelligence.

Examples of models related to intelligence applications are as follows:[10]

- The purchase of chemicals, building materials, hardware that, by itself, would not be unusual, but taken together and related by a common transaction mechanism (e.g. a credit card);
- An attempt to gain a common set of rare or unusual skills (such as the attempt by the World Trade Center attackers to gain pilot certification);
- The theft of particular classes of vehicles – a mixture of UPS or Federal Express trucks as an example. These vehicles are so ubiquitous in our society that they are effectively invisible in large metropolitan areas.

8.8.2 Robot Soldiers

The Pentagon predicts that robots will be a major fighting force in the U.S. military in less than a decade, hunting and killing enemies in combat. Robots are a crucial part of the Army's effort to rebuild itself as a 21st-century fighting force, and a $127 billion project called Future Combat Systems is the biggest military contract in U.S. history.[11] Pentagon officials and military contractors say the ultimate ideal of unmanned warfare is combat without casualties. Failing that, their goal is to give as many dirty, difficult, dull or dangerous missions as possible to robots, conserving soldiers' bodies in battle.

However there are financial considerations in addition to the military impact of implementing robotic soldiers. The Pentagon today owes its soldiers $653 billion in future retirement benefits that it cannot pay. Robots, unlike old soldiers, do not fade away. The cost of a soldier from enlistment to interment is about $4 million today and growing, according to a Pentagon study. Robot soldiers are supposed to cost a tenth of that or less.

---

[10]E. Cox, Computational intelligence and threat assessment, *PC AI*, *15*, October, pp. 1620, 2001.

[11]T. Wiener, Pentagon has sights on robot soldiers, *New York Times*, 02/16/2005.

There is another issue of a moral character, though. Decades ago, Isaac Asimov posited three rules for robots: Do not hurt humans; obey humans unless it violates Rule 1; and, defend yourself unless it violates Rules 1 and 2. Obviously, robot soldiers violate Asimov's rules.

8.8.3 Smart Weapons

One of the key application areas of smart weapons is automatic target recognition, which uses computational intelligence. The famous application is the Predator with more than 1200 drones in the U.S. military arsenal in 2005. Today drones as small as a crow and as big as a Cessna are searching for roadside bombs, seeking out insurgents, and watching the backs of the troops. They are cheap, they can stay in the air longer than any manned aircraft, and they can see a battlefield better – all without risking a pilot. In 2006, an unmanned aircraft made military history by hitting a ground target with a small smart bomb in a test from 35,000 feet, flying 442 miles an hour.

Computational intelligence is also the key technology for the next generation of smart weapons – the swarms of unmanned, unattended and untethered drones on the ground, in the air and underwater. These machines would be capable of independently handling events in a hostile combat zone, such as surveillance, strike and even capture and detention.

Another possible application is a smart dust of micro-robots based on swarm intelligence. They could be used in searching for mines, reconnaissance buildings, and destroying enemies in caves.

8.8.4 Cyber Wars

Computer simulation of military activity is of critical importance for analyzing combat strategy and planning. Most of the available simulation environments include computational intelligence methods. One of the popular cyber war packages used worldwide by the military operations research community is EINSTein.[12] It has pioneered the simulation of combat on a small-to-medium scale by using autonomous agents to model individual behaviors and personalities rather than hardware. EINSTein offers a bottom-up, generative approach to modeling combat and stands in stark contrast to the top-down, or reductionist philosophy that still underlies most conventional military models and illustrates how many aspects of land combat may be understood as self-organized, emergent phenomena. The range

---

[12]A. Ilachinski, *Artificial War: Multiagent-Based Simulation of Combat,* World Scientific Publishing, Singapore, 2004.

of available cyber war games is also very broad. On the one end are games like NERO, which allows the use of different tactics to defeat the enemy by using intelligent agents, based on neuroevolution. On the other end are the complex strategic war games, such as Cyberwar XXI, based on different machine learning techniques, which can simulate the consequences of strategic decisions.

8.9 Protect Health

In a domain like healthcare, where medical knowledge grows at an exponentially growing pace, the only option is improving the decision-making process. Computational intelligence is very well positioned in this expanding industry with well-educated key customers open to new technologies, like most of the baby boomers. Many personal health-protection needs, from early symptom diagnostics to comprehensive health monitoring systems, can benefit from the capabilities of computational intelligence. The key areas of potential value creation in healthcare through computational intelligence are shown in Fig. 8.11 and discussed below.

8.9.1 Medical Diagnosis

It is a well-known secret that medical doctors look at the task of diagnosing disease as the most important part of their job and are resistant to any attempts to delegate it to a computer. Unfortunately, some of the initial applications in the late 1970s and early 1980s of applied AI directly challenged doctors' competency with the message of replacement of their expertise by a piece of software. As a result, attempts to

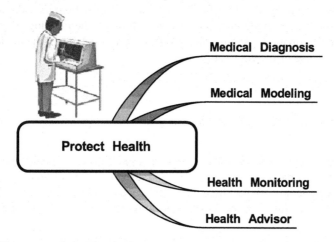

Fig. 8.11 Necessary technical features for successful health protection

improve the medical diagnosis capabilities by expert systems were coldly interpreted as competition and job threat.

However, nowadays, medical doctors understand much better the place of the computer in the diagnosis process and they are much more open to use computer systems as an aid in their practice. The main reasons are improved accuracy of computer systems, which implement multiple components of applied computational intelligence.

In addition, applied computational intelligence, with its philosophy of not replacing but enhancing human intelligence, can contribute significant improvements in the speed and quality of medical diagnosis done by medical doctors. Of special appeal to medical doctors is the combination of fuzzy systems, which can quantify and numerically process the verbal nature of health information, with the pattern-recognition capabilities of machine learning techniques. Typical examples are the diagnosis systems for early detection of breast and lung cancer, based on neural networks or support vector machines. There are several computer-aided detection (CAD) systems on the market, such as R2 ImageChecker, which are used for screening mammography. It is claimed that the ImageChecker could result in earlier detection of up to 23.4% of the cancers currently detected with screening mammography in women who had a mammogram – 24 months earlier.[13] There are other Federal Drug Administration (FDA)-approved CAD systems for detection of lung cancer on 3D computer tomography images as well as chest X-rays. There are also a number of CAD systems for detection of colon cancer.

A major problem for CAD systems is the generation of false positive detections. The radiologist needs to examine carefully every CAD detection, which can be time consuming and can also lead to incorrect increase in biopsies of benign lesions. A major effort is underway currently to improve the performance of CAD systems by reducing the false positive detections and at the same time to have the sensitivity of the system (the ratio of the detected cancer to existing cancers) high enough.

Advances in computer voice recognition made possible the use of voice recognition systems in many radiology facilities to transfer the radiologist's dictated case report to text. This resulted in faster automated filing of the radiologist's report in the hospital information system, leading to faster determination of the final diagnosis and the course of patient treatment.

8.9.2   Personal Health Modeling

Another big potential market for computational intelligence is the development of personal medical models. Due to the tremendous complexity of human organisms and the uniqueness of each individual, it is unrealistic to expect models based on fundamental principles. The personal health model will include a set of empirical

---

[13] According to the website: http://www.pamf.org/radiology/imagechecker.html

and semi-empirical models, which represent different dependencies related to a specific health component. For example, type 2 diabetic patients may have an empirical model which predicts the glucose level based on the different types of food and physical exercises. Type 1 diabetic patients may have an optimal insulin control system, based on the previous model. Neural networks, support vector machines, and symbolic regression via genetic programming could be the key building blocks for this type of medical modeling. An example is a neural network model for assessment of the absence of coronary artery stenosis, based on 19 factors, such as age, blood pressure, creatinine content, dialysis, etc. After each laboratory analysis or measurement, the model can predict the necessity of coronary angiography.

8.9.3 Health Monitoring

A typical health monitoring application is monitoring the depth of anesthesia of a patient during surgery. Currently anesthesiologists are adjusting the anesthetic dosage with heuristic rules based on their experience and traditional signs, such as spontaneous breathing, muscular movement, and changes in blood pressure or heart rate. However, these subjective adjustments depend on experience and cannot always account for the wide range of patient responses to anesthesia. As a result, inadequate dosage may cause severe complications during surgery and is one of the leading sources of lawsuits.

An alternative monitoring system for depth of anesthesia is based on a combination of neural networks and fuzzy logic.[14] The patient model uses as input the electroencephalograms (EEG) from the brain and quantifies the depth of anesthesia in a scale between 0 (fully awake) and 1 (complete EEG suppression). The experimental results on 15 dogs demonstrate the advantages of the neuro-fuzzy monitoring system, such as high accuracy (90%), robustness towards different patients and anesthetic techniques, prediction of the appearance of clinical signs of an inadequate anesthesia, and no need for calibration.

8.9.4 Personal Health Advisor

One of the expected tendencies in handling our health is to concentrate all available information and personal health models in our possession. Our personal computerized doctor will also include some decision-making functions based on the available data and models. It will communicate and adjust its recommendations with

---

[14]J. Huang *et al.*, Monitoring depth of anesthesia, *Computational Intelligence Processing in Medical Diagnosis*, M. Schmitt *et al.* (Eds), Physica-Verlag, 2002.

your medical doctor when it is necessary. Intelligent agents is the technology that can gradually make this necessity a reality.

8.10 Have Fun

The entertainment software industry has shown a tremendous expansion recently, with sales up to $25.4 billion dollars in 2004.[15] The key areas of development of this industry, linked to computational intelligence, are shown on the mind-map in Fig. 8.12 and discussed below.

8.10.1 Intelligent Games

Computational intelligence can be a useful technology in developing different types of games, such as checkers, chess, bridge, poker, backgammon, cribbage, etc. One of the first examples is Blondie24 developed by a team led by David B. Fogel[16] and described briefly in Chap. 5.

An amazing feature of Blondie24 is that its ability to play checkers does not depend on the human expertise of the game but only on the evolutionary process itself. The evolving players did not even know how individual games ended.

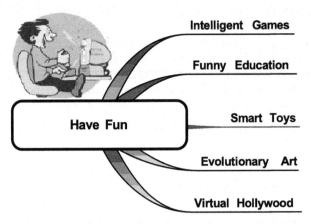

Fig. 8.12 Key directions of intelligent entertainment

---

[15]R. Miikkulainen *et al.*, Computational intelligence in games, *Computational Intelligence: Principles and Practice*, G. Yen and D. Fogel (Eds), IEEE Press, 2006.

[16]D. Fogel, *Blondie24: Playing at the Edge of AI*, Morgan Kaufmann, 2002.

8.10.2 Funny Education

Educational games are fundamentally different from the prevalent boring instructional mode of education. They are based on punishment, reward, and learning through doing and guided discovery, in contrast to the "tell and test" methods of traditional instruction. However, effective use of games is likely to be limited unless educational institutions are willing to consider significant changes in pedagogy and content, and rethink the role of teachers and other educational professionals.[17]

Some commercial computer games are already being used in high-school classrooms. For example, *Civilization III* – a computer game about the development and growth of world civilizations – is used in classrooms around the world to teach history, geography, science, and other subjects. Other commercial video games that have been used in high schools include: *SimCity*, a city planning game, and Roller Coaster Tycoon, a game used in classrooms to teach physics concepts such as gravity and velocity.

Another three games aim to address three different themes, namely strategies for responding to terrorist attacks (Mass Casualty Incident Response, developed in coordination with the New York City Fire Department), the science of immunology (Immune Attack), and the mathematics of ancient Mesopotamia (Discover Babylon).

The objective of these instructional games is making the subject matter more appealing to the three target groups: elementary school students for Discover Babylon, adolescents for Immune Attack, and adult workers for Mass Casualty Incident Response.

8.10.3 Smart Toys

Another application area for computational intelligence is toys with some level of intelligence. An interesting example is a modern version of an old traditional guessing game, called 20 questions (20Q), played on most long journeys, school trips and holidays with the parents. The idea of the game is to answer the questions the 20Q throws at you as accurately as possible. The twenty questions can be answered with either "yes", "no", "sometimes" or "don't know".

How does the 20Q guess what you are thinking[18]? The computational intelligence method behind the game is a neural network. The 20Q.net online version has about ten million neurons, and the pocket version has about 250,000 neurons. The game uses the neural network to choose the next question as well as deciding what to guess.

---

[17]*Harnessing the Power of Video Games for Learning*, Summit on Education Games 2006, Federation of American Scientists, 2006.

[18]http://www.radicagames.com/20q.php

8.10.4   Evolutionary Art

Using evolutionary computation in painting was discussed earlier in this chapter. Creating digital music is another potential application area of computational intelligence. Several implementations of neural networks and evolutionary computation methods include, but are not limited to: virtual orchestra, robotic drummers, and even artificial composers.

Another interesting area is creating interactive drama, such as in the project *Façade*,[19] where a real-time 3D animated experience is created of being with two "live" actors who are motivated to make a dramatic situation happen. The author actively participates in the generated scenes and directs the action as she likes it.

Façade is a computational intelligence-based art/research experiment. In *Façade*, you, the player, using your own name and gender, play the character of a longtime friend of Grace and Trip, an attractive and materially successful couple in their early thirties. During an evening get-together at their apartment that quickly turns ugly, you become entangled in the high-conflict dissolution of Grace and Trip's marriage. No one is safe as the accusations fly, sides are taken and irreversible decisions are forced to be made. By the end of this intense one-act play you will have changed the course of Grace and Trip's lives, motivating you to replay the drama to find out how your interaction could make things turn out differently the next time.

8.10.5   Virtual Hollywood

It is possible that with improving graphical capabilities and intelligent agents, computational intelligence builds the basis for creating movies, entirely based on virtual reality. An example is the virtual-reality drama, The Trial, The Trail.[20] It is a brand new type of dramatic entertainment, where instead of identifying with the protagonist, the audience becomes the protagonist. Human audience members in this interactive drama sink virtually into the world of the characters on the screen.

Instead of using a joystick to compete against virtual characters as in a video game, the actions of human users determine how the virtual characters respond, based on an ever-growing "library" of actions and verbal communications with which the virtual reality characters are endowed.

So, as the human user proceeds through the drama, his/her actions are being recorded computationally over the Internet, interpreted psychologically and used to prompt the responses by the virtual characters.

In a sense, the computational agents in the drama must improvise around the human user, who is acting spontaneously without a script.

---

[19]J. Rauch, Sex, lies, and video games, *The Atlantic Monthly*, November 2006, pp. 76–87.

[20]J. Anstey, *et al.*, Psycho-drama in virtual reality, *Proceedings of the 4th Conference on Computation Semiotics*, 2004.

8.11   Summary

Key messages:

Computational intelligence improves business competitive capabilities by accelerating innovations, impressing customers, and attracting investors.

Computational intelligence increases productivity by enhancing creativity, improving manufacturing efficiency, and optimizing the supply chain.

Computational intelligence significantly contributes to modern national defense.

On a personal level, computational intelligence has the capabilities to protect our health and make our leisure more fun.

The Bottom Line

Applied computational intelligence has the potential to create value in almost any area of human activity.

Suggested Reading

A book with a good summary of effective innovation strategies:
M. George, J. Works, and K. Watson-Hemphill, *Fast Innovation*, McGraw-Hill, 2005.

A good summary of the state of the art of applied computational intelligence in medical diagnostics is the book:
M. Schmitt *et al* (Eds), *Computational Intelligence Processing in Medical Diagnosis*, Physica-Verlag, 2002.

Two important references in the area of defense applications of CI:

E. Cox, Computational intelligence and threat assessment, *PC AI*, *15*, October, pp. 16–20, 2001.
A. Ilachinski, *Artificial War: Multiagent-Based Simulation of Combat*, World Scientific Publishing, Singapore, 2004.

Chapter 9
Competitive Advantages of Computational Intelligence

> *If you don't have competitive advantage, don't compete.*
> Jack Welch

Competitive advantage is on the top ten list of the most used (and abused) phrases by managers. Even more, very often it is the winner of the game "Bullshit Bingo" popular in the corporate world.[1] Fortunately, competitive advantage is beyond the hype and has a very clear business meaning and as such is at the basis of almost any economic analysis and strategic decision-making. It is not entirely the case when this phrase is used to evaluate the economic impact of a given research approach relative to others. Usually the focus is on academic comparative studies of the technical supremacy or limitations between methods. However, the transformation of an undisputed scientific superiority of a given approach into improved business performance with clear impact in the competitive market is still a vast unknown exploratory effort.

In reality, defining competitive advantage is the first question we need to address when applying new emerging technologies. The objective of this chapter is to give some answers in the case of computational intelligence.

9.1 Competitive Advantage of a Research Approach

Webster's Dictionary defines the term "advantage" as the superiority of position or condition, or a benefit resulting from some course of action. "Competitive" is defined in *Webster's* as relating to, characterized by, or based on competition (rivalry). A competitive advantage can result either from implementing a value-creating strategy not simultaneously being employed by current or prospective competitors or through superior execution of the same strategy relative to

---

[1]The interesting reader can enjoy the game from the website http://www.bullshitbingo.net/cards/bullshit/

A.K. Kordon, *Applying Computational Intelligence*,
DOI 10.1007/978-3-540-69913-2_9, © Springer-Verlag Berlin Heidelberg 2010

competitors. Based on this economic characterization, we define the competitive advantage of a research approach as technical superiority that cannot be reproduced by other technologies and can be translated with minimal efforts into a position of competitive advantage in the marketplace. The definition includes three components. The first component requires clarifying the technical superiority (for example, better predictive accuracy) over alternative methods. The second component assumes minimal total cost of ownership and requires assessment of the potential implementation efforts of the approach and the competitive technologies. The third component is based on the hypothesis that the technical gains can improve the business performance and contribute to economic competitive advantage.

The practical procedure for evaluation of competitive advantage of an emerging technology includes three generic steps, shown in Fig. 9.1 and described below.

9.1.1 Step 1: Clarify Technical Superiority

The objective of this initial step of competitive analysis is to define and quantify the specific technical advantages of the research method relative to its key competitive

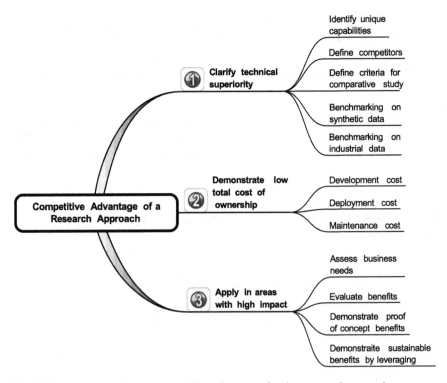

Fig. 9.1 Key steps of evaluating competitive advantage of a given research approach

approaches. One of the possible ways to accomplish this goal is by the following sequence:

- *Identify Unique Capabilities* – Focus on the features where the evaluated approach *shines*. Here are some examples from computational intelligence technologies: fuzzy systems can handle imprecise information and represent phenomena by a level of degree; neural networks can identify hidden patterns and relationships by learning from data; and evolutionary computation can generate systems with novel features.
- *Define Competitive Research Approaches* – Select the most relevant known methods closely related to the evaluated approach. An analysis of the key scientific approaches competing directly with computational intelligence is given in the next section.
- *Define Criteria for Technical Comparison* – Specify the list of criteria based on statistical characteristics or expert knowledge for relative performance evaluation. On data-related comparative studies it is strongly recommended to validate if the difference in the results is statistically significant. Without it, the conclusions lack credibility.
- *Benchmark on Synthetic Data* – Apply defined criteria on well-accepted standard benchmarks, typical for any research community. They are well-known and can be downloaded from the Web.[2] The most widely used sites for each computational intelligence approach are given in Part I of this book. In addition, there are many academic-driven comparative studies between almost all computational intelligence techniques that could be used in the analysis.[3]
- *Benchmark on Real-World Data* – The final technical test is on a case-study maximum close to the class of problems targeted for application. It is recommended that the benchmark includes all the nastiness of the real-world applications like missing and erroneously entered data, inconsistent knowledge, and vague requirements.

9.1.2 Step 2: Demonstrate Low Total Cost of Ownership

The goal of this step in competitive advantage analysis is to assess the application cost. Different aspects of the cost have to be taken into account, such as capital cost (the need for computers, clusters, laboratory equipment, very expensive software packages, etc.), software licenses, and, above all, labor cost. The level of detail of the evaluation is flexible and it is not definitely necessary to count the pennies. Very often a guess of the expense in an order of magnitude is sufficient. Of big

---

[2]http://ieee-cis.org/technical/standards/benchmarks/

[3]L. Jain and N. Martin (Eds), *Fusion of Neural Networks, Fuzzy Sets, and Genetic Algorithms: Industrial Applications*, CRC Press, 1999.

importance is defining the cost savings in the three main components of the total cost of ownership: development, deployment, and maintenance costs.

- *Development Cost* – Usually it includes: the necessary hardware cost (especially if more powerful computational resources than PCs are needed), the development software licenses, and above all, the labor efforts to introduce, improve (internally or with external collaboration), maintain, and apply the technology.
- *Deployment Cost* – The estimate takes into account the following: the hardware cost for running the application, the run-time license fees (sometimes they are higher than the software development fees), the labor efforts to integrate the solution into the existing work processes (or to create a new work procedure), to train the final user, and to deploy the model.
- *Maintenance Cost* – Involves assessment of the long-term support efforts of the application for at least five years. Some types of applications, such as emissions monitoring systems, require substantial hardware expenses for periodic data collection and model validation to pass quarterly or annual regulatory tests. Very often, in supporting complex solutions, high-skilled labor at a Ph.D. level is needed and that significantly increases the maintenance cost.

9.1.3   Step 3: Apply in Areas with High Impact

This is the most difficult part of the competitive advantage evaluation process since there is no established methodology and not enough publicly available test cases and data. At the same time, it is the most important step of the analysis as the potential for transforming the technical advantages into value has to be assessed. The following sequence of actions is recommended:

- *Assess Business Needs* – Identify and prioritize unresolved problems in a targeted business. Understand the key obstacles for preventing a potential solution – lack of support, insufficient data, limited resources, no urgency, etc. Another important task is to estimate the existing experience with modeling and statistics and assess the willingness of the potential user to take the risk and apply the new technology.
- *Evaluate Benefits of Applied Approach* – Includes the critical task of mapping the technical superiority features of the method on the business needs. For example, if there is a need for improved optimal scheduling of the existing supply-chain systems, the benefits of using the unique capabilities of several computational intelligence methods, such as genetic algorithms, evolutionary strategies, ant colony optimization, and particle swarm optimizers, has to be assessed. The ideal final result of this action item will be if an estimate of the potential value is given, in addition to the technical benefits.
- *Demonstrate Proof of Concept Benefits* – Following the generic rule that "the proof is in the pudding", it is necessary to validate the business hypothesis for competitive advantage with selected applications. It is preferable to illustrate the

technical and business advantages relative to the current industrial standard solutions for the chosen class of problems even if they are based on different methods. Very often an issue in this action item is the higher-than-expected development cost of computational intelligence. One of the ways to overcome this problem is by compensating the increased development cost with the potential of leveraging the technology in similar future applications.

- *Demonstrate Sustainable Benefits by Leveraging* – A successful transformation of the technical superiority of a given research approach into business competitive advantage assumes growing application opportunities. Demonstrating benefits on one application case is insufficient if it is not supported by an implementation strategy for how the proof-of-concept solution will be leveraged in similar areas. In addition, identifying new application areas is always a plus. Introduction of a new emerging technology in industry becomes a very expensive exercise. Only techniques with clear technical superiority and capability of delivering sustainable benefits in a broad class of applications can win the competition. Fortunately, computational intelligence is such a type of emerging technology.

9.2   Key Competitive Approaches to Computational Intelligence

One of the most important steps in defining competitive advantages of computational intelligence is identifying the key technological rivals in the race. A very generic competitive approaches selection combined with high-level comparison with computational intelligence is proposed in this section. One of the comparative analysis issues is that the specific computational intelligence technologies have a very broad range of features. For example, the advantages of fuzzy systems are significantly different from the benefits of swarm intelligence. In such high-level analysis, however, we compare the competitors with the wide spectrum of features that all computational intelligence technologies, such as fuzzy systems, neural networks, support vector machines, evolutionary computation, swarm intelligence, and intelligent agents, offer for practical applications. It is possible to implement the same methodology for each specific computational intelligence technique at a more detailed level.

From the numerous techniques used for solving practical problems, the following approaches have been selected (shown in a field of track runners analogy in Fig. 9.2): first-principles modeling, statistics, heuristics, and classical optimization.

In selecting the competitors we took into account not only the tremendous value that has been delivered by each approach in many applications, but also the almost religious support it has from different researchers and practitioners. Behind each method, there is a large community of gurus, research groups, vendors, consultants, and diverse types of users. The probability of interacting with some of these communities in promoting or applying computational intelligence is very high.

Computational intelligence

First-principles modeling

Statistics

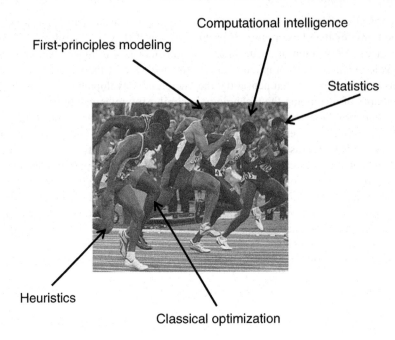

Heuristics

Classical optimization

Fig. 9.2 Main competitors to computational intelligence

On the surface, such a generic comparison looks mundane and the needed arguments seem trivial. Surprisingly, our practical experience from introducing emerging technologies shows that the common user (engineer, economist, or manager) is not aware of the practical implications of all of these well-known methods. Very often we observed a bias toward specific approaches with exaggeration of their advantages and neglecting the inevitable limitations. Based on this experience, we recommend beginning the comparative analysis with a clear answer about the advantages and disadvantages of each competitor relative to computational intelligence. In principle, the list of pros and cons could be very long. However, for the purposes of the high-level analysis and delivering a simple message, we will focus on the top three positive/negative features of each method. The final objective of the analysis includes also identified areas where computational intelligence can improve the corresponding competitive approach.

9.2.1   Competitor #1: First-Principles Modeling

First-principles model building represents the classical notion of modeling. It is the backbone of numerous successful applications in many industries. An army of researchers, engineers, and managers has the firm conviction that this is the only

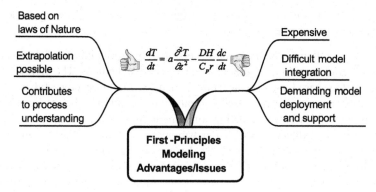

Fig. 9.3 Pros and cons of first-principles modeling

credible way of modeling. However, the growing complexity of manufacturing processes and the high dynamics of the market require solutions beyond the capabilities of the known laws of Nature.[4] Let us look at the key pros and cons of this well-known competitor, as shown in Fig. 9.3.

Key advantages of first-principles modeling:

- *Based on Laws of Nature* – This undisputed feature of first-principles modeling allow us to develop solutions on a solid theoretical framework. It automatically introduces high credibility of the model and opens the door for better acceptance from potential users while reducing the risk-taking threshold in their decision for practical implementation. In many cases, such as chemical reactors, heat-exchangers, and distillation towers, analytical models are a must. They are accepted by the majority of process engineers as regular tools that require only model parameter fit.
- *Extrapolation Possible* – Maybe the most important advantage of first-principles models from a practical point of view is their ability to predict correctly beyond the ranges of development data. In most real-world applications there is a high probability that they will operate in new process conditions which have not been taken into account or tested during model development. The robustness of the solution towards unknown operating regimes is critical for model credibility and long-term use. First-principles models are one of the few approaches that have the theoretical basis to claim such extrapolation capabilities.
- *Contributes to Process Understanding* – A significant side-effect of first-principles modeling is the ability to increase process knowledge and understanding. Very often this leads to discovery of new unknown properties. Process understanding, gained by intensive simulations of first-principles models, also reduces the risk of scale-up of the application, which is one of the most expensive and risky operations in industry.

---

[4]Basic laws of natural sciences, such as physics and chemistry.

The discussed impressive pros are balanced with the following three key disadvantages of first principles modeling:

- *Expensive* – Often the total cost of ownership of first-principles models is at least of an order of magnitude higher than the alternative methods. The high development cost is based on several causes, such as: (a) more powerful hardware needed because of the increased computational requirements, (b) license fees for first-principles-based modeling software are one of the most expensive, and (c) costly data collection for model validation. The lion's share, however, is taken by the high-skilled labor cost. Even with the most powerful and user-friendly simulation software, first-principles modeling takes much longer than alternative methods. It requires time-consuming hypothesis search through many possible physical mechanisms and very exhaustive model validation with increased number of parameters (hundreds, even thousands). The trend, however, is towards reduction of first-principles modeling cost due to increasing capabilities of developed commercial software.
- *Difficult Model Integration* – In many applications the adequate modeling of the process requires several different first-principles modeling techniques. For example, a common case in the chemical industry is to combine material-balances with computational fluid dynamics and thermodynamics models. Usually the individual models are developed in separate software with entirely different interfaces and data structure. Even more important than the software incompatibility are the differences in the principles and assumptions of modeling methods. As a result, integrating all first-principles models, necessary for a given application, is very difficult. It significantly slows down the model development process.
- *Demanding Model Deployment and Support* – Another issue in deploying first-principles models is their relatively slow execution speed even on the most advanced computer systems. This creates problems, especially if frequent model-based control and optimization is needed in real time. First-principles models also require very highly qualified support, since model understanding assumes specialized knowledge about the core science, e.g., physics, and the corresponding software packages. Especially tricky is the periodic model validation. It requires very intensive data collection to verify the model assumption space. A well-recognized issue is the lack of confidence measure of the predictions if the model operates outside the postulated space and is theoretically invalid.

One of the objectives of the competitive analysis is to identify opportunities for the computational intelligence to resolve some of the issues of the specific competitor while keeping the advantages of the compared approach. In the case of first-principle modeling, the key opportunities for computational intelligence can be defined as:

- *Reduce Development Cost, Decrease the Hypothesis Search Space* – One possible way to shorten first-principles model development time is by reducing the hypothesis search process for finding appropriate physical/chemical mechanisms. Using the methods of evolutionary computation and swarm intelligence can significantly accelerate this process. An example of combining genetic programming and fundamental model building with substantial reduction of new product development time is given in Chap. 14.

- *Integration of Different First-Principles Models by Emulators* – Emulators are empirical models that represent the key functionality of first principles models. The data for the emulators are generated by different fundamental models. The final integrated empirical model is derived by neural networks or genetic programming using the combined data, generated from all necessary models. The extremely difficult concatenation on the first-principles modeling level can be easily accomplished on the data-driven modeling level.
- *Reduce Deployment and Maintenance Costs with Emulators* – The empirical proxies are fast and easy to deploy. Their execution time is milliseconds, not minutes or hours. There is no particular need for specialized hardware and high-skilled support. Tracking the performance of empirical models is also much cheaper than looking at the specific requirements at each different first-principles model.

9.2.2   Competitor #2: Statistical Modeling

Statistics is probably the most widespread modeling approach in practice. Most engineers, biologists, and economists have some statistical background. Recently, even managers began to use statistical methods in their decision-making, not to mention that Six Sigma became one of the necessary buzzwords in the top management mantra.

Very often, however, this popularity is at the expense of mass-scale misuse and abuse of the approach. It also shapes a mindset that all the problems related to data must be resolved by statistics and all other methods are statistically not credible. The comparative analysis has to be done very carefully and in a balanced way. The top three advantages/disadvantages of statistics are shown in Fig. 9.4.

First, we will focus on the well-known advantages of statistics.

- *Solid Theoretical Basis* – If the mechanistic models are founded on the laws of Nature, statistics is based on the laws of Numbers (especially if the number of observations is big). The statistical concepts and principles are well accepted by practitioners and the credibility of the statistically based models is as high as that of their fundamental counterparts. A clear advantage of statistics is that its key theoretical foundation is known to a broader audience of practitioners than the much more specific first-principles laws. Another significant influence of statistics in the area of real-world applications is the defined metrics for performance evaluation. The theoretical statistical measures of model quality are at the basis of all contemporary process quality systems and the Six Sigma work process.

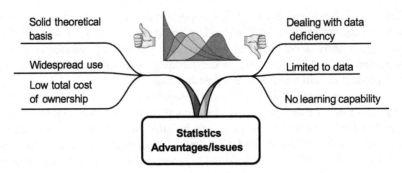

Fig. 9.4 Pros and cons of statistics

Statistical theory on design of experiments is the key guideline in planning and evaluating scientific and practical experiments.

- *Widespread Use* – From all the discussed competitive approaches, statistics has the broadest user audience. There are several sources of this popularity – available university courses, many training opportunities, and several good introductory books, to name a few. An additional factor is the variety of available software options, from Excel add-ins to professional packages like JMP (the SAS Institute), MINITAB (Minitab, Inc), or SYSTAT (Systat Software, Inc). The acceptance of Six Sigma methodology in industry created tens of thousands of black and green belts, actively using statistical methods in their projects. It is not an exaggeration to claim that statistics is the accepted *lingua franca* in dealing with data.

- *Low Total Cost of Ownership* – For data-driven solutions, statistical modeling is at the economic minimum. All three components of the total cost of ownership (development, deployment, and maintenance costs) are at their lowest values. Development cost is low because: (a) there is no need for specialized hardware, (b) the software licenses even for the professional packages are minimal, (c) training cost is also minimal, and (d) development time is short. Deployment cost is low since the final models can be implemented on almost any software environment. Maintenance cost is low because there is no need for specialized training of the support teams.

Meanwhile statistics, as any approach, has its own problems. The top three issues, related to practical implementation of statistical modeling are discussed below:

- *Dealing with Data Deficiency* – Statistics has difficulties in cases when the number of variables is bigger than the number of data records (the so-called "fat" data sets). Most of the statistics algorithms for model parameter calculations require at least the number of records to be equal to the number of variables (for a linear model). For polynomial models the number of required records is larger. For some industrial applications, especially in microarray processing, this requirement is difficult to satisfy.

- *Limited to Data* – The key strength of statistics – dealing with data – is also a limitation. Statistics does not offer conceptual model development based on the laws of Nature, as mechanistic models do, nor knowledge representation and handling vague information, as heuristics and fuzzy logic do. As we know, intelligence is not based on data only, i.e. we need complementary methods to capture the full complexity of real-world applications.
- *No Learning Capabilities* – Statistical models are static and are based on the available data during model development. In the case of changes in process conditions, the models cannot adapt to the new environment by learning. This is especially difficult if the new conditions are far away from the assumption space of initial model development data.

Fortunately, most of the limitations of statistics can be solved by different computational intelligence methods. Even more, there are many ways of integrating both approaches. Here are some examples:

- *Using Statistics in Computational Intelligence* – Several computational intelligence approaches could benefit from more systematic use of statistics. Of special importance is the effective integration of statistical analysis in evolutionary computation, since this approach is based on a large number of populations. An example of the benefits of such integration is given in Chap. 11.
- *Broadening Empirical Modeling with Computational Intelligence Techniques* – Several computational intelligence techniques, such as symbolic regression via genetic programming, neural networks, and support vector machines for regression, can deliver high-quality empirical models. They can complement the models, delivered by statistics. Another option is to use computational intelligence methods to derive linearizing transforms of input variables and allow a credible statistical solution in dealing with nonlinear problems.
- *Promoting Computational Intelligence Through Statistical Software and Work Processes* – Due to the vast popularity of statistical modeling, the best strategy for mass-scale promotion of computational intelligence is by integration with the most widespread statistical software. Several vendors (like the SAS Institute) already offer neural network capability in their statistical products. Another option for effective promotion is to integrate computational intelligence into statistically based work processes, such as Six Sigma. The efficiency of this widespread industrial quality improvement program will grow significantly by broadening the existing statistical modeling capabilities with high-quality empirical solutions, generated by computational intelligence. An example of integrating computational intelligence into Six Sigma is given in Chap. 12.

9.2.3   Competitor #3: Heuristics

Heuristics modeling represents human knowledge as defined by experts. The assumption is that this knowledge has not been derived by data or machine learning

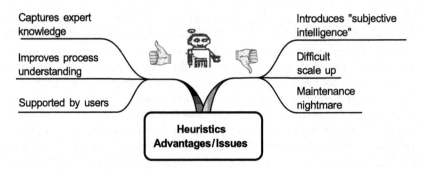

Fig. 9.5 Pros and cons of heuristics

techniques. In most cases, heuristics is implemented in the computer as rules. In fact, it is based on classical artificial intelligence and illustrates the early phases of introducing experts' knowledge in computers. The pros and cons of heuristics are shown in the mind-map in Fig. 9.5.

The top three identified advantages of heuristics are the following:

- *Captures Expert Knowledge* – The existing knowledge of a practical problem is not limited to first-principles or data-driven models. There are many "rules of thumb", guesses based on unique experience, and recommended actions in the case of system failure in the heads of the experts. Heuristics allows the transfer of this special experience into the computer system in the form of defined rules. Very often this knowledge is a substantial component in the final solution and leads to cost reduction. The savings are due to the smaller number of numerical models, optimal conditions defined by the experts, and suggested reliable operation limits.
- *Improves Process Understanding* – An inevitable effect of capturing and refining the knowledge from the best experts in the field is an enriched view of the process. In many cases it triggers additional ideas for process improvements and helps the first-principles model building by reducing the hypothesis search space with correct initial guesses of potential mechanisms.
- *Supported by Users* – Introducing experts' knowledge into the final solution has a very important psychological effect in favor of the future use of the developed system. By recognizing the experts, who are often the final users, we automatically introduce credibility into the suggested solution. When the experts see that the system includes their rules, they look at it as their "baby". From that moment they are personally interested in the success of the application and take care of the growth of the "baby".

Let us now switch to the key issues with heuristics:

- *Introduces "Subjective Intelligence"* – With all due respect to the experts' knowledge, it is based on their personal perceptions, qualifications, and beliefs. Biases, individual preferences, even personal attitudes toward other experts, are more the rule than the exception. Since at the center of the heuristics-related rules is the expert with his subjective thinking, we call this form of knowledge representation "subjective intelligence". A key feature of subjective intelligence is that the defined knowledge is based on human assessment only. It is not derived from "objective" sources, such as the laws of Nature or empirical dependencies, supported by data. "Subjective intelligence" could be very dangerous if the expertise in a field is scarce. The combination of limited numbers of experts with insufficient knowledge may transfer the expected application from a problem solver into a "subjective intelligence" disaster.

- *Difficult Scale-up* – Usually the defined rules of thumb capture the local experience of the problem. It is extremely difficult to transfer heuristics to another similar problem. The generalization capability of "subjective intelligence" is very limited. Especially difficult is scaling-up the solution to problems with a large number of variables.
- *Maintenance Nightmare* – In addition to the local nature of heuristic rules, they are static. This feature may create significant problems in the case of dynamic changes in the system since many related rules have to be recoded. If the system has high dimensionality and complexity (interdependent rules), it may become difficult and costly to maintain and support. An additional detail is the objective ignorance in new process conditions by the experts themselves, i.e., there is no longer a reliable source of rule definition.

There are many opportunities to improve heuristics by computational intelligence, mostly by replacing "subjective intelligence" with "objective intelligence", described in the next section. In fact, this is the process of transferring the classical artificial intelligence solutions into the new approaches of computational intelligence, which are discussed in many places in the book.

9.2.4   Competitor #4: Classical Optimization

Optimizing manufacturing processes, supply-chains, or new products is of ultimate interest to industry. Classical optimization uses numerous linear and nonlinear methods. The linear methods are based on least-squares techniques that guarantee finding analytically the global optimum. Classical nonlinear optimization techniques are based on either direct-search methods (simplex method and Hooke–Jeeves) or gradient-based methods (steepest descent, conjugate gradient, sequential quadratic programming, etc.) which can find local optima. Usually classical optimization in industry is multicriteria and includes many constraints.

Classical optimization is widely used by different types of users on various software packages – from the Excel solver to specialized optimizers in complex modeling programs. The top three pros and cons of this key competitor are shown in Fig. 9.6 and discussed below.

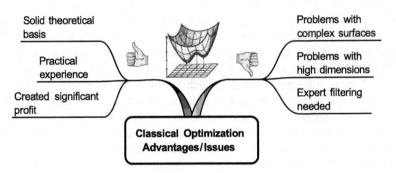

Solid theoretical basis		Problems with complex surfaces
Practical experience		Problems with high dimensions
Created significant profit		Expert filtering needed

Classical Optimization Advantages/Issues

Fig. 9.6 Pros and cons of classical optimization

The key advantages of classical optimization can be defined as:

- *Solid Theoretical Basis* – Most classical optimization methods are mathematically derived with proven convergence either to a global or to a local optimum. The users trust the theoretical basis and accept the mathematical credibility of the recommended optimal solutions even without deep knowledge of the specific optimization methods.
- *Practical Experience* – Classical optimizers are used in almost any type of industrial applications in manufacturing, supply-chain, or new product development. The requirements for applying the different optimization techniques are not significantly different and the training cost of the user is relatively low. The experience in optimizing one process can easily be transferred to other different processes.
- *Created Significant Profit* – The key advantage of classical optimization is that it already creates tens of billions of dollars profit to industry. The main sources of profit as a result of successful optimization are as follows: (a) operating plants with minimal waste of energy and raw materials, (b) distributing products with minimal transport expenses, and (c) designing new products with maximal quality and minimal material losses. An additional factor in favor of classical optimizers is that in most cases the profit gain does not require capital investment.

On the negative side, the top three issues of classical optimizers are as follows:

- *Problems with Complex Surfaces* – Most real-world optimization problems are based on noisy data and complex landscapes. In these conditions, classical optimizers get stuck in local minima and finding a global solution is difficult and time-consuming.
- *Problems with High Dimensionality* – Some complex industrial problems require optimization of hundreds, even thousands of parameters. Classical optimizers have limited capabilities for handling these types of tasks or the delivered solutions require specialized computer clusters and are usually very slow. Both of these factors raise the cost and reduce the efficiency of optimization due to the lower speed of execution.

- *Expert Filtering Needed* – Running complex industrial optimizers requires some expert knowledge on defining the initial conditions and the constraints. During the first runs it is also necessary to verify the physical meaning of the recommended optimal values. Very often adding feasibility constraints to protect the system from mathematically correct but practically unacceptable solutions is needed. In fact, almost any real-world optimizer requires a set-up period of a couple of months before operating fully optimally.

Computational intelligence methods, such as evolutionary computation and swarm intelligence, offer many opportunities to handle the issues of classical optimizers. Above all, they increase the probability of finding global optima even in very noisy and high-dimensional surfaces.

9.3   How Computational Intelligence Beats the Competition

The key competitors' analysis demonstrates that computational intelligence has unique capabilities to improve the performance of each discussed approach. The issue is, that in contrast to the competition, computational intelligence is virtually unknown to industry. Therefore, it is so important to define as broadly as possible the main competitive advantages of this new emerging technology. The selected pros must capture the essence of the strengths of computational intelligence and help the potential user to draw a clear demarcation line between this technology and the competitors.

In order to help the future user, we define the following key advantages of computational intelligence, shown in the mind-map in Fig. 9.7 and discussed in detail in this section.

9.3.1   Creating "Objective Intelligence"

The most important feature that boosts computational intelligence ahead of the competition is the "objective" nature of the delivered "smart" solutions. "Objective intelligence" is similar to first-principles and statistical models in that they are also "objective", since they are based on the laws of Nature and laws of Numbers. However, "objective intelligence" differentiates itself with its capability to automatically extract solutions through machine learning, simulated evolution, or emerging phenomena. The difference with "subjective intelligence" has already been discussed in the previous section.

The advantages of "objective intelligence" have a significant impact on the application potential of computational intelligence. The most important advantages are shown in Fig. 9.8:

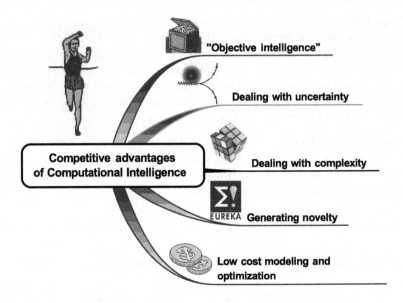

Fig. 9.7 Key competitive advantages of computational intelligence

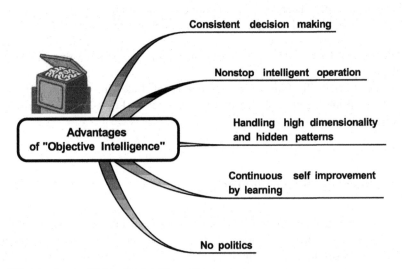

Fig. 9.8 Advantages of "objective intelligence"

- *Consistent Decision-Making* – In contrast to classical expert systems, the decisions suggested by "objective intelligence" are derived from and supported by data. As a result, the defined rules are closer to reality and the influence of the subjective biases and individual preferences is significantly reduced.

Another advantage of "objective intelligence" is that its decisions are not static but adapt to the changes of the environment.

- *Nonstop Intelligent Operation* – "Smart" devices, based on computational intelligence, operate continuously and reliably for long periods of time in a wide range of process conditions. As we all know, human intelligence cannot endure a 24/7 mode of intellectual activity. Even the collective intelligence of rotating shifts, typical in manufacturing, has significant fluctuations due to the wide differences in operators' expertise and their attention to the process at a given moment. In contrast, "objective intelligence" continuously refreshes itself by learning from data and knowledge streams. This is one of the key differences between computational intelligence and the competition. The competitive solutions can also operate nonstop, but they cannot continuously, without human interference, maintain, update, and enhance their own intelligence.

- *Handling High Dimensionality and Hidden Patterns* – Computational intelligence can infer solutions from multidimensional spaces with thousands of factors (variables) and millions of records. This feature is beyond the capabilities of human intelligence. Another advantage of "objective intelligence" is its ability to capture unknown complex patterns from available data. It is extremely difficult for human intelligence to detect patterns with many variables and in different time scales.

- *Continuous Self-improvement by Learning* – Several learning approaches, such as neural networks, statistical learning theory, and reinforced learning, are the engines of an almost perpetual progress in the capabilities of "objective intelligence". The competitive methods, even human intelligence, lack this unique feature.

- *No Politics* – One can look at "objective intelligence" as an honest and loyal "employee" who works tirelessly to fulfill her/his duties while continuously improving her/his qualifications. Political maneuvering, growing pretensions, or flip-flopping, so typical in the behavior of human intelligence, is unknown. It is not a surprise why this feature sounds very appealing to management.

9.3.2 Dealing with Uncertainty

The only certain thing in real-world applications is uncertainty. There are technical and natural sources of uncertainty, such as measurement errors, the stochastic nature of phenomena (weather is a typical example), or unknown process behavior, to name a few. The biggest problems, however, are generated by the human origins of uncertainty. Here the list is much longer and includes sources such as vague expressions, flip-flopping in application support, unpredictable organizational changes, etc. It is extremely difficult to address the second type of uncertainties with any available method, although fuzzy logic and intelligent agents have the capabilities to reduce to some extent even this form of ambivalence by simulating social systems.

The key strength of computational intelligence, however, is in handling technical uncertainty. The economic benefits of this competitive advantage are substantial. Reduced technical uncertainty leads to tighter control around process quality, faster new product design, and less-frequent incidents. All of these benefits explicitly translate the technical advantages into value.

One of the advantages of statistics is that uncertainty is built into its foundation. Of special practical importance are the statistical estimates of the uncertainty of model predictions, represented by their confidence limits. The different ways computational intelligence handles uncertainty are shown in Fig. 9.9 and discussed below.

- *Minimum A Priori Assumptions* – Fundamental modeling deals with uncertainty only within strictly defined *a priori* assumptions, dictated by the validity regions of the laws of Nature; statistics handles uncertainty by calculating confidence limits within the ranges of available data and heuristics explicitly builds the fences of validity of the rules. All of these options significantly narrow down the assumption space of the potential solutions and make it very sensitive toward changing operating conditions. As a result, their performance lacks robustness and leads to gradually evaporating credibility and imminent death of the application outside the assumption space. In contrast, computational intelligence has a very open assumption space and can operate with almost any starting data or pieces of knowledge. The methods that allow computational intelligence to operate with minimum *a priori* information are highlighted below.
- *Handling Imprecise Information* – Computational intelligence, and especially fuzzy systems, can capture vague expressions and process them numerically

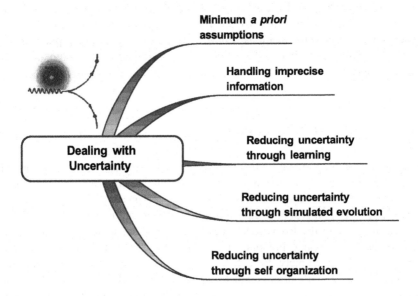

Fig. 9.9 How computational intelligence reduces uncertainty

with high accuracy and specifics. This capability of fuzzy logic is a clear technical superiority.

- *Reducing Uncertainty Through Learning* – One of the possible ways to deal with unknown operating conditions is through continuous learning. By using several learning methods, computational intelligence can handle and gradually reduce wide uncertainty. This allows adaptive behavior and low cost.
- *Reducing Uncertainty Through Simulated Evolution* – Another approach to fight the unknown conditions is by evolutionary computation. This technology is one of the rare cases when modeling can begin with no *a priori* assumptions at all. Uncertainty is gradually reduced by the evolving population and the fittest winners in this process are the final result of this fight with the unknown.
- *Handling Uncertainty Through Self-organization* – In self-organizing systems, such as intelligent agents, new patterns occur spontaneously by interactions, which are internal to the system. As in simulated evolution, this approach operates with no *a priori* assumptions. Uncertainty is reduced by the new emerging solutions.

9.3.3 Dealing with Complexity

Modern technologies and globalization have pushed the complexity of real-world applications to levels that were unimaginable even a couple of years ago. A short list includes the following changes: (i) the number of interactive components has risen by several orders of magnitude; (ii) the dynamic environment requires solutions with both continuous adaptation and capable of abrupt transformations; and (iii) the nature of interactions depends more and more on time-critical relationships between the components. Another factor that has to be taken into account in dealing with the increased complexity of practical applications is the required simple dialog with the final user. The growing complexity of the problem and the solution must be transparent to the user.

The competitive approaches face significant problems in dealing with complexity. First-principles models have relatively low dimensionality; even statistics has difficulties in dealing with thousands of variables and millions of records; heuristics is very limited in representing large numbers of rules; and classical optimization has computational and convergence problems with complex search spaces of many variables.

The different ways in which computational intelligence handles complexity better than the competition are shown in Fig. 9.10.

- *Reducing Dimensionality Through Learning* – Computational intelligence can cluster the data by learning automatically how they are related. This condensed form of information significantly reduces the number of entities representing the system.

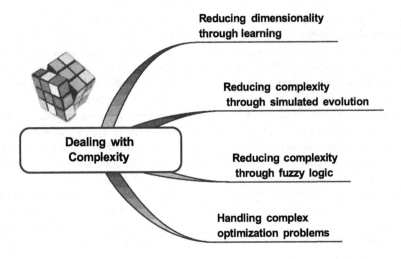

Fig. 9.10 How computational intelligence reduces complexity

- *Reducing Complexity Through Simulated Evolution* – Evolutionary computation delivers distilled solutions with low complexity (especially when a complexity measure is included in the fitness function). One side-effect during the simulated evolution is that the unimportant variables are gradually removed from the final solutions, which leads to automatic variable selection and dimensionality reduction.
- *Reducing Complexity Through Fuzzy Logic* – The most complicated task in solving practical problems is representing the complexity of human knowledge. Fuzzy logic and the new related technologies of granular computing and computing with words give us the tools to condense the linguistic-based knowledge. The knowledge is transferred into mathematical expressions that could adapt with the changing environment. This leads to another reduction in complexity due to the universal interaction between linguistic-based and numerical-based entities.
- *Handling Complex Optimization Problems* – Evolutionary computation and swarm intelligence may converge and find optimal solutions in noisier and more complex search spaces than the classical approaches.

9.3.4 Generating Novelty

Probably the most valuable competitive advantage of computational intelligence is its unique capability to automatically create innovative solutions. In the classical way, before shouting "Eureka", the inventor goes through a broad hypothesis search, and trials of many combinations of different factors. Since the number of hypotheses and factors is close to infinity, the expert also needs "help" from nonscientific forces like luck, inspiration, "divine" spark, even a bathtub or falling apple. As a result, classical discovery of novelty is an unpredictable process.

Computational intelligence can increase the chances of success and reduce the overall efforts for innovation discovery. Since generating intellectual property is one of the key components in economic competitive advantage, this unique strength of computational intelligence may have an enormous economic impact.

The three main ways of generating novelty by computational intelligence are shown in the mind-map in Fig. 9.11 and discussed next.

- *Capturing Emerging Phenomena from Complex Behavior* – Self-organized complex adaptive systems mimic the novelty discovery process by its property of *emergence*. This property is a result of coupled interactions between the parts in a system. As a result of these complex interactions, new unknown patterns emerge. The features of these novel patterns are not inherited or directly derived from any of the parts. Since the emerging phenomena are unpredictable discoveries, they require being captured, interpreted and defined by an expert with high imagination.
- *Extracting New Structures by Simulated Evolution* – One specific method in evolutionary computation, genetic programming, can generate almost any types of new structures based on a small number of given building blocks. This feature was discussed in detail in Chap. 5.
- *Finding New Relationships* – The most widespread use of computational intelligence, however, is in capturing unknown relationships between variables. Of special importance are the derived complex dependencies, which are difficult to reproduce using classical statistics. The development time for finding these relationships is significantly shorter than building first-principles or statistical models. In the case of simulated evolution, even these dependencies are derived automatically and the role of the expert is reduced to selection of the most appropriate solutions based on performance and interpretability.

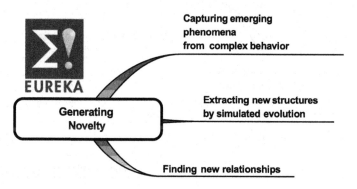

Fig. 9.11 How computational intelligence generates novelty

9.3.5   Low-Cost Modeling and Optimization

Finally, what really matters for practical applications is that all the discussed technical competitive advantages of computational intelligence lead to costs of modeling and optimization that is lower than the competition. The key ways in which computational intelligence accomplishes this important advantage are shown in Fig. 9.12 and the details are given below.

- *High-Quality Empirical Models* – The models, derived by computational intelligence, especially by symbolic regression via genetic programming, have optimal accuracy and complexity. On the one hand, they represent adequately the complex dependencies among the influential process variables and deliver accurate predictions. On the *other* hand, their relatively low complexity allows robust performance in minor process changes when most competitive approaches collapse. In principle, empirical models have minimal development cost. In addition, the high-quality symbolic regression models with their improved robustness, significantly reduce deployment and maintenance cost.
- *Optimization for a Broad Range of Operating Conditions* – As we discussed earlier, evolutionary computation and swarm intelligence broaden the capabilities of classical optimization in conditions with complex surfaces and high dimensionality. As a result, computational intelligence gives more technical opportunities to operate with minimal cost in new, previously not-optimized areas. Of special importance are the dynamic optimization options of swarm intelligence where the process could track continuously, in real time, the economic optimum.
- *Easy Integration into Existing Work Processes* – Another factor that contributes to the low cost of computational intelligence is the relatively obvious way it could be introduced within the established work processes in industry. It fits

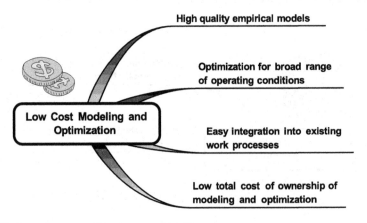

Fig. 9.12 Cost advantages of computational intelligence

very well with Six Sigma, which is the most popular work process in practice. What still needs to be done is including the full variety of computational intelligence technologies into the most popular Six Sigma software packages.

- *Low Total Cost of Ownership of Modeling and Optimization* – All discussed advantages contribute to the overall reduction of the total cost of ownership of the combined modeling and optimization efforts, driven by computational intelligence. Some competitors may have lower components in the cost. For example, the development and deployment cost of statistics is much lower. However, taking into account all the components, especially the growing share of maintenance costs in modeling and optimization, computational intelligence is a clear winner. The more complex the problem, the bigger the advantages of using this emerging technology. All known technical competitors have very limited capabilities to handle imprecision, uncertainty, complexity, and to generate novelty. As a result, they function inadequately in new operating condition, reducing the profit and pushing maintenance costs through the roof.

Still, the biggest issue is the high introductory cost of the technology. Since computational intelligence is virtually unknown to industry at large, significant marketing efforts are needed. One of the purposes of this book is to suggest solutions to reduce this cost.

9.4   Summary

Key messages:

Defining competitive advantages is a necessary prerequisite before introducing a new technology, such as computational intelligence, in practice.

Analyzing pros and cons of key competitive approaches, such as first-principles modeling, statistics, heuristics, and classical optimization builds the basis for comparison.

Computational intelligence beats the competition technically with its unique capabilities to handle uncertainty and complexity, and to generate innovative solutions, based on objective intelligence.

Computational intelligence beats the competition economically with lower development, deployment, and maintenance cost.

The Bottom Line

With the growing complexity of global markets it will become more and more difficult to be competitive without using computational intelligence.

Suggested Reading

This is the classical book for an introduction into the area of economic competitive advantage:
M. Porter, *Competitive Advantage: Creating and Sustaining Superior Performance*, Free Press, 1998.

This book includes a good technical comparative analysis between neural networks, fuzzy systems, and evolutionary computation:
L. Jain and N. Martin (Editors), *Fusion of Neural Networks, Fuzzy Sets, and Genetic Algorithms: Industrial Applications*, CRC Press, 1999.

This is the author's favorite book for building statistical models:
D. Montgomery, E. Peck, and G. Vining, *Introduction to Linear Regression Analysis*, 4th edition, Wiley, 2006.

The book is a popular introduction to both classical artificial intelligence and some computational intelligence methods:
M. Negnevitsky, *Artificial Intelligence: A Guide to Intelligent Systems*, Addison-Wesley, 2002.

This is the author's favorite book on optimization:
Z. Michalewicz and D. Fogel, *How to Solve It: Modern Heuristics*, 2nd edition, Springer, 2004.

This book introduces the key concepts of self-organization and emergence:
J. Holland, *Emergence: From Chaos to Order*, Oxford University Press, 1998.

This book is one of the best sources for an introduction to complexity theory:
R. Lewin, *Complexity: Life at the Edge of Chaos*, 2nd edition, Phoenix, 1999.

Chapter 10
Issues in Applying Computational Intelligence

> *Nothing is more difficult than to introduce a new order.*
> *Because the innovator has for enemies all those who have*
> *done well under the old conditions and lukewarm defen-*
> *ders in those who may do well under the new.*
>
> Niccolò Machiavelli

Applying any emerging technology is not trivial and requires some level of risk-taking even when the competitive advantage is clear. In the case of computational intelligence the process is even harder due to the different nature of the comprising methods, the lack of marketing, affordable professional tools and application methodology. Another important factor slowing down computational intelligence applications is the wrong perception of the technology. To many potential users it looks like it's either too expensive or it's rocket science. The pendulum of expectations also swings from one extreme of anticipating a silver bullet to all problems to the other extreme of awaiting the next technology fiasco.

The topic of this chapter is to focus on the most important application issues of computational intelligence, shown in the mind-map in Fig. 10.1. Their understanding and resolution is critical for the success of applied computational intelligence.

Suggestions for how to resolve these issues are given in the third part of the book.

10.1 Technology Risks

The first application issue is related to the generic factors which lead to unsuccessful introduction of technology. From the multitude of potential causes of technology failure we'll focus on four: (1) balance between the user crisis pushing for introduction of technology and the total perceived pain of adaptation of the new technology; (2) the prevailing technocentric culture; (3) increased complexity as a result of the introduction of technology; and (4) technology hype. These factors are shown in the mind-map in Fig. 10.2.

A.K. Kordon, *Applying Computational Intelligence*,
DOI 10.1007/978-3-540-69913-2_10, © Springer-Verlag Berlin Heidelberg 2010

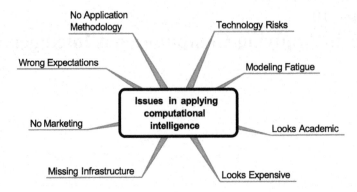

Fig. 10.1 Key issues in applying computational intelligence

Fig. 10.2 Key factors for failure of new technology

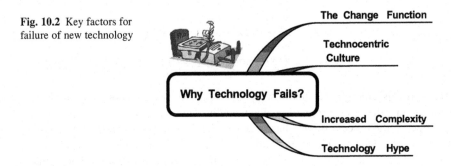

10.1.1   The Change Function

In a recent book about introducing new technologies, Pip Coburn has proposed the idea about The Change Function as the key driving force for opening the door to novel solutions. According to The Change Function, people are only willing to change and accept new technologies when the pain of their current situation out-weighs the perceived pain of trying something new,[1] i.e.

The Change Function

$= \mathbf{f}($ user crisis vs. total perceived pain of adoption)

The arguments for The Change Function are as follows. Potential users have some level of reaction from indifference to crisis whenever they encounter a new technology product. People are more willing to change the higher the level of crisis that they have in their current situation.

The total perceived pain of adaptation is the perception of how painful it will be to actually adopt the product and especially change a habit. The Change Function looks at both sides of the issue of emerging technologies. On the one hand, it defines

---

[1]P. Coburn, *The Change Function*, Penguin Group, 2006.

the critical level of the services a new technology might offer. On the other hand, it evaluates the total perceived pain of adoption associated with that new service.

Applying The Change Function to computational intelligence suggests that the technology must minimize the total perceived pain of adoption of potential users by offering a user-friendly environment, simple solutions, and easy integration into the existing work processes. The other factor for success is the matching of the unique capabilities of computational intelligence to a critical problem of the user. Usually these are the cases when computational intelligence has a clear competitive advantage, i.e. a novel solution is needed in a dynamic environment of high complexity or uncertainty.

10.1.2 Technocentric Culture

According to Clayton Christensen – one of the leading gurus in technology innovations – three quarters of the money spent on product development investment results in products that do not succeed commercially.[2] It is observed that most new technologies are hated by users due to the high total pain of adoption. A possible cause for this unpleasant result is the prevailing technocentric culture in industry. It could be defined as an obsession with the critical role of the technology while neglecting the real customer needs, i.e., the famous Levitt's Law is totally ignored:

Levitt's Law

When people buy quarter-inch drill bits, it's not because they want the bits themselves. People don't want quarter-inch drill bits - they want quarter-inch holes.

Ted Levitt

A typical result of the technocentric culture is pushing technology improvement at any cost by management. Often introducing new emerging technologies is part of this process and as such may become a potential issue for applied computational intelligence. Imposing the technology for purely technology's sake may lead to lost credibility, as we know from applied AI (see Chap. 1). In principle, computational intelligence development requires higher implementation efforts, i.e. high total perceived pain of adoption. It is critically important to justify the user needs before suggesting the introduction of technology. The new technology has to be transparent and the life of the user must be easier not harder after the technology is applied. An example of what the user does not need is the Microsoft Office advisor, based on computational intelligence and introduced in the late 1990s. The initial expectations of Microsoft to deliver a killer computational intelligence application evaporated soon through the unanimous cold response from the users. Absolutely boring and useless, it was removed by almost all users of the popular product.

---

[2]C. Christensen and M. Raynor, *The Innovator's Solution*, Harvard Business School Press, 2003.

10.1.3   Increased Complexity

Increased complexity of applied new solutions is the key root cause for the high total perceived pain of adoption. It is especially painful when the application requires changing a habit. Modifying an established work process is a potential disaster as well. Adding new equipment and pushing to learn new software helps the negative response of the user.

In many cases the imposed complexity is well balanced with the benefits of the applied computational intelligence system. A typical example is the nonlinear control systems based on neural networks. The advantages of using this type of system to control processes with difficult nonlinear behavior partially overcomes the pain of complex tuning. It has to be taken into account that tuning the neural network-based controllers requires 10-12 parameters in comparison to three in the prevailing PID controllers.

10.1.4   Technology Hype

Another factor that contributes to technology failure is eroding the credibility by overselling its capabilities. Unfortunately, technology hype is an inevitable evil in the introduction of emerging technologies which needs to be taken into account and counteracted with realistic expectations. We will focus on the following key sources of technology hype:

- *Technology Hype from Vendors* – Usually exaggerates the capabilities for direct value creation while it hides the application issues, especially the potential for growing maintenance cost. The ultimate example is one of the early slogans "We will turn your data into gold" of one of the leading vendors of neural networks-based systems.
- *Technology Hype from Management* – It is like the regular top-down push of corporate initiatives. In principle, it is slightly more realistic, since it is based on some practical experience. The potential topic of exaggeration is mostly the leveraging capacity of the technology.
- *Technology Hype from R&D* – It may come from the top academics in the field, which usually are entirely convinced that their methods are The Ultimate Solution Machine to almost any problem in the Universe. Another source could be industrial R&D which compensates for the lack of professional marketing of the technology with typical oversell.
- *Technology Hype from the Media* – Applied AI in the past and recently different computational intelligence methods are periodically the focus of attention from the media. Sometimes the media-generated hype goes beyond any sensible limits. A typical case was the absurd story of using all the power of computational intelligence to create an intelligent litter box for cats.

10.2 Modeling Fatigue

Using modeling for new product design and process improvement has a long history in industry. Many profitable businesses did significant investments in developing and deploying different types of models for their critical operations. In some cases, the modeling efforts have already reached the saturation state when almost everything necessary for process optimization and advanced monitoring and control has been accomplished. As a result, the opportunities for introducing and applying new modeling approaches are limited.

The key features that are at the basis of this critical challenge for successful application of computational intelligence are captured in the mind-map in Fig. 10.3 and discussed below.

10.2.1 The Invasion of First-Principles Models

The clearest feature of modeling saturation is the case of mass applications of models based on the laws of Nature. Here is a short list for when a business needs first-principles models:

- Understand process physics and chemistry;
- Design model-based optimization and control systems;
- Expect long-term profit from manufacturing;
- Process has manageable complexity and dimensionality;
- Expected return from modeling investment is high.

At the basis of the decision to initiate a costly program of fundamental modeling is the assumption that the future productivity gains will justify the expense. Another factor in support of first-principles modeling is the increased efficiency of the recent software environments from vendors like Aspen Technologies, Fluent, and Comsol Inc. In addition, fundamental modeling includes the best available domain experts in the specific modeling areas and that builds credibility in the implemented models.

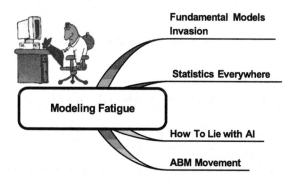

Fig. 10.3 Key symptoms of modeling fatigue

On the negative side, however, first-principles modeling creates a barrier to other modeling approaches. One factor is the reluctance of management to invest more money after the high cost already spent on fundamental model development and deployment. The key factor, however, is the united front from the developers of mechanistic models against empirical solutions. While the principal argument of the advantages of the laws of Nature over empiricism is difficult to argue, going to the other extreme of ignoring data-driven solutions is also not acceptable.

10.2.2   Statistical Models Everywhere

The biggest generator of models in industry, however, is not the laws of Nature but the laws of Numbers. Thanks to Six Sigma and the increased role and availability of data the number of applied statistical models in industry is probably orders of magnitude higher than the number of first-principles models. While the advantages of statistical models are clear and were discussed in the previous chapter, mass-scale applications create some issues that deserve attention.

The key issue is creating the image of statistics as a universal solution to all data-related problems. Part of the problem is the opposition of many professional statisticians to some empirical methods that are not statistically blessed, such as neural networks and symbolic regression. As a result, the empirical modeling opportunities are implemented almost entirely as statistical models. This is a big challenge to computational intelligence since significant marketing efforts are needed to promote the technology, especially to Six Sigma black belts and professional statisticians. A good starting point could be a realistic assessment of the performance of existing statistical models. There is a high probability that at least a part of them will need significant improvement. One of the potential alternative solutions is using symbolic regression models.

10.2.3   How to Lie with AI

A special category of models is heuristic-based applications, which emphasize the expert role at the expense of either first-principles or empirical approaches. Unfortunately, the average credibility of some of these models in industry is not very high. The reasons for this negative image were discussed in Chap. 1.

10.2.4   Anything but Model (ABM) Movement

An unexpected result of modeling fatigue due to the mass-scale modeling efforts is the appearance of the so-called Anything But Model (ABM) movement. It includes the model users who are disappointed by the poor performance of applied models and tired of the administrative modeling campaigns. They are strong opponents to introducing new models and prefer direct human actions, based on their experience. Several factors contribute to this attitude. One of the factors is that the excess of models leads to confusion. At least some of the models do not predict reliably and the users gradually lose patience and confidence in using them. Another factor is the growing maintenance cost with the increased number and complexity of models. There is also a critical threshold of modeling intensity beyond which the saturation of the users becomes obvious.

10.3 Looks Too Academic

Another issue of applying computational intelligence is its academic image in industry. Part of the problem is the lack of popular references that would help the potential users to understand the diverse computational intelligence approaches. The dynamic growth of the technology makes it difficult to track the state of the art even for researchers, not to mention practitioners. These key factors are shown in the mind-map in Fig. 10.4 and discussed in this section.

10.3.1 Difficult to Understand

One of the reasons for the academic image is the lack of understanding of the technology outside several closed research communities. A number of issues contribute to this situation. First, undergraduate and graduate courses on computational intelligence are offered in very few universities. The technology is virtually unknown to most students in related technical disciplines and this significantly narrows the application opportunities. It is extremely difficult for potential users to

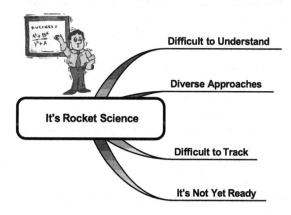

Fig. 10.4 A mind-map with the main reasons why computational intelligence looks too academic

find popular explanation of the key computational intelligence approaches without scientific jargon and heavy math.[3] It has to be taken into account that some of the methods, especially support vector machines, are not easy to translate into plain English. As was discussed several times, one of the key objectives of the book is to fill this obvious need.

The other option of understanding and learning the capabilities of computational intelligence by playing with popular and user-friendly software is also virtually non-existent.

10.3.2   Diverse Approaches

An additional factor contributing to the academic image of computational intelligence is the wide diversity of the comprising approaches. They differ in scientific principles, mathematical basis, and user interaction. It is extremely challenging even for a person with a sound technical and mathematical background to be easily introduced to all approaches. The confusion is also increased by the still divisive environment among the research communities developing the different methods. Usually each community tries to glorify the role of the specific approach as the ultimate technology, very often at the expense of the others. The spirit of competition prevails over the efforts of pursuing the synergetic benefits. There are very few popular references and practical guidelines on integration of these approaches, which is critical for the success of real-world applications.

10.3.3   Difficult to Track

The fast speed of growth of the existing computational intelligence methods and the emergence of numerous new approaches contributes to the academic image of the technology as well. This makes the task of tracking the scientific progress of computational intelligence very time-consuming and more complex than looking at state of the art of an average technology. For example, to cover the whole field it is necessary to keep abreast of at least five scientific journals, such as the *IEEE Transaction on Neural Networks, IEEE Transaction on Fuzzy Systems, IEEE Transaction on Evolutionary Computation, IEEE Transaction on Systems, Man, and Cybernetics,* and *IEEE Intelligent Systems Journal,* and the yearly proceedings of several top conferences, organized by the different scientific communities.

Even more difficult is keeping track of available software, and especially of successful real-world applications. The potential sources of information are very limited. Fortunately, there is a trend of a more sizable industrial presence in some of

---

[3]One of the few exceptions is the book by V. Dhar and R. Stein, *Seven Methods for Transforming Corporate Data into Business Intelligence*, Prentice Hall, 1997.

the conferences in special sessions and workshops devoted to practical issues of computational intelligence.

10.3.4   It's Not Yet Ready for Industry

A negative consequence from the fast scientific progress of computational intelligence is the perception of technology and incompleteness it creates among practitioners. In principle, it is much more difficult to convince a potential user to apply a technology that is still in the high-risk dynamic development phase. In addition, management is reluctant to give support for a technology which may require continuous internal R&D development.

The limited knowledge about industrial success stories based on computational intelligence and the lack of affordable professional software contribute to the image of technology immaturity as well.

10.4   Perception of High Cost

The prevailing academic impression of computational intelligence in industry is combined with the perception of high total cost of ownership of the applied solutions. It is expected that research-intensive technology will require higher-than-average development cost. The investment in new infrastructure and the necessary training will increase the deployment cost as well. To potential users, it seems that the maintenance cost will also be high due to the specialized skills of the support teams. All of these application issues are presented in the mind-map in Fig. 10.5 and discussed briefly below.

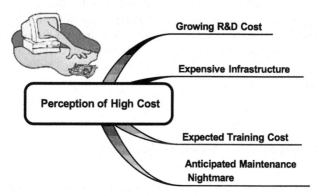

Fig. 10.5 Key factors that contribute to the image of high total cost of ownership of computational intelligence

10.4.1   Growing R&D Cost

Unfortunately, the scientific expansion of computational intelligence creates the logical estimate not only of high development cost but even expectation for growing R&D expenses in the future. The technology is still treated by the management as research-intensive and the diverse methods on which it is based create the impression of a steep learning curve. The estimated high development cost is based on the assumption that probably external resources from academia or vendors will be needed. In some cases the development cost will be even higher if internal R&D efforts are necessary to improve the technology for important specific classes of applications. An additional factor in the high development cost of computational intelligence is that the opportunity analysis for potential applications is not trivial and requires more than average efforts for exploratory analysis.

10.4.2   Expensive Infrastructure

The assessment of high deployment cost is based on the assumption of necessary investment for the computational intelligence infrastructure. In some cases, such as intensive use of evolutionary computation, investment in more powerful computer hardware, such as clusters, is recommended. Unfortunately, the cost of available professional software is relatively high. Very often, it is necessary to allocate resources for internal software development and support, which also significantly increases the cost.

10.4.3   Expected Training Cost

A significant share of the expected high total cost of ownership of applied computational intelligence is related to training. Firstly, it is a nonstandard training for developers with an expected steep curve of understanding for the key approaches. Secondly, it could be very difficult training for the users, since most of them do not have the technical and mathematical skills of the developers. Thirdly, training tools explaining the computational intelligence approaches in plain English are practically unavailable.

10.4.4   Anticipated Maintenance Nightmare

The fear of difficult maintenance and gradually reducing performance of applied computational intelligence solutions is the key concern to most potential users. The little experience from the known industrial applications of computational intelligence and the maintenance lessons from applied AI contribute to this perception. It is unclear to potential industrial users if it is possible to support the applied computational intelligence systems without Ph.D.-level skills. An additional source of concern is the uncertain long-term future of most of the computational intelligence vendors and their limited capabilities for global support.

10.5 Missing Infrastructure

Clarifying the requirements for the necessary infrastructure to support applied computational intelligence is of key concern to potential users of this emerging technology. Unfortunately, there are very few available sources to discuss this issue of critical importance for the final decision of whether or not to give the green light for the promotion of technology.

The objective of this section is to address the key concerns related to the computational intelligence infrastructure, represented in the mind-map in Fig. 10.6.

10.5.1 Specialized Hardware

The first infrastructural question that needs to be clarified to the potential users of computational intelligence is the possible need for more powerful hardware than high-end personal computers. Fortunately, with the exception of evolutionary

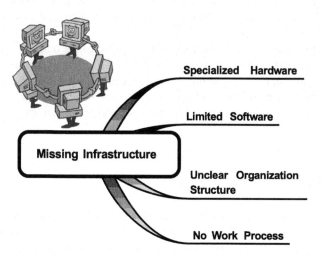

Fig. 10.6 A mind-map representing the key issues related to the required infrastructure for applied computational intelligence

computation, there are no special requirements for high computational power. The needs for evolutionary computation are problem-dependent. In the case of high-dimensionality data or use of complex simulation packages (as is the situation with applying genetic programming for inventing electronic circuits) computer clusters or grid computing is a must. Most evolutionary computation algorithms are inherently parallel and benefit from distributed computing.

10.5.2   Limited Software

The situation is much more challenging with the other component of the applied computational intelligence infrastructure – the available software. Unfortunately, an affordable software platform which integrates all key approaches discussed in this book is not available. Different approaches have been integrated in some commercial packages, such as Intelligent Business Engines™ (NuTech Solutions), SAS Enterprise Miner (SAS Institute), and Gensym G2 (Gensym Corporation). However, the price of the software is relatively high and affordable for big corporate clients only.

A more realistic option is to build a software infrastructure with commercial products. Several vendors offer professional user-friendly software based on specific methods, mostly neural networks. Typical examples of off-line solutions are Neurodynamics (NeuroSolutions) and Neuroshell (Ward Systems Group). The leading products for on-line neural network applications in advanced process control and environmental compliance are ValueFirst™ (Pavilion Technologies) and Aspen IQModel (Aspen Technologies).

An interesting option is the commercial software which is embedded in popular products, such as Excel. The advantage of this solution is the minimal training for development and support. Examples are products like NeuralWorks Predict (Neuralware) and NeuralTools (Palisade).

The most realistic option to find software for most of the discussed approaches is on universal modeling platforms such as MATLAB (Mathworks) and Mathematica (Wolfram Research). In addition to commercially available toolboxes on these platforms, there are many free packages developed from academia. However, their user-friendliness is far from that desired for real-world applications.

10.5.3   Unclear Organization Structure

The third infrastructural question of interest to potential users of computational intelligence is the possible need for organizational changes. The answer is not trivial, especially for large corporations. In the case of expected high return from applying the technology, a special group of developers can be established. The objective of an applied computational intelligence group is to own the technology inside the company. In this most favorable scenario for introducing the technology, there is a clear focal point for internal marketing, opportunity analysis, potential technology improvement, application development, deployment and maintenance.

Another possible organizational option is to establish a small group within a modeling-type of R&D department with more limited tasks of focusing on opportunity analysis for specific computational intelligence applications with minimal development efforts. The third organizational scenario is when the individual computational intelligence experts are spread out in different groups and promote the technology through professional networks.

10.5.4   Work Process Not Defined

The fourth infrastructural question of interest to potential users of computational intelligence is the possible need to change existing work processes or define new ones, related to the specifics of the technology. In principle, solution development, implementation, and support is not significantly different from other technologies and probably doesn't require the introduction of new work processes. However, the need for nontrivial internal marketing of the unique computational intelligence capabilities will demand adequate work process. The ideal scenario is to link the computational intelligence application methodology into the existing standard work processes, like Six Sigma, and this is discussed in Chap. 12.

10.6   No Marketing

The most critical issue of applied computational intelligence from a practical point of view is the lack of professional marketing of this emerging technology. As a result, we have the paradox when the technology is introduced to potential users and markets through the "back door" by individual efforts of R&D enthusiasts. The need to open the field to industry through the "front door" with professional marketing is discussed briefly in this section. The key issues that need to be resolved are captured in the mind-map in Fig. 10.7 and addressed below. A more comprehensive discussion is given in Chap. 13.

10.6.1   Product Not Clearly Defined

The first challenge in professional marketing of applied computational intelligence is the nontrivial definition of the final product. There are several sources of confusion, such as the wide diversity of methods and the broad application areas, to name a few. The deliverables are also extremely different. Here are some obvious

Fig. 10.7 A mind-map of the key issues to be resolved for professional promotion of computational intelligence in industry

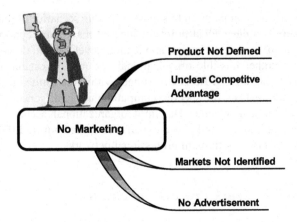

Product Not Defined

Unclear Competitve Advantage

No Marketing

Markets Not Identified

No Advertisement

examples of the main types of products, derived from applied computational intelligence:

- Predictive models;
- Problem classifiers;
- Complex optimizers;
- System simulators;
- Search engines.

They could be the basis of a broad definition of the expected products from this emerging technology. For marketing purposes it could be advertised that applied computational intelligence enhances the productivity of human intelligence through complex predictive models, problem classifiers, optimizers, system simulators, and search engines.

10.6.2 Unclear Competitive Advantages

The previous chapter was an attempt to fill this gap.

10.6.3 Key Markets Not Identified

The broad view of the markets with most perspective is given in Chap. 2. However, more specific efforts for identifying the opportunities and the size of individual markets are needed.

10.6.4   No Advertisement

A significant part of the marketing efforts is advertising the technology to a very broad nontechnical audience of potential users. The advertisement objective is to define a clear message for representing the technology and capturing attention. The key principle is focusing on the unique deliverables which give competitive advantage. Specific marketing slides and examples of elevator pitches are given in the corresponding chapters for each computational intelligence approach. In addition, examples of advertising applied computational intelligence to nontechnical and technical audiences are given in Chap. 13.

10.7   Wrong Expectations

Probably the most difficult issue of applied computational intelligence is to help the final user in defining the proper expectations from the technology. Very often the dangerous combination of lack of knowledge, technology hype, and negative reception from some applied research communities creates incorrect anticipation about the real capabilities of computational intelligence. The two extremes of wrong expectations either by exaggeration or by underestimation of the computational intelligence capabilities cause almost equal damage to promotion of the technology in industry.

The key ways of expressing wrong expectations about applied computational intelligence are shown in the mind-map in Fig. 10.8 and discussed below.

10.7.1   Magic Bullet

The expectation for technical magic is based on the unique capabilities of applied computational intelligence to handle uncertainty, complexity, and to generate novelty. The impressive features of the broad diversity of methods, such as fuzzy logic,

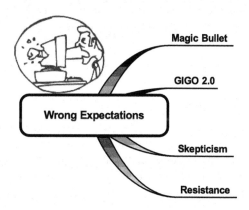

Fig. 10.8 Key ways of expressing wrong expectations from applied computational intelligence

machine learning, evolutionary computation, and swarm intelligence, contribute to such a Harry Potter-like image even when most of the users do not understand the principles behind them. Another factor adding to the silver bullet perception of computational intelligence is the technology hype from the vendors, the media, and some high-ranking managers.

As a result, potential users look at applied computational intelligence as the last hope to resolve very complex and difficult problems. Often, they begin looking at the technology after several failed attempts of using other methods. In some cases, however, the problems are ill-defined and not supported by data and expertise. In order to avoid the magic bullet trap, it is strongly recommended to identify the requirements, communicate the limitations of the appropriate methods, and to define realistic expectations in the very early phase of potential computational intelligence applications.

10.7.2 GIGO 2.0

The worst-case scenario of the magic bullet image is the GIGO 2.0 effect. In contrast to the classical meaning of GIGO 1.0 (Garbage-In-Garbage-Out), which represents the ignorant expectations of a potential solution, GIGO 2.0 embodies the next level of arrogant expectations defined as Garbage-In-Gold-Out. In essence, this is the false belief that low-quality data can be compensated for with sophisticated data analysis. Unfortunately, computational intelligence with its diverse capabilities to analyze data is one of the top-ranking technologies that create GIGO 2.0 arrogant expectations. It is observed that the bigger the disarray with data the higher the hope of exotic unknown technologies to clean up the mess. Usually this behavior is initiated by top management who are unaware of the nasty reality of the mess.

It is strongly recommended to protect potential computational intelligence applications from the negative consequences of the GIGO 2.0 effect. The best winning strategy is to define the requirements and the expectations in advance and to communicate clearly to the user the limitations of the methods. Better to reject an impossible implementation than to poison the soil for many feasible computational intelligence applications in the future.

10.7.3 Skepticism

In contrast to the magic bullet optimistic euphoria, disbelief and lack of trust in the capabilities of applied computational intelligence is the other extreme of wrong expectation, but in this case on the negative side. Usually skepticism is the initial response of the final users of the technology on the business side. Several factors contribute to this behavior, such as lack of awareness of the technical capabilities and the application potential of computational intelligence, lessons from other over-hyped technology fiascos in the past, and caution from ambitious R&D initiatives pushed by management.

Skepticism is a normal attitude if risk is not rewarded. Introducing emerging technologies, like computational intelligence, requires a risk-taking culture from all participants in this difficult process. The recommended strategy for success and reducing skepticism is to offer incentives to the developers and the users of the technology.

10.7.4 Resistance

The most difficult form of wrong expectations from computational intelligence is actively opposing the technology due to scientific or political biases. In most cases the driving forces of resistance are part of the industrial research community which either does not accept the scientific basis of computational intelligence or feels threatened by its application potential.

Usually the resistance camp against computational intelligence is led by the first-principles modelers. They challenge almost any empirical method in general, and have severe criticisms against black-box models in particular. Their opposition strategy is to emphasize the advantages of first-principles models, which are well known, and to ignore the benefits of computational intelligence, which unfortunately are not familiar to a broad audience. Often first-principles modelers are very aggressive in convincing management to question the capabilities and the total-cost-of-ownership of computational intelligence. In most of the cases they silently reject the opportunities for collaboration with computational intelligence developers and to deliver an integrated solution.

The other part of the resistance movement against computational intelligence includes the professional statisticians and a fraction of the Six Sigma community. Most professional statisticians do not accept the statistical validity of some of the computational intelligence methods, especially neural networks-based models. The key argument of these fighters for the purity of the statistical theory is the lack of a statistically sound confidence metric of the nonlinear empirical solutions derived by computational intelligence. They have support in the large application base of the Six Sigma community, since classical statistics is the key application method.

The third part of the resistance camp against computational intelligence includes the active members of the Anything But Model (ABM) movement who energetically oppose any attempt to introduce the technology. In this category we can also add the users expecting high perceived pain of adoption from computational intelligence.

10.8   No Application Methodology

Another application issue of computational intelligence is the confusion of potential users about how to implement the technology in practice. In contrast to the key competitive approaches, such as first-principles modeling, statistics, heuristics, and optimization, there is no well-known application methodology for computational intelligence. Potential users have difficulties selecting the proper methods for solving their problems. They don't know how to take advantage of integrating the different approaches. The specific application sequence is also unclear to users and there are very few references to answer the practical application questions related to computational intelligence.

These key issues related to the lack of computational intelligence application methodology are represented in the mind-map in Fig. 10.9 and discussed briefly below.

10.8.1   Method Selection

The users' confusion begins with the difficult task of understanding the technical advantages of the different computational intelligence approaches and linking them to the specific application problem. Unfortunately, it is hard to find simple practical advice on this subject in the available literature. However, the reader is provided with useful recommendations which can answer this question in several places in this book. A general roadmap of all discussed methods and the related application areas is given in Chap. 2 followed by analysis of specific methods in Chaps. 3–7. A specific table for the selection of computational intelligence methods is given in Chap. 12.

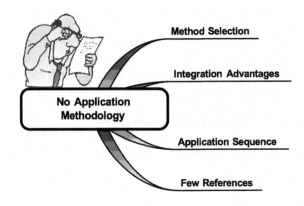

Fig. 10.9 A mind-map with the gaps in computational intelligence application methodology

10.8.2 Integration Advantages

Unfortunately, most potential users are unaware of the big benefits of integrating the different computational intelligence approaches. This gap is partially filled with the different integration options discussed in Chap. 11. An integrated methodology based on statistics, neural networks, support vector machines, genetic programming, and particle swarm optimization is presented, and illustrated with industrial examples.

10.8.3 Application Sequence

Potential computational intelligence users also need more clarity on the specific steps in applying the technology. This issue is covered in Chap. 12 with two options for potential users. The first option is a generic computational intelligence application methodology as a separate work process at the application organization. The second option is directed to the potential users of Six Sigma and includes practical suggestions on how to integrate computational intelligence with this established work process.

10.8.4 Few References

Potential users of computational intelligence face another challenge due to the very few sources with practical information about the applicability of the different methods. Some vendors offer a development methodology in their manuals for a specific approach, mostly related to neural networks or genetic algorithms. However, this information is very limited and usually requires specialized training. In addition, parameter settings for the different approaches are not easy to find, which is very important from a practical point of view.

Fortunately, this gap is filled in the different chapters of this book. The necessary information about the application sequence and parameter settings of the different approaches, as well as the relevant references of practical interest, are given in the corresponding chapters.

10.9 Summary

Key messages:

A necessary condition for successful application of computational intelligence is that the benefits of the solution will outweigh the perceived pain of technology adoption.

The academic image of computational intelligence creates a perception of high total-cost-of-ownership and alienates potential users.

The missing computational intelligence infrastructure, due to limited professional software, unclear organizational structure, and undefined work processes, reduces the application opportunities of the technology.

Marketing the competitive advantages and application capabilities of computational intelligence is the key step to open the door into industry.

Wrong expectations about computational intelligence can either destroy the credibility of the technology or prevent its introduction.

The Bottom Line

The key application issues of computational intelligence are mostly related to insufficient knowledge of the real capabilities of the technology.

Suggested Reading

Three books with a good summary of effective innovation strategies are:
C. Christensen and M. Raynor, *The Innovator's Solution*, Harvard Business School Press, 2003.
P. Coburn, *The Change Function*, Penguin Group, 2006.
M. George, J. Works, and K.Watson-Hemphill, *Fast Innovation*Innovation, McGraw-Hill, 2005

Part III
Computational Intelligence
Application Strategy

Chapter 11
Integrate and Conquer

Practical sciences proceed by building up; theoretical sciences by resolving into components.

St. Thomas Aquinas

The objective of the third part of this book is to present a path for transferring computational intelligence from the realm of exciting research ideas into a highly competitive generator of valuable solutions to real-world problems. The proposed application strategy begins in this chapter by emphasizing the importance of integrating various approaches for resolving real-world problems. Various ways to introduce, apply, and leverage computational intelligence in a business are discussed in Chap. 12. Chapter 13 presents different techniques for addressing one of the key issues in the application strategy of computational intelligence – marketing the technology to technical and nontechnical audiences. The last chapter in the third part, Chap. 14, includes examples of successful and failed industrial applications.

Since the value creation capability is the key driving force in real-world applications, the strategy for applying computational intelligence is based on factors like minimizing modeling cost and maximizing model performance under a broad range of operating conditions. An obvious result of this strategy is the increased efforts in robust empirical model building, which is very often at an economic optimum. Unfortunately, the robustness of the empirical solutions, i.e. their capability to operate reliably during minor process changes, is difficult to accomplish with one modeling method only. Very often the nasty reality of real-world problems requires the joint work of several modeling techniques. In order to meet this need, it is necessary to develop a consistent methodology that effectively combines different modeling approaches to deliver high quality models with minimal efforts and maintenance. It is based on the almost infinite number of ways to explore the synergetic benefits between the modeling methods. The result is an integrated methodology which significantly improves the performance and compensates the disadvantages of the corresponding individual methods.

A.K. Kordon, *Applying Computational Intelligence*,
DOI 10.1007/978-3-540-69913-2_11, © Springer-Verlag Berlin Heidelberg 2010

The motto of the chapter is clear: Integrate the modeling methods and conquer the real world. From our experience it is the winning strategy which opens the door of industry to computational intelligence.

11.1   The Nasty Reality of Real-World Applications

In the same way as an abstract spherical cow differs from the real animal, the purely academic versions of a modeling approach are far away from the challenges of industrial reality. Unfortunately, the reality corrections cannot be done only through intensive computer simulations, as is usually the case. The big issue is that the nasty reality of industrial applications requires the joint solution of the technical, infra-structural, and people-related components of a given problem. The lion's share of the application effort is focused on handling the technical issues, due to the lack of a systematic approach or a lack of information about the other two components. Ignoring the infrastructural and people-related issues, however, may lead to an implementation fiasco even if the technical issues are resolved.

In order to avoid this common mistake, a broader view of the main issues in real-world applications, which includes technical, infrastructural, and people-related components, is presented in Fig. 11.1 and discussed below.

- *Unclean, Incomplete, and Noisy Data* – Data quality and availability is at the top of the list of technical real-world application issues. Usually two extreme situations need to be handled. In the first situation, typical for manufacturing, the available data has high dimensionality (up to thousands of variables and millions of records). In the other extreme is the second situation, typical of new product development, when the available data are limited to a few records and the cost of each new record is very high. In both cases, the data may contain gaps, erroneously entered values, units of measurement conversion chaos, and different levels of noise. Handling all of this data mess with effective data preparation is the prerequisite for the success of any data-driven approach, including most computational intelligence methods.

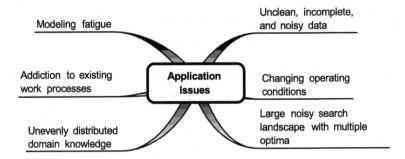

Fig. 11.1 Main issues in real-world applications

- *Changing Operating Conditions* – The real world operates in a dynamic environment. The pace and amplitude of changes, however, varies significantly from microseconds to decades and from small variations to deviations in orders of magnitude. It is extremely difficult to represent this broad dynamic range with one modeling technique only. Even the meaning of a steady-state model has to be interpreted with relation to the current business needs. The use of the model depends strongly on the current demand of the product or service it is linked to. The demand varies with the economic environment and at some point the steady-state conditions on which the model was developed become invalid and the model's credibility evaporates. Either parameter readjustments or complete redesign is needed. In both cases, the maintenance cost grows significantly.

- *Large Noisy Search Solution Landscape with Multiple Optima* – The potential solutions to real-world problems depend on a multitude of factors which are related in complex ways and often represented by messy data. The high dimensionality, as well as the complex search space with multiple optima and noisy surface, is a tough challenge for any individual optimization technique. In addition, the solutions of real-world problems are inherently multiobjective. This includes active participation of the final user in the trade-off process of solution selection.

- *Unevenly Distributed Domain Knowledge* – Domain knowledge is critical not only for expert systems or first-principles models, but for any real-world application. The presence of subject matter experts is decisive in all phases of model development and deployment – from project specification to solution selection and use. The popular phrase that the model is as good as the experts involved in its development is not an exaggeration. Unfortunately, domain knowledge has high dynamics (experts change organizations frequently) and is unevenly distributed in an organization. Involving the subject matter expert in driving the real-world application is a challenging but critical task for the final success. It has to be taken into account that some top-level experts have no habits of sharing their knowledge easily.

- *Addiction to Existing Work Processes* – The working environment in a business is based on the established infrastructure, work processes, and behavior patterns (or culture). Any successful real-world application must adapt to this environment and not try to push the opposite, i.e. to adapt the existing work processes to the technical needs of the application. Most people are addicted to their working environment and hate significant changes, especially in well-established work processes. Understanding how the solution of the real-world problem will fit into the existing working infrastructure is one of the most critical (and most ignored) factors for success.

- *Modeling Fatigue* – As was already discussed in Chap. 10, many businesses have experienced several invasions of different modeling approaches, some of them pushed with a high level of hype. As a result, the credibility of modeling, as a solution to real-world problems, has been eroded. It is extremely difficult to promote new "exotic" technologies in such an environment. If a similar attitude is identified, it is recommended not to waste time changing the mindset but to look for other opportunities. The best way to fight a modeling credibility crisis is by demonstrating long-term success in another place.

In summary, when we solve real-world problems, we realize that such systems are typically ill-defined, difficult to model, and possess large solution spaces with multiple optima. The data we may use are messy and noisy, the operating conditions change, and the domain knowledge is difficult to extract. In addition, we need to understand and integrate the solution into the existing work infrastructure and to find support, not opposition to the application.

11.2   Requirements for Successful Real-World Applications

The best way to capture the reality correction is by defining and satisfying the specified features of the applied systems. The most important requirements that may lead to successful real-world applications are shown in the mind-map in Fig. 11.2.

- *Credibility* – At the top of the list is the requirement that the derived solution is trustworthy in a wide range of operating conditions. Usually model credibility is based on its principles, performance, and transparency. First-principles-based models have built-in trust by the laws of Nature. Statistical models with their confidence metrics are also accepted as credible by the majority of users, especially if they are involved in data collection and model development. All other solutions, built on either empirical or expert knowledge methods are treated as lacking a solid theoretical basis and must fight for their credibility with almost flawless performance. An additional factor that may increase an empirical model's credibility is the level of its transparency to the user. On the one extreme are complex black boxes, like neural networks or support vector machines, which are among the least trusted models. On the other extreme are simple symbolic regression equations, especially if they could be physically interpreted. Our experience shows that users have high confidence in this type of empirical solution.

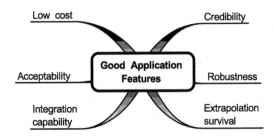

Fig. 11.2 Key features of a successful real-world application

The key credibility factor, however, is model performance according to expected behavior. The credibility erosion begins with questioning some model predictions and grows into raising doubts about model validity after a series of erroneous behaviors. The final phase of lost credibility is the silent death of the applied solution, which is gradually eliminated from use. Unfortunately, it's almost impossible to recover an application with lost credibility. Even more, an accidental credibility fiasco in a specific application may impose a bad image for the whole technology and poison the soil for future implementation for a very, very long time.

- *Robustness* – Performance credibility depends on two key features of applied models: (1) the ability to handle minor process changes and (2) the ability to avoid erratic behavior when moving to the unknown extrapolation areas. Process changes driven by different operating regimes, equipment upgrades, or product demand fluctuations are more the rule than the exception in industry. It is unrealistic to expect that all the variety of process conditions will be captured by the training data and reflected in the developed empirical or fundamental models. One of the potential solutions with increased robustness is deriving models with an optimal balance between accuracy and complexity. A current survey from several industrial applications in The Dow Chemical Company demonstrates that it is possible to generate high-quality symbolic regression models with very low level of complexity by Pareto-front GP.[1] The derived symbolic regression models show improved robustness during process changes relative to conventional GP as well as neural network-based models.

- *Extrapolation Survival* – Robust models have their limits (usually within 10% outside the original ranges of model development), and when they are crossed, the behavior in the extrapolation mode becomes critical for the fate of the application. It is expected that the potential solution can manage the performance degradation in a gradual fashion until at least 20% outside the training range. Switching from normal to erratic predictions due to model limitation within the range of training data (as is usually the case with neural networks), can destroy the application credibility in minutes. A potential empirical solution to the problem is selecting statistical or symbolic regression models with a low level of nonlinearity. Another technique that may help models to avoid an extrapolation disaster is adding a self-assessment capability using built-in performance indicators or by using combined predictors and their statistics as a confidence indicator of the model's performance.

- *Integration Capability* – Since industry has already invested in developing and supporting the infrastructure of the existing work processes, the integration efforts of any new technology become a critical issue. The applied solution must be integrated within the existing infrastructure with minimal efforts. The best-case

---

[1] A. Kordon, F. Castillo, G. Smits, and M. Kotanchek, Application issues of genetic programming in industry, *Genetic Programming Theory and Practice III*, T. Yu, R. Riolo, and B. Worzel (eds), Springer, pp. 241–258, 2006.

scenario is to use the well-established dialog with the user and the existing software for deployment and maintenance. It is assumed that different tools and software environments can be used during model development, but the final runtime solution has to be integrated into the known environment of the user. Very often, it could be the ubiquitous Excel.

- *Acceptability* – A very important requirement for the success of real-world applications is the support from all stakeholders. Several factors contribute to the model's acceptability. The first factor is the credibility of the modeling principles, which was already discussed. Another critical factor is making the development process user-friendly with minimal tuning parameters and specialized knowledge. Evolutionary computation models are developed with a minimal number of assumptions unlike, for example, first-principles models that have many assumptions stemming from physical considerations or by statistical considerations. Another factor in favor of model acceptance is a known real-world solution to a similar problem. In the real world, proven practical results speak louder than any purely technical arguments.

  Two factors are critical for transferring acceptability into real projects, though – buy-in from users and a powerful internal champion. Without support from users the proposed solution probably will not be accepted and it means that users must be involved at every step of the project. It is also true that without the blessing of a powerful advocate most projects cannot be funded or succeed.

- *Low Total Cost of Ownership* – The sum of development, deployment, and expected maintenance cost of the real-world application must be competitive versus the other alternative solutions. In the case of new technology introduction, the potentially higher development cost must be justified with expected benefits, such as competitive advantage or expected solution leverage that will reduce the future cost.

11.3   Why Integration Is Critical for Real-World Applications

The challenges of industrial reality and the tough criteria for success of real-world applications raises a very high standard for any modeling method. It is much more difficult for a new emerging technology, such as computational intelligence, to satisfy this standard. The comparative analysis of the strengths and weaknesses of the different approaches, presented in Chap. 2, clearly defines the limitations of each method. Independently of the demonstrated industrial successes by the individual methods, described in the related chapters, their direct application potential is very limited and the probability of a credibility fiasco is high.

At the same time, the synergetic potential between computational intelligence methods is enormous. Almost any weakness of a given approach can be compensated for by another method. A clear example is when a neural network or a fuzzy system model optimizes its structure using a genetic algorithm. Exploring the integration potential of computational intelligence gives tremendous application opportunities at a relatively low price and only a slightly complicated model development process.

11.3.1 Benefits of Integration

First, we'll focus on the key benefits from the synergetic effects between the various computational intelligence methods, shown in Fig. 11.3.

- *Improved Model Development* – The combination of different methods leads to the following gains during model development:
    - *Condensed Data* – The information content of the available data can be increased by nonlinear variable selection (performed by neural networks) and important records selection (performed by SVM). The informationally rich and condensed data are supplied for model generation to GP.
    - *Optimal Parameters* – A typical case is using GA or PSO for parametric or structural optimization of neural networks, fuzzy systems, or agent-based models.
    - *Fast Model Selection* – It is based on the condensed data and optimal parameters which generate high-quality models in fewer iterations.

- *Increased Credibility* – An obvious benefit from the synergy between computational intelligence methods is increasing the critical factor for application success - its credibility. It is based on the following specific factors:
    - *Improved Robustness* – The final result of the integrated model development is generating models with the optimal trade-off between accuracy and complexity. These relatively simple solutions are more suitable to operate reliably during minor process changes.

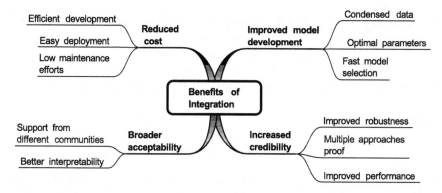

Fig. 11.3 Key benefits of integrating different modeling approaches for successful real-world applications

- *Multiple Approaches Proof* – The fact that solutions with similar performance are generated by methods with an entirely different scientific basis increases the level of trust of the proposed solution. The first-time users of the technology are really impressed by its capability to deliver multiple proofs of the expected results in different algorithmic forms without substantial increase of the development cost.
- *Improved Performance* – The final result of the integrated methodology is an application with the best trade-off of accurate prediction, robustness, and extrapolation crisis handling.

- *Broader Acceptability* – Using several methods in model development increases the application support, especially if the synergetic opportunities with first-principles and statistics are explored. The sources for this broader acceptability are as follows:

  - *Support from Different Communities* – Integration not only optimizes the technical capabilities between methods but also builds bridges between different research communities. Of critical importance for introducing computational intelligence in large organizations with advanced R&D activities is to have on board the first-principles modelers and the statisticians. The best way to accomplish this is by integrating computational intelligence with their approaches.
  - *Better Interpretability* – As a result of integration, the methods, which are more appropriate for interpretation, such as symbolic regression or fuzzy rules, can be used as final delivery. However, the other methods are used in improving the performance of fuzzy rules and symbolic regression during the integrated model development.

- *Reduced Cost* – All total cost of ownership components are positively affected by the integration, as is discussed briefly below:

  - *Efficient Development* – The optimal development process significantly reduces the development time and increases the quality of the developed models.
  - *Easy Deployment* – The final models are from technologies such as symbolic regression or statistics that have minimal run-time software requirements and can easily be integrated into the existing infrastructure.
  - *Low Maintenance Efforts* – The increased robustness and potential for extrapolation crisis handling reduce significantly the maintenance cost.

11.3.2   The Price of Integration

Integration adds another level of complexity over the existing methods. Having in mind that the theoretical foundations of most computational intelligence approaches are still under development, the analytical basis of an integrated system is available only in several specific cases, such as neuro-fuzzy or genetic-fuzzy systems. However, the practical benefits of integration have already been demonstrated and explored before developing a solid theory.

Integration also requires more complex software and broader knowledge about the methods discussed in this book. In comparison to the benefits, however, the price for integrating computational intelligence methods is relatively low.

11.4 Integration Opportunities

The good news about building an integrated system is that the ways to explore the synergistic capabilities of its components are practically unlimited. The bad news is that it is practically unrealistic to devote theoretical efforts to analyze all of them. As a result, most of the designed and used integrated systems lack a solid theoretical basis. However, the experience from several big multinational companies, such as GE,[2] Ford,[3] and Dow Chemical,[4] demonstrate the broad application of integrated systems in different areas of manufacturing, new product design, and financial operations.

We'll focus on three layers of integration. The first layer includes the synergy between the computational intelligence methods themselves, the second layer integrates computational intelligence methods with first-principles models, and the third layer explores the bridges between computational intelligence and statistics.

11.4.1 Hybrid Intelligent Systems

Hybrid intelligent systems combine at least two computational intelligent technologies, for example neural networks and fuzzy systems. The key topic of hybrid intelligent systems is the methodology of integration. The best way to illustrate the direction of potential integration is by borrowing a social analogy from the famous quote of Prof. Zadeh about a good and a bad hybrid system: "A good hybrid system is when we have British Police, German Mechanics, French Cuisine, Swiss Banking and Italian Love. But a hybrid of British Cuisine, German Police, French Mechanics, Italian Banking, and Swiss Love is a disaster". In the same way, a hybrid intelligent system could be a tremendous success or a complete fiasco depending on the effectiveness of a component's integration. As discussed in Chap. 2, each

[2]P. Bonissone, Y. Chen, K. Goebel, and P. Khedkar, Hybrid soft computing systems: industrial and commercial applications, *Proceedings of the IEEE*, *87*, no. 9, pp. 1641–1667, 1999.

[3]O. Gusikhin, N. Rychtyckyj, and D. Filev, Intelligent systems in the automotive industry: Applications and trends, *Knowl. Inf. Syst.*, *12*, 2, pp. 147–168, 2007.

[4]A. Kordon, Hybrid intelligent systems for industrial data analysis, *International Journal of Intelligent Systems, 19*, pp. 367–383, 2004.

Fig. 11.4 Ways of integrating computational intelligent systems

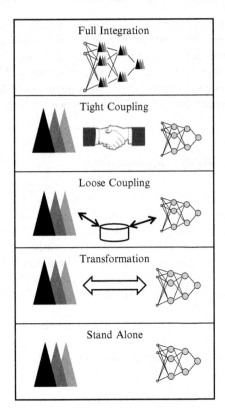

computational intelligent technology has its strengths and weaknesses. Obviously, the synergistic potential is based on the compensation of the weaknesses of one approach with the strengths of another. The key ways of integration are shown in Fig. 11.4 illustrated by integrating neural networks and fuzzy systems.[5]

- *Stand Alone* – The different technologies do not interact in any way and can be used in parallel. The derived models can be compared and even used in an ensemble. As we already discussed, the independent model generation from approaches with different scientific basis increases significantly the application credibility.
- *Transformation* – One technology is transformed into a different technology, which interacts with the final user. An example is the transformation of the classes, captured by an unsupervised neural network, into rules and membership functions, implemented as a fuzzy system.

[5]L. Medsker, *Hybrid Intelligent Systems*, Kluwer, 1995.

- *Loose Coupling* – The different technologies communicate indirectly through data files. An example is when neural networks are used for variable selection, which generates a data file with reduced number of variables. This file is later used by SVM to extract the support vectors in the data and to reduce the number of records only to data with high informational content. The latter condensed data file can be used by different techniques for high-quality data-driven model generation.
- *Tight Coupling* – The different technologies communicate directly within a common software environment. For example, parameters of fuzzy rules are updated by neural networks in real time.
- *Full Integration* – The different technologies can be embedded in each other as a unified system. For example, the neurons of a neural network can be represented by fuzzy rules.

  Different combinations of the discussed types of integration and the computational intelligence approaches have been explored and applied. The most frequently used are as follows:

- *Neuro-fuzzy Systems* – These are neural networks that are trained to develop If-Then fuzzy rules and determine their membership functions. The synergy between both computational intelligence approaches makes neural networks interpretable and fuzzy systems capable of learning and adapting in a changing environment. The generic structure of a neuro-fuzzy system with multiple inputs and multiple outputs is shown in Fig. 11.5.

  Layer 1 and output Layer 5, three hidden layers represent membership functions and fuzzy rules. Layer 2 includes neurons that capture the input membership functions used in the antecedents of fuzzy rules. It is also called the fuzzifier. The fuzzy rules are implemented by the corresponding neurons in

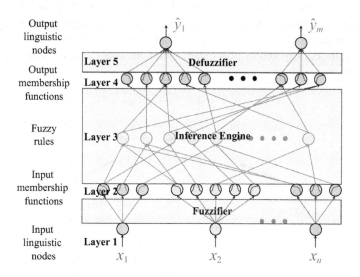

Fig. 11.5 A generic architecture of a neuro-fuzzy system

Layer 3, called the fuzzy rule layer. Layer 4 includes neurons that represent fuzzy sets used in the consequent of the fuzzy rules in Layer 3. It is also called the output membership layer and receives inputs from the fuzzy rule neurons and combines them using fuzzy operations. Both Layers 3 and 4 represent the inference engine of the neuro-fuzzy system. The top Layer 5 of the neural network combines the outputs from the different rules in Layer 4 and applies some defuzzification method, i.e. the final results of the neuro-fuzzy system are linguistic variables.

The famous neuro-fuzzy systems application is the popular rice cooker, discussed briefly in Chap. 8.

- *Genetic Fuzzy Systems* – A genetic fuzzy system is basically a fuzzy system augmented by a learning process based on a genetic algorithm. Genetic tuning assumes an existing rule base and aims to find the optimal parameters for the membership functions. A chromosome encodes the parameterized membership functions associated with the linguistic terms of all the variables. Genetic learning processes cover different levels of applications: from the simplest case of parameter optimization (adaptation or tuning) to the highest level of complexity of learning the rule set of the fuzzy system (learning).

  Genetic fuzzy *systems* have been successfully applied in control, manufacturing, and transportation by companies like GE.
- *Evolutionary Neural Networks* – The marriage between evolutionary computation and neural networks can lead to structures with optimal topology and parameters. An example of an evolving neural network using genetic algorithm is shown in Fig. 11.6.

The principle of integration between genetic algorithms and neural networks is simple. The problem domain of the neural network that needs to be optimized – represented by its structure or weights – is encoded as a genotype chromosome. Then a population of genotypes evolves using genetic operators until the performance of the corresponding phenotype neural networks reaches some specified optimal fitness.

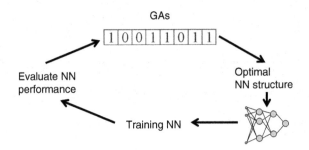

Fig. 11.6 An example of an evolving neural network

Fig. 11.7 An example of joint integration of computational intelligence methods and first-principles models

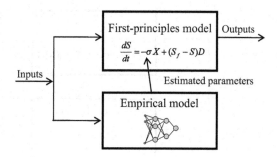

A big application area of evolving neural networks is computing games. It has been used in checkers, chess, and complex games like Nero.[6]

11.4.2 Integration with First-Principles Models

One of the key issues in industrial model development, especially in the chemical industry and biotechnology, is reducing the total cost of ownership of first-principles modeling. This is a niche that computational intelligence can fill very effectively by various ways of integration of its methods into the first-principles model building process. Unfortunately, the academic exploration of this area is orders of magnitude less intensive relative to hybrid intelligence systems even though the demand is higher. We'll focus on four potential ways of integrating computational intelligence methods and mechanistic models: (1) joint integration, (2) parallel operation, (3) accelerating first-principles model development by empirical proto-models, and (4) representing fundamental models by their empirical substitutes for on-line optimization.

The first integration scheme of fundamental and empirical models, generated by computational intelligence, is shown in Fig. 11.7. The idea is to combine a simplified parametric first-principles model with some empirical estimator of the unknown parameters. A typical case is a mass balance hybrid model of a fed-batch bioreactor, in which an artificial neural network estimates the specific kinetic rates, such as biomass growth and substrate consumption.

In the second integration scheme, shown in Fig. 11.8, the simple fundamental model and the empirical model (a neural network in this example) operate in parallel to determine the model output. The first-principles model serves as a physically sound but not detailed representation of the process. The model deficiency, however, is compensated for by the parallel empirical counterpart. The neural network is trained on the residual between the real response and the fundamental model-based estimates. As a result, the model disagreement is minimized and the combined prediction is more accurate.

---

[6]http://www.nerogame.org

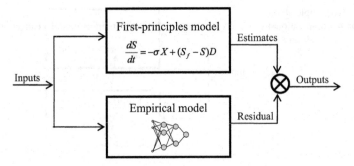

Fig. 11.8 An example of parallel operation of computational intelligence methods and first-principles models

Fig. 11.9 A generic scheme of using empirical proto-models in first-principles model development

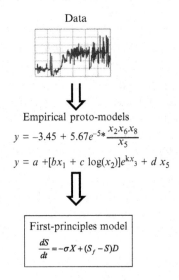

The third scheme of integrating fundamental and computational intelligence models, shown in Fig. 11.9, is not based on direct or indirect model interaction but on using empirical methods during first-principles modeling.

The idea is to reduce the cost of one of the most expensive and unpredictable phases in fundamental model development – hypothesis search. Usually this phase is based on a limited amount of data and a large number of potential factors and physical mechanisms. The current process of finding the winning hypothesis and mechanisms only based on the available data could be time-consuming and costly. However, if the data are used first to generate symbolic regression models, the process could be significantly shortened. In this case, the starting position for hypothesis search is based on a set of empirical proto-models, which may help the first-principles modeler with ideas for proper mechanisms. The benefits of this form of integration for development of structure-properties relationships are illustrated in Chap. 14.

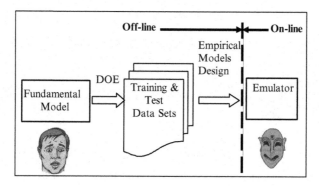

Fig. 11.10 Representing first-principles models as empirical models or emulators for on-line optimization

The fourth scheme of integrating fundamental and computational intelligence models, shown in Fig. 11.10, links both paradigms with the data, generated by Design Of Experiments (DOE).

The first-principles model (the face in Fig. 11.10) generates high-quality data off-line, which are the basis for deriving an empirical model that emulates the high-fidelity model's performance on-line (the mask in Fig. 11.10). This form of integration opens the door using complex first-principles models for on-line optimization and control. Very often, the execution speed of fundamental models is too slow for real-time operations and this limitation reduces the frequency and efficiency of process optimization.

Empirical emulators are one of the rare cases when neural networks can be reliably used, since we have full control over the ranges of data generation and the potential for extrapolation is minimal. An example of a successful application of emulators in the chemical industry is given in Chap. 14.

11.4.3 Integration with Statistical Models

Surprisingly, the least-explored area of integration is between computational intelligence and statistics even though the benefits could be used by a large army of statisticians. Of special interest is the synergy between evolutionary computation, which is population-based, and statistical modeling. We'll focus on the integration opportunities between genetic programming (GP), as one of the most applicable evolutionary computation methods, and statistics.[7] The main advantages of integration are captured in the mind-map in Fig. 11.11 and discussed below.

---

[7] F. Castillo, A. Kordon, J. Sweeney, and W. Zirk, Using genetic programming in industrial statistical model building, *Genetic Programming Theory and Practice II.*, U.-M. O'Reilly, T. Yu, R. Riolo, and B. Worzel (eds), Springer, pp. 31–48, 2004.

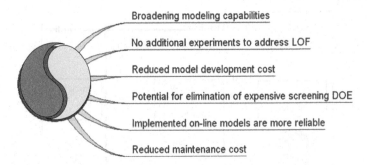

Broadening modeling capabilities

No additional experiments to address LOF

Reduced model development cost

Potential for elimination of expensive screening DOE

Implemented on-line models are more reliable

Reduced maintenance cost

Fig. 11.11 Benefits from integration of computational intelligence and statistics

The key benefit from the synergy of GP and statistical model building in industrial model development is broadening the modeling capabilities of both approaches. On the one hand, GP allows model building in cases where it would be very costly or physically unrealistic to develop a linear model. On the other hand, statistical modeling with its well-established metrics gives GP models all the necessary measures of statistical performance. Some of these measures, like the confidence limits of model parameters and responses are of critical importance for model implementation in an industrial environment.

There are several economic benefits from the synergy between GP and statistical model building. The most obvious is the elimination of additional experimental runs to address model Lack Of Fit (LOF).[8] Another case of economic benefit is the potential for the elimination of expensive screening Design Of Experiments (DOE). Since the dimensionality of real industrial problems can be high (very often the numbers of inputs is 30–50), the screening process is often very time consuming and costly. Input screening can be addressed using the GP algorithm. An additional potential benefit from the synergy between GP and statistical model building is that the applied models may have higher reliability (due to the confidence limits and reduced instability) and require less maintenance in comparison to the nonlinear models generated by GP alone.

Examples of successful integration of GP and statistics for developing models based on designed and undesigned data are shown in the next sections.

11.5 Integrated Methodology for Robust Empirical Modeling

As we already discussed, robust empirical models are the golden eggs of contemporary industrial modeling. They are based on two distinct types of data – (1) undesigned data or data that could be collected without any additional process intervention; and (2) designed data that must be extracted after carefully planned

---

[8]A statistical measure that indicates that the model does not fit the data properly.

experiments. The usual case in empirical modeling is a data collection of undesigned data, which does not require any systematic changes in the process, according to a well-defined plan. In the case of designed data, the process is analyzed in advance and the data collection is driven by a sequence of statistically designed experiments. The difference in data collection leads to a more significant contrast in the nature of the derived empirical models. While undesigned data are open to any data-driven method and the resulting models do not claim causality, designed data are limited to specific statistical representations, which may deliver causal relationships between the factors of interest and the response. Fortunately, both modeling strategies can benefit from integration with computational intelligence, and the separate methodologies are presented below and illustrated with specific applications in the next section.

11.5.1 Integrated Methodology for Undesigned Data[9]

While much cheaper than designed data, undesigned data provides many challenges, among them: data collinearity, inability to draw cause-and-effect conclusions, and limitations due to narrow variable ranges. Very often applying classical statistics under these circumstances either delivers low-quality solutions or models with limited operation ranges. Exploring the synergetic effects of several computational intelligence technologies allows us to resolve some of the undesigned data issues and generate high-quality robust empirical models.

An integrated methodology that explores the synergies is proposed and described in this section, and its main blocks are shown in Fig. 11.12.

The key objectives of the integrated methodology modeling process is to reduce data dimensionality, perform automatic generation of models, choose the models that exhibit the best trade-off between accuracy and complexity, and translate the selected models into a statistically sound linear model. It is assumed that the input of the integrated methodology is a full-size data set of undesigned data. The data dimensionality of the input is first reduced by variable selection techniques including GP or analytical neural networks (to be described in the next section). Once the variable selection has been performed, records are identified and selected that are found to contribute the most substantial information about the model, further condensing the data dimensionality via support vector machines techniques. The resulting prepared data set is supplied for model generation performed by Pareto-front genetic programming. Pareto-front genetic programming produces candidate nonlinear symbolic regression models. The produced symbolic regression models that are found to have the best trade-off between accuracy and complexity are then transformed (when possible) into linear statistically correct solutions that can be

---

[9]The initial version of the methodology is published in: A. Kordon, G. Smits, E. Jordaan and E. Rightor, Robust soft sensors based on integration of genetic programming, analytical neural networks, and support vector machines, *Proceedings of WCCI 2002*, Honolulu, pp. 896–901, 2002.

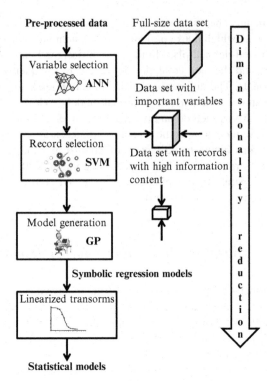

Fig. 11.12 Integrated methodology for robust empirical model development based on undesigned data

further explored. The transformed models exhibit all known statistical metrics, are statistically credible (therefore widely acceptable), and can be implemented on most of the available software environments.

11.5.1.1 Variable Selection

It is assumed that at the input of this block we have the full-sized data set, which is representative of the problem and clean of obvious outliers. It is also recommended to divide the data set for training, validation, and test purposes. The dimensionality of the full data set could vary from 10 to 20 variables and hundreds of records to thousands of variables and millions of records. Obviously, directly using such large data sets for empirical modeling could be very inefficient, especially if GP is used for automatic model generation. The first step in decreasing the size of the full data set is selecting only the most influential variables. A popular approach in this case is dimensionality reduction using Principal Component Analysis (PCA) and building linear models with projections to latent structures by means of Partial Least Squares (PLS).[10] This approach, however, has two key issues: (1) the model interpretation is

---

[10]L. Eriksson, E. Johansson, N. Wold, and S. Wold, *Multi and Megavariate Data Analysis: Principles and Applications*, Umeå, Sweden, Umetrics Academy, 2003.

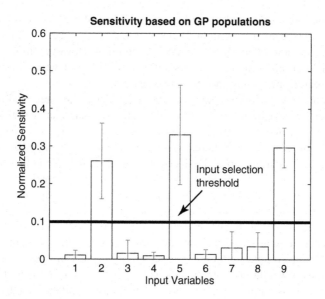

Fig. 11.13 Variable selection based on GP

difficult and (2) it is limited to linear systems. In the proposed integrated methodology, two other technologies are used for nonlinear variable selections – GP and analytic neural networks.

The first method is limited to data sets with modest dimensionality of up to 50 variables.[11] It is based on one of the unique features of GP to select the variables that are related to the solutions with high fitness during the simulated evolution and to gradually ignore variables that are not. An example is given in Fig. 11.13.

The figure represents a metric, called nonlinear sensitivity analysis of all input variables, participating in the simulated evolution. The reasoning behind the high nonlinear sensitivity of a specific input to the output is that important input variables will be used in equations that have a relatively high fitness. So the fitness of input variables is related to the fitness of the equations they are used in.[12] The example, shown in Fig. 11.13, demonstrates the final sensitivities (normalized between 0 and 1) after a simulated evolution of 300 generations of nine inputs, used for automatic model generation by GP. The final sensitivities include the average sensitivities with their standard deviations for each input x. It is clearly seen that only three inputs (x_2, x_5, and x_9) have been consistently selected in the generated functions during the simulated evolution. The variable selection can be

---

[11]The number is suggested from practical experience.

[12]The method is described in: G. Smits, A. Kordon, E. Jordaan, C. Vladislavleva, and M. Kotanchek, Variable selection in industrial data sets using Pareto genetic programming, *Genetic Programming Theory and Practice III*, T. Yu, R. Riolo, and B. Worzel (eds), Springer, pp. 79–92, 2006.

done after defining a threshold based on the statistics of the low sensitive inputs. The key advantage of this method is that it is built in the GP model generation process and is a side effect of the simulated evolution. However, it is not recommended for high-dimensional data sets since the validity of variable selection in such large search spaces has not been investigated.

The second method for nonlinear variable selection in the integrated methodology is based on sensitivity analysis of the stacked analytical neural networks. Analytical neural networks are based on a collection of individual, feedforward, single-layer neural networks where the weights of the input to hidden layer have been initialized according to a fixed distribution such that all hidden nodes are active. The weights of the hidden to output layer can then be calculated directly using least squares. The advantages of this method are: it is fast, and each neural network has a well defined, single, global optimum.

The variable selection begins with the most complex structure of all possible inputs. During the sensitivity analysis, decreasing the number of inputs gradually reduces the initial complex structure. The sensitivity of each structure is the average of the calculated weight derivatives on every one of the stacked neural nets. The procedure performs automatic elimination of the least-significant inputs and generates a matrix of input sensitivity vs. input elimination. An example of nonlinear sensitivity analysis by analytical neural networks is given in Sect. 11.6.

11.5.1.2 Data Record Selection

The purpose of the next block, based on support vector machines (SVM), is to further reduce the size of the data set to only those data points that represent the substantial information about the model. One of the main advantages of using SVM as a modeling method is that the user has direct control over the complexity of the model (i.e. the number of support vectors). The complexity can be controlled implicitly or explicitly. The implicit method controls the number of support vectors by controlling the acceptable noise level. To explicitly control the number of support vectors, one can either control the ratio of support vectors or the percentage of nonsupport vectors. In both cases, a condensed data set that represents the appropriate level of complexity is extracted for effective symbolic regression.

An additional option instead of using symbolic regression is to deliver a model based solely on SVM. As was already discussed in Chap. 4, SVM models based on mixed global and local kernels have very good extrapolation features. In the case that a symbolic regression model, generated by GP, does not have acceptable performance outside the range of training data, an SVM-based model is a viable robust solution.

11.5.1.3 Model Generation

The next block of the integrated methodology uses the multiobjective GP approach of symbolic regression to search for potential analytical relationships in a

Fig. 11.14 Explored
symbolic regression models
of interest on the Pareto front

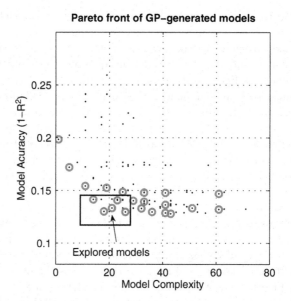

condensed data set of the most sensitive inputs. The search space is significantly reduced through the previous steps, and the effectiveness of GP is considerably improved.

A typical example of symbolic regression model generation results from Pareto-front GP is shown in Fig. 11.14 where the interesting nondominated solutions on the Pareto front are shown with encircled dots.

The explored models of interest occupy the area with the best trade-off between accuracy (measured by $1-R^2$) and expressional complexity (measured by the total number of nodes of the symbolic expression). Usually the explored area for model selection is narrowed down to the section on the Pareto front with the biggest gain in accuracy for the smallest expressional complexity (see the corresponding area in Fig. 11.11). The performance of the individual models is explored as well as the interpretability of the derived functional forms from physical consideration. The final model selection is always done with the blessing of the users. In some cases, several models with different inputs are selected in an ensemble to improve the robustness in the case of input failure.

In many cases the integrated technology delivers the final solutions as symbolic regression models at this step. However, the derived solutions, even if accepted by the users, are statistically "incorrect" and lack a statistical metric.

11.5.1.5 Model Linearization

The objective of the final step of the integrated methodology is to tailor a statistical dress to the selected symbolic regression models. The idea is very simple, and, as such, very effective: decompose the nonlinear symbolic regression model into linearized transforms and compose a statistical model, which is linear in the transformed variables.[13] In this case we have the best of both worlds. On the one hand, the nonlinear behavior is captured by the transformed variables, derived automatically from the selected GP equation. On the other hand, the final linearized solution is linear in the parameters, statistically correct, and has all related measures, such as confidence limits and variance inflation factors (VIF).[14] Recent analysis demonstrates that these types of linearized models, with transforms derived by GP, have reduced multicollinearity and do not introduce bias in the parameter estimates.[15] It has to be taken into account that the presence of severe multicollinearity can seriously affect the precision of the estimated regression coefficients, making them quite sensitive to the data collected and producing models with poor precision. The most important effect, however, is that linearized models give the so-needed statistical credibility and open the door to symbolic regression modeling and GP as a technology for the broad statistical community.

However, selecting linearized transforms is not automatic and trivial. It requires experience in statistical interpretation of the derived models. For example, looking at the error structure of the linearized models is a must. If the error structure does not have constant variance and shows patterns, models with other transforms need to be selected. Another test for acceptance of the derived linearized models is calculating the variance inflation factors (VIF) and validating the lack of multi-collinearity between the transformed variables. As the statisticians like to say, the linearized model must make "statistical sense".

11.5.2   Integrated Methodology for Designed Data[16]

The same idea of using nonlinear transforms, generated by GP, is used in developing statistical models for designed data. At the basis of building statistical models on DOE is the underlying assumption that for any system there is a fundamental relationship between the inputs and the outputs that can be locally approximated over a limited range of experimental conditions by a polynomial or a linear regression model. The capability of the model to represent the data can often be

---

[13]The idea was proposed by Flor Castillo initially for designed data and is fully described in F. Castillo, K. Marshall, J. Green, and A. Kordon, Symbolic regression in design of experiments: A case study with linearizing transformations, *Proceedings of GECCO 2002*, New York, pp. 1043–1048, 2002.

[14]Variance Inflation Factor (VIF) is a statistical measure of collinearity between input variables.

[15]F. Castillo and C. Villa, Symbolic regression in multicollinearity problems, *Proceedings of GECCO 2005*, Washington, D.C., pp. 2207–2208, 2005.

[16]The material in this section was originally published in F. Castillo, A. Kordon, J. Sweeney, and W. Zirk, Using genetic programming in industrial statistical model building, *Genetic Programming Theory and Practice II*, U.-M. O'Reilly, T. Yu, R. Riolo, and B. Worzel (eds), Springer, pp. 31–48, 2004. With kind permission of Springer-Verlag.

assessed through a formal Lack Of Fit (LOF) test when experimental replicates are available. Significant LOF in the model indicates a regression function that is not linear in the inputs; i.e. the polynomial initially considered is not adequate. A polynomial of higher order that fits the data better may be constructed by augmenting the original design with additional experimental runs. However, in many situations a second-order polynomial has already been fit and LOF is still present. In other cases the fit of a higher-order polynomial is impractical because experiments are very expensive or technically infeasible due to extreme experimental conditions. This problem can be handled if appropriate input transformations are used, provided that the basic statistical assumption that errors are uncorrelated and normally distributed with mean zero and constant variance are satisfied.

Some useful transformations have been previously published.[17] Unfortunately, transformations that linearize the response without affecting the error structure are often unknown, at times they are based on experience and frequently they become at best a guessing game. This process is time consuming and often inefficient in solving LOF situations.

Fortunately, GP-generated symbolic regression provides a unique opportunity to rapidly develop and test these transformations. This multiplicity of models, derived by GP, with different analytical expressions, provides a rich set of possible transformations of the inputs, otherwise unknown, which have the potential to solve the LOF issue.

Therefore, once LOF is confirmed with a statistical test, and transformations of the inputs seems to be the most practical approach to address this situation, GP-generated symbolic regression can be used. The key steps of the integrated methodology of statistics and GP for designed data are illustrated in Fig. 11.15 and consist of selecting equations with correlation coefficients larger than a threshold level.

These equations are analyzed in terms of the R^2. The original variables are then transformed according to the functional form of these equations. Then a linear regression model, defined as Transformed Linear Model (TLM), is fit to the data using the transformed variables. The adequacy of the transformed model is initially analyzed considering Lack Of Fit and R^2. Then the error structure of the models not showing significant LOF is considered and the correlation among model parameters is evaluated. This process ensures that the transformations given by GP not only remove LOF but also produce the adequate error structure needed for least-square estimations with no significant correlations among the model parameters.

11.6 Integrated Methodology in Action

The integrated methodology will be illustrated with an industrial application for an emissions estimation inferential sensor based on undesigned data. The curious

---

[17]G. Box and N. Draper, *Empirical Model Building and Response Surfaces*, Wiley, 1987.

Fig. 11.15 Key steps of
applying GP on statistical
modeling of designed data

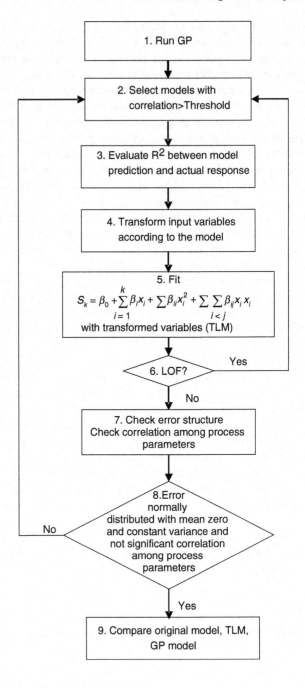

Fig. 11.16 Nonlinear
sensitivity of emissions
estimator inputs from GP
model generation

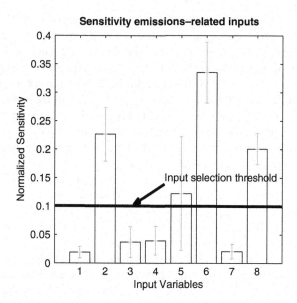

reader can find an example of how the methodology for designed data was deployed for particle size distribution of a chemical compound in Castillo *et al.*[18]

Soft sensors for emission estimation are one of the most popular application areas and a viable alternative to hardware analyzers. Usually an intensive data collection campaign is required for empirical model development. However, during on-line operation the output measurement is not available and some form of inferential sensor performance self-assessment is highly desirable. Since it is unrealistic to expect that all possible process variations will be captured during the data collection campaign, an inferential sensor with increased robustness is required. This type of inferential sensor, based on the proposed integrated methodology for undesigned data, was developed and implemented in one of The Dow Chemical Company plants in Freeport, Texas. The key results from implementation of the main blocks, shown in Fig. 11.12, are as follows.

A representative data set from eight potential process input variables and the measured emissions as output included 251 data points for training and 115 data points for testing. The test data is 20% outside the range of the training data which by itself is a severe challenge for the extrapolation capability of the model.

Both methods for nonlinear variable selection – GP and analytic neural networks – were explored. The results from the GP sensitivity analysis are shown in Fig. 11.16 and demonstrate that only four inputs, x_2, x_5, x_6, and x_8, are strongly related to emissions. Of special interest is the production rate (input x_2), called the key input.

[18]F. Castillo, A. Kordon, J. Sweeney, and W. Zirk, Using genetic programming in industrial statistical model building, *Genetic Programming Theory and Practice II*, U.-M. O'Reilly, T. Yu, R. Riolo, and B. Worzel (eds), Springer, pp. 31–48, 2004.

This result is also confirmed from the analytic neural network variable selection. The sequence of input elimination based on the sensitivity of the corresponding input to the emissions is shown in Fig. 11.17, where the dark color represents low sensitivity. The competition begins with an analytic neural network structure of 30 stacked networks with 10 neurons in the hidden layer and all eight inputs. At the end of the first modeling sequence, the input with the lowest sensitivity (input x_3) is eliminated. The second sequence begins with seven inputs, and at the end of model generation the next laggard in sensitivity (input x_7) is eliminated. The recommended neural network structure includes only four inputs (x_2, x_5, x_6, and x_8), the same as the structure according to the GP sensitivity analysis, shown in Fig. 11.18. It is selected based on the performance of the test data as the most parsimonious structure before the performance drops (see Fig. 11.18).

As a result of the nonlinear sensitivity analysis based on both GP and analytical neural networks, the number of variables was reduced by half, starting from eight and finishing with four relevant inputs.

The next step of the methodology, using SVM for records reduction, delivers a model with only 34 data points (support vectors) and extraordinary extrapolation capability, shown in Fig. 11.19. The figure shows the measured and predicted emissions relative to the key input (production rate). The model is based on a mixture of a second-order polynomial global kernel and an RBF local kernel with a width of 0.5 in a ratio of 0.95. It is clearly seen that the model reliably predicts in the interesting area with high production rate above 40,000.

As a result of variable selection and data record reduction, the representative data set for deriving the final symbolic regression model is drastically decreased to only 8.4% of the original training data set.

The initial functional set for GP model generation includes: {addition, subtraction, multiplication, division, square, change sign, square root, natural logarithm,

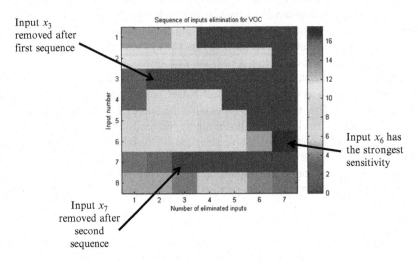

Fig. 11.17 Sequence of input elimination by analytic neural networks

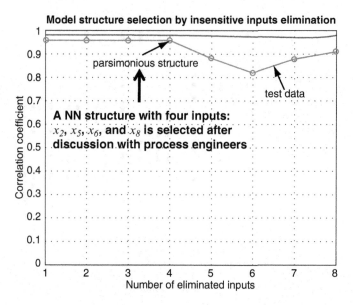

Fig. 11.18 Variable selection based on analytic neural network performance on test data

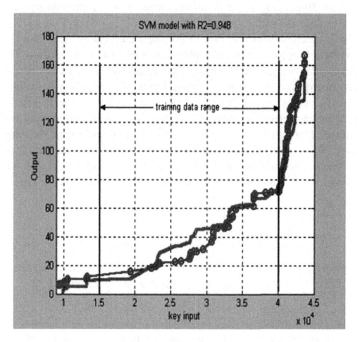

Fig. 11.19 Performance of a SVM model of emissions with four inputs and 34 support vectors

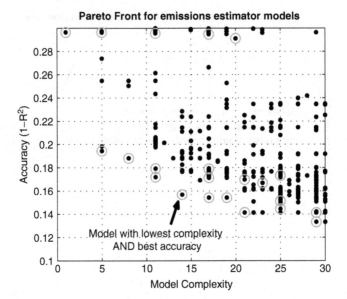

Fig. 11.20 Model selection on the Pareto front for emissions estimator

exponential, and power}. Function generation takes 20 independent runs of 30 cascades with 30 generations, a population size of 300, a probability of function selection 0.6, and size of the archive 75%.[19]

The results are represented in Fig. 11.20 where each dot is a generated symbolic regression model with corresponding model accuracy or error (measured by $1-R^2$) and model complexity (measured by the total number of nodes in the corresponding mathematical expression). There are several potential models with optimal trade-off between accuracy and complexity, which are lying on the Pareto front and represented by encircled dots. The final selected model, shown with an arrow, has the advantage of having the lowest complexity (14 total nodes) and the biggest gain in accuracy (R^2 is ~ 0.85). The other models with better accuracy are much more complex. The functional form of the model is given in Equation (11.1), where y is the emissions variable and x_2, x_5, x_6, and x_8 are the selected process inputs.

$$y = -3.45 + 5.67e^{-5} * \frac{x_2 x_6 x_8}{x_5}. \tag{11.1}$$

The performance of this solution has been fully explored. Examples of a response surface and contour plots are given in Figs. 11.21 and 11.22.

---

[19]The parameters are selected according to the values recommended in: F. Castillo, A. Kordon, and G. Smits, Robust Pareto front genetic programming parameter selection based on design of experiments and industrial data, In: R. Riolo and B. Worzel (eds): *Genetic Programming Theory and Practice IV*, Springer, pp. 149–166, 2007.

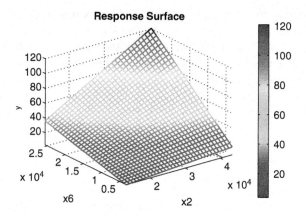

Fig. 11.21 Response surface of emissions variable when inputs x_2 and x_6 are changed by 10% outside the full ranges and inputs x_5 and x_8 are fixed

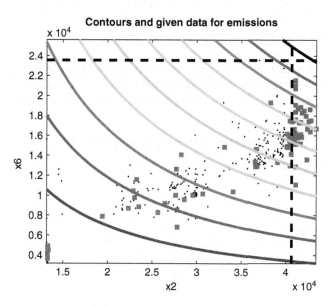

Fig. 11.22 Contour plot of the emissions variable when inputs x_2 and x_6 are changed by 10% outside the full ranges and inputs x_5 and x_8 are fixed. The extrapolation area is to the right of the dotted line. Test data are represented with small squares

Both plots give the opportunity for testing the model performance using different What-If scenarios. Usually this assessment is done jointly with the users who make the final decision of selecting or rejecting the model. In this particular case, the selected nonlinear model behaved according to process experience, gave sound predictions, and was blessed by the production engineers.

The last step of the methodology explores the opportunity to make this model statistically correct by defining linearized transforms and using them in a statistical model. At the basis of defining transforms is the symbolic expression of the selected GP solution. For example, for the expression in equation (11.1), the following linearized transforms can be defined:

$$z_1 = x_2 * x_6 \tag{11.2}$$

$$z_2 = 1/x5 \tag{11.3}$$

$$z3 = x8, \tag{11.4}$$

where x_i are different process inputs, related to the emissions variable. The following multiple linear regression model with R^2 of 0.85 (the same as the original nonlinear model) on training data was used:

$$y = -238.6 + 1.13e^{-7} * z_1 + 41,066 * z_2 + 0.92 * z_3. \tag{11.5}$$

The confidence limits, calculated using the statistical software package JMP®[20] are shown in Fig. 11.23.

A clear advantage of this type of model is that it is based on the theory of multiple linear regression. As a result, all the different statistics, including the confidence limits of prediction and parameters, outlier detection, and detection of

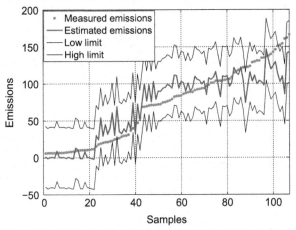

Fig. 11.23 Statistical confidence limits of the selected linearized model

---

[20]JMP is a registered trademark of SAS Institute Inc., Cary, NC, USA.

significant observations apply. In addition, good practices of model fitting, including checking of multicollinearity, can be performed.

11.7   Summary

Key messages:

The nasty reality of industrial applications requires the joint solution of the technical, infrastructural, and people-related components of a given problem.

The key factors for success of a real-world application are credibility, acceptability, robustness, extrapolation survival, integration capability, and, of course, low cost.

Integration of modeling approaches improves model development, reduces cost, and increases the credibility and acceptability of the derived solutions.

It is possible to integrate successfully not only different computational intelligence approaches but to broaden the synergy with first-principles models and statistics.

Integrated methods have been successfully used in industry by companies like GE, Ford and Dow Chemical.

The Bottom Line

Integrate the modeling methods and conquer the real world.

Suggested Reading

The following books give detailed technical descriptions of the key integration methods of computational intelligence techniques:

L. Jain and N. Martin, Fusion of Neural *Networks, Fuzzy Sets, and Genetic Algorithm: Industrial Applications*, CRC Press, 1999.

R. Khosla and T. Dillon, *Engineering Intelligent Hybrid Multi-Agent System*, Kluwer, 1995.

L. Medsker, *Hybrid Intelligent Systems*, Kluwer, 1995.

Z. Michalewicz, M. Schmidt, M. Michalewicz, and C. Chiriac, *Adaptive Business Intelligence*, Springer, 2007.

D. Ruan (Editor), *Intelligent Hybrid Systems: Fuzzy Logic, Neural Networks, and Genetic Algorithm*, Kluwer, 1997.

Chapter 12
How to Apply Computational Intelligence

There is only one valid definition of business purpose: to create a customer.

Peter F. Drucker

Implementing successfully any new emerging technology, such as computational intelligence, requires consistent support of many stakeholders, appropriate infrastructure, and readiness to accept change. The objectives of this very important chapter in the book is to propose some guidelines for applying computational intelligence in a business, to give some flavor of the application pains, and to illustrate the application methodology with real examples.

12.1 When Is Computational Intelligence the Right Solution?

Some answers to this question that may open or close the door to potential computational intelligence applications are captured in the mind-map in Fig. 12.1.

- *Competitive Advantage* – The most important criterion for applying computational intelligence is the competitive advantage it gives to the user. It will be extremely difficult to sustain long-term success if there is no clear vision of the sources of competitive advantage in implementing this emerging technology in specific business areas. The three generic sources of computational intelligence competitive advantage are highlighted below and were already discussed in Chap. 3.
- *Complex Problems* – Often practitioners look at computational intelligence after exhausting the capabilities of other methods. The key reason is the high level of complexity of the problem, especially related to nonlinear interactions. This category also includes problems with very high dimensionality and any type of social systems modeling. The methods that contribute mostly to complexity

A.K. Kordon, *Applying Computational Intelligence*,
DOI 10.1007/978-3-540-69913-2_12, © Springer-Verlag Berlin Heidelberg 2010

Fig. 12.1 Criteria for evaluating potential computational intelligence applications

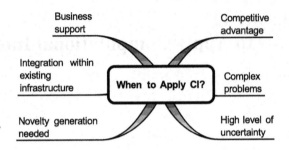

modeling are intelligent agents, neural networks, support vector machines, and evolutionary computation.

- *High Level of Uncertainty* – Another issue that could be resolved by using computational intelligence is the imprecise nature of available expertise combined with inaccurate data. Fuzzy systems, especially in combination with neural networks and intelligent agents can reduce this type of uncertainty and build reliable models. Another form of uncertainty is the high level of *a priori* assumptions required for model development. In contrast to first-principles and statistical models, neural networks, support vector machines, or evolutionary computation do not need strict assumptions for building models. In addition, machine learning allows reducing uncertainty through continuously learning about the changes in the environment and updating model parameters accordingly.

- *Novelty Generation* – Computational intelligence is very effective for businesses where innovation is critical. Intelligent agents can capture emergent behavior from local interactions, which could be identified as novelty. Evolutionary computation automatically generates novel structures, such as electric circuits, optical lenses, and control systems. Neural networks, support vector machines, and evolutionary computation capture unknown dependencies between variables that could be used for process monitoring and new product design.

- *Integration Within Existing Infrastructure* – The level of modeling experience and the established work processes in the targeted business are very important factors to consider. The efforts of integrating computational intelligence into the existing software environments and operating processes, such as Six Sigma, need to be evaluated very carefully. It is also recommended to pay attention to factors like training of key developers and users and maintenance of the derived solutions.

- *Business Support* – Demonstrating the technical advantages of applying computational intelligence to a specific business is one side of the equation. On the other side is the clear commitment of the business to allocate the necessary resources for the required period of time. The support must be formally communicated to the organization or the key stakeholders in a short document, like a statement of direction. It is preferable to clarify the application funding in advance as well.

12.2   Obstacles in Applying Computational Intelligence

The second step in the application strategy of computational intelligence is avoiding potential pitfalls of a technical and nontechnical nature. From the detailed analysis of computational intelligence issues, discussed in Chap. 10, we'll focus on the following specific obstacles that need to be assessed.

12.2.1   Technical Obstacles in Applying CI

The important technical issues that may reduce the efficiency of the computational intelligence applications are shown in Fig.12.2 and discussed next.

- *Data Quality* – Since most computational intelligence methods are data-driven, data quality becomes a critical factor for success. Firstly, data availability must be checked very carefully. It is possible that the historical records are too short to capture seasonal effects or trends. Secondly, the ranges of the most important factors in the data have to be as broad as possible to represent the nonlinear behavior. Data-driven models developed on narrow data ranges have low robustness and require frequent readjustment. Thirdly, the frequency of data collection must be adequate to the nature of the modeling. For example, dynamic modeling requires more frequent collection and data sampling. Steady-state models, on the other hand, assume slow data collection frequency that filters the dynamic effects. Fourthly, the noise level has to be at acceptable limits to avoid the classical Garbage-In-Garbage-Out (GIGO) effect. In the case when some of these requirements are not met, it is recommended to create an adequate data collection infrastructure and to begin the application only after collecting the right data. Making a compromise with data quality is one of the most frequent

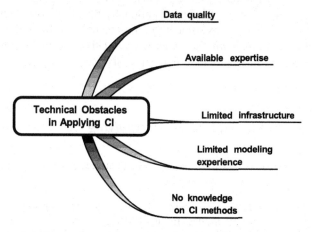

Fig. 12.2 Key technical issues in applying computational intelligence

mistakes in applying computational intelligence. The perception that poor data can be compensated for with the "magic" and sophistication of advanced modeling methods is one of the leading causes of an application fiasco.

- *Available Expertise* – The access to and the quality of domain knowledge is of great importance for applying computational intelligence. However, allocating the necessary resources and winning their support for the application is not trivial. Most subject-matter experts are excited to participate in projects with advanced technologies, especially if their role is well recognized. It is possible, however, that some domain gurus feel challenged by the potential smart computerized "competitor" and may not cooperate enthusiastically. Having the key experts "on board" the application efforts is time-consuming and may require management intervention as well.

- *Limited Infrastructure* – The success of computational intelligence application depends on the integration capabilities of the existing hardware, software, and work processes infrastructure. Usually most of the off-line applications do not require significant changes in the existing infrastructure. A special case is users' addiction to Excel and the requirement to interact only within its environment. Fortunately, this limitation is either resolved by using Excel add-ins in some of the professional software products, such as SAS Enterprise Miner, or with additional software efforts. For on-line implementation of computational intelligence, however, a careful analysis of the software limitations and the maintenance infrastructure is a must.

- *Limited or No Modeling Experience* – The success and the speed of implementing computational intelligence depends also on the previous record of modeling applications. Even the lessons from applying simple statistical models are of help since a modeling culture has been introduced. As a result, users have some experience in using and maintaining models as well as assessment of the created value. On the other hand, one of the issues of limited modeling experience is defining unrealistic expectations, which paves the way to an application fiasco. Usually the lack of modeling culture is combined with limited infrastructure for implementation and support, which additionally raises the total cost of ownership due to the necessary investment in infrastructure and training.

- *No Knowledge on Computational Intelligence* – This is the common case in most businesses and the necessary tools to resolve this issue are discussed in this book.

12.2.2 Nontechnical Obstacles in Applying CI

The key nontechnical issues that may lead to unsuccessful computational intelligence applications are shown in Fig. 12.3.

- *No Management Support* – As with any emerging technology, computational intelligence needs initial management blessing before it can begin to deliver sustainable value. Of critical importance is also the consistency of the support

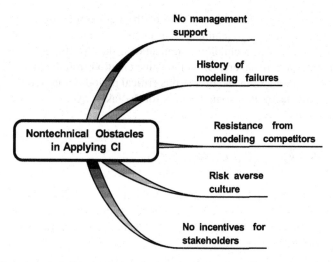

Fig. 12.3 Key nontechnical issues in applying computational intelligence

for a period of at least three years. Unfortunately, this requirement may be unrealistic in businesses with frequent restructuring and management changes. The best strategy to address this issue is to find application areas with fast demonstration of value creation and promote computational intelligence with effective marketing.

- *History of Modeling Failures* – Nothing can impede a computational intelligence application more than a previous modeling fiasco, independent of the applied methods. Unfortunately, it takes years to recover the destroyed credibility from this type of failure. Without firm management support and enthusiastic users (most of them not related to the past bad experience), it is too risky to start a computational intelligence application in such a poisonous environment. If that happens, it is recommended as a precondition for the application to analyze the causes of the previous modeling failures in a short document and to define the differences with the new computational intelligence methods. Defining clear deliverables upfront in a document is a must.
- *Resistance from Modeling Competitors* – The different types of resistance have been already discussed in Chap. 10. The best way to address this issue is by exploring the benefits from the synergies between computational intelligence and the different competitive approaches. The comparative analysis in Chap. 9 and the examples in the book are a good starting point in this process.
- *Risk-Averse Culture* – It is difficult to apply computational intelligence in organizations with a low level of taking risks. Unfortunately, it is not trivial to assess in advance the potential for this problem. Some patterns for detecting technological conservatism can be defined as: no investment in high-tech in the last five years, no work process to bring in and support innovation and process improvements, and no incentives for taking risk.

- *No Incentives for Stakeholders* – One of the key lessons from the numerous applications of emerging technologies is the need for an adequate reward system for all participants in technology applications: developers, users, support and maintenance, and managers. Often the value creation from computational intelligence takes a long time and during this critical period it is preferable to support the innovative team with some benefits for the taken risk.

12.2.3   Checklist "Are we Ready?"

To summarize the factors that may influence the decision-making process of initiating a computational intelligence application, a generic checklist is shown below. Examples of typical answers are given in brackets, having in mind the inferential sensor application:

- Define appropriate application (on-line inferential sensors).
- Define competitive advantage (lower cost than hardware analyzers).
- Get management support (decision to fund the projects).
- Allocate available stakeholders (modelers, process engineers, operators).
- Check data quality (process and laboratory measured data availability and ranges).
- Identify infrastructure needs (new software for development and run-time inferential sensor execution is needed; project will be developed within Six Sigma work process; maintenance infrastructure must be established).
- Estimate users' attitudes (skepticism over inferential estimates credibility, concern about model support, positive experience from using first-principles models for physical properties estimation).
- Estimate training needs (full training for developers, short training for users).
- Propose incentives for all stakeholders (offer management specific incentives within the business reward practices).

12.3   Methodology for Applying CI in a Business

If the checklist shows that an organization has the capacity and is willing to explore the benefits of computational intelligence, we move to the critical application step. We'll separate this step into two parts. Firstly, the sequence of applying CI in a business will be discussed in this section. Secondly, the key steps of CI project management will be described in the next section.

There are many ways to introduce and integrate a new technology, such as CI, in a business.[1] They also depend on the size of the business and its capacity to allocate

---

[1]T. Davenport and J. Harris, *Competing on Analytics: The New Science of Winning*, Harvard Business School Press, 2007.

resources for internal development of the technology. At one end are the small businesses that cannot afford any development efforts and are interested in cheap and simple solutions with minimal capital and training cost. At the other extreme are the large international corporations with tremendous capabilities to use internal and external resources for development and deployment of the technology across the globe. This applies in the case of The Dow Chemical Company, as illustrated in this book. The generic phases for applying CI, however, are similar, and are shown in Fig. 12.4. Each specific business can adjust the efforts according to its size and available capacity.

The methodology separates the application sequence into three distinct phases with growing impact: (1) introduction phase, (2) application phase, and (3) leveraging phase.

The purpose of the introduction phase is to validate the technology potential based on well-selected pilot projects. Since there is a deficit of knowledge about computational intelligence, external resources from the universities, vendors and consultants are used in marketing, training, and project development. The objective of the application phase is to validate the value creation potential of the technology in several real-world projects. Very often, external resources from vendors are used in project development and support. The goal of the leveraging phase is to maximize the benefits of computational intelligence through its integration into existing work processes. A more detailed description of each phase follows next.

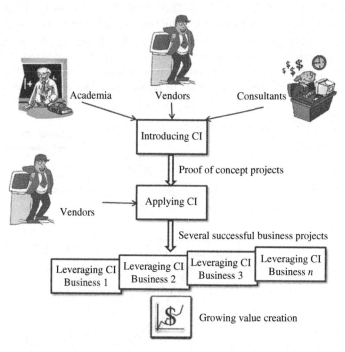

Fig. 12.4 Key phases of methodology for applying computational intelligence in a business

12.3.1 Steps for Introducing CI in a Business

The key steps for introducing CI in a business are shown in Fig. 12.5.

1. *Marketing CI* – The obvious first step in introducing computational intelligence in an organization with no or minimal knowledge of the technology is marketing. The objective is to demonstrate the competitive advantage of computational intelligence to a broad audience of potential users. Some guidelines for organizing the marketing efforts are given in the next chapter, and examples for presentation slides and elevator pitches are given in each chapter for the key CI methods. It is possible to include external resources from the leading researchers, consultants, or vendors. The ideal case is a presentation that illustrates the benefits of applying computational intelligence in areas similar to those in the targeted business.

2. *Educate Key Stakeholders* – It is good practice to introduce the technology with more details to future developers and users. The resources given in this book are a good starting point. Many introductory presentations for each of the computational intelligence technologies are available on the Web as well. Most of the leading computational intelligence conferences also offer tutorials on the technologies. It is strongly recommended to attend these conferences and to begin building a network with the research community. Computational intelligence is a very dynamic research area and needs to be continuously on the radar screen.

3. *Evaluate Business Opportunities* – The driving force for using computational intelligence is some business needs that are difficult to solve using existing methods. At this early stage it is necessary to identify and prioritize the generic needs. Examples of such generic needs are: improved process monitoring systems, process control based on inferential sensors, and developing robust empirical models for new products. In many cases, the existing solutions for

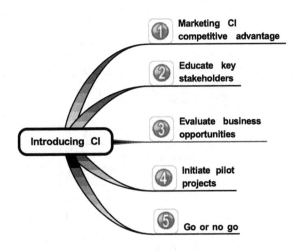

Fig. 12.5 Key steps for introducing computational intelligence in a business

each of these needs, based either on first-principles models or on statistical models, have exhausted their potential. Most computational intelligence technologies offer capabilities, such as low-cost data-driven models or pattern-recognition algorithms, which may deliver the desired solutions.

4. *Initiate Pilot Projects* – The most important step in the introduction stage is the problem of identifying the top priority business needs, to demonstrate the competitive advantage over the existing solutions, and to implement this in a relatively short period of time, preferably less than six months. An example of a pilot project is development and deployment of a specific inferential sensor in a manufacturing process. At least four computational intelligence technologies can be used: neural networks, symbolic regression, support vector machines, and fuzzy systems. During the pilot project development and deployment it is normal practice to transfer knowledge from external resources.

5. *Go or No Go* – The results from the pilot projects are the litmus test for triggering the next phase of the application strategy — applying computational intelligence in different large-scale projects. The decision has to be made based on the pilot project performance metric and acceptance of the solution by the users.

12.3.2 Steps for Applying CI in a Business

The key steps for applying CI in a business are shown in Fig. 12.6. It is assumed that the pilot projects have raised interest in and demonstrated the unique capabilities of computational intelligence. The objective of this step is to demonstrate the value creation potential of the technology by effective applications for solving appropriate business problems.

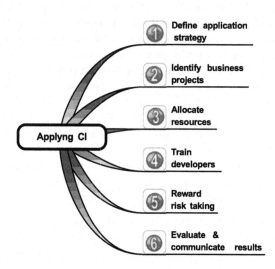

Fig. 12.6 Key steps for applying computational intelligence in a business

1. *Define Application Strategy* – It is strongly recommended to direct the computational intelligence application efforts in a systematic way. As a good starting point, a more detailed survey of the current business needs than in the introductory phase and their value assessment is recommended. Often, some of the selected application areas are linked to the pilot application (for example, focusing on development of inferential sensors). In this case, it is preferable to gradually increase the number of applications based on the delivered value. Identifying other application areas is always a plus, for example, new product development.
2. *Identify Business Projects* – Selecting appropriate projects must follow the application strategy. However, it is recommended not to jump immediately from pilot to large-scale projects. For example, an ambitious project to develop 50 inferential sensors in a wide range of manufacturing processes creates a significant risk for failure due to the limited experience and insufficient internal resources.
3. *Allocate Resources* – The biggest challenge in the application phase is the available capacity for development, deployment, and support of the potential projects. On the one hand, it is too early to estimate the long-term demand and to commit substantial internal resources. On the other hand, relying only on external resources doesn't lead to a strategic presence of computational intelligence in the business. The key in the decision-making process on this topic is the assessment of the need for internal development. If the results from the first few applications demonstrate value creation, it is recommended to evaluate the option of gradually allocating resources for internal development of the projects. For example, in the case of growing demand for different inferential sensors, it is better to transfer the model development from the vendor to several internal experts in the area of process monitoring.
4. *Train Developers* – An inevitable step in the case of allocating resources for internal model development is a comprehensive training on the computational intelligence methods and corresponding tools. Due to the rapid growth of the technology, the training mode is almost continuous. The training could be done by the vendors and at the key conferences related to computational intelligence.
5. *Reward Risk-Taking* – One of the best practices to encourage new technology development is to link a specified percentage of the bonuses of all stakeholders in the application projects to the level of risk-taking.
6. *Evaluate and Communicate Results* – Nothing speaks louder in support of computational intelligence than a series of successful applications with documented value creation. Of special importance is the demonstration of competitive advantage over established solutions. For example, the overall savings from several applied inferential sensors may exceed millions of dollars relative to the established but expensive hardware analyzers.

In order to build credibility, it is recommended to communicate also the issues and the lessons learned from the failed applications. For example, some concerns about inferential sensor support have to be addressed and illustrated with comments from final users. Cases of unsuccessful inferential sensor development due to insufficient data have to be given as examples of the limitations of the proposed technology.

12.3.3 Steps for Leveraging CI in a Business

The best-case scenario of the application methodology is when the business finds opportunities for sustainable value creation by using computational intelligence in different areas. The key steps for leveraging the technology are shown in the mind-map in Fig. 12.7.

1. *Mandate from Management* – The critical step in moving ahead with computational intelligence across the business must be clearly supported by the top management. Ideally, a top manager takes the lead and forms a steering committee with key experts and executives. The mandate for leveraging the technology assumes strategic support for several years and corresponding organizational decisions. The most important factor is the vision and commitment of the top manager in charge. Unfortunately, there is always a risk of management changes and the replacement of the visionary sponsor with a narrow-minded bureaucrat. In this case, there are no guarantees that the mandate will be fulfilled with the previous enthusiasm and support. The best protection is the sustainable value creation from the existing applications.

2. *Define Strategy of Direction* – One of the important documents that navigates the leveraging efforts is the statement of direction. It explicitly defines the appropriate uses of computational intelligence in the business (for example, in process monitoring and control, data mining, and new product development). Another important section of the document is the scope of applicability (for

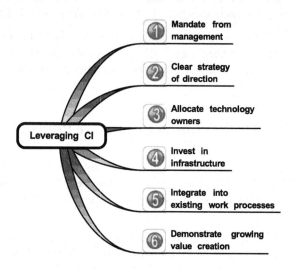

Fig. 12.7 Key steps for leveraging computational intelligence in a business

example, the list of businesses, software environments, data warehouses, and control systems). The statement of direction also includes: the appropriate actions that should be taken by the different stakeholders; the next steps to be taken to further leverage the technology in a specific period of time; the key milestones with specific dates, and the owners, stakeholders, and contacts.

3. *Allocate Technology Owners* – If the previous two phases could be done by individuals or a network of technology users and supporters, leveraging requires allocating an organization or group that takes ownership of computational intelligence. Some big companies, like Microsoft, GE, and Google, have specialized groups in R&D for internal development of the technology as well. The more common situation is when the computational intelligence owner is some generic modeling or data mining group. The specialized resources are part of these groups and work on other modeling projects as well. The technology owner takes responsibility for of being the center of computational intelligence expertise, manages the strategy of direction, leads the implementation efforts, and takes care of technology training.

4. *Invest in Infrastructure* – Of special importance is resolving the issue of mainte-nance and support of existing applications. In addition to the support from the vendors, internal resources are needed and must be allocated.

5. *Integrate into Existing Work Processes* – The best way to leverage the technol-ogy with minimum infrastructural investment is by using the existing work processes. An example of how to integrate computational intelligence into one of the most popular work processes in industry, Six Sigma, is given in this chapter.

6. *Demonstrate Growing Value Creation* – The definition of leveraging success is self-sustainability, i.e. the created value from implemented projects fully covers the total cost of ownership. At that point the technology growth is based on project demand not on management support. In our example with inferential sensors, after several applications in different manufacturing processes, the demand and, respectively, the generated value and support began to grow.

In addition to the economic benefits, successful computational intelligence leveraging creates a growing network of users and technology supporters who initiate the next application wave in a natural way.

12.4 Computational Intelligence Project Management

One can manage computational intelligence projects in many different ways.[2] We'll focus on a very generic project sequence, shown in Fig. 12.8, where in addition to the key steps of project development, the expected deliverables are

---

[2]A very systematic and detailed methodology for developing models for real-world applications is given in the book of D. Pyle, *Business Modeling and Data Mining*, Morgan Kaufmann, 2003.

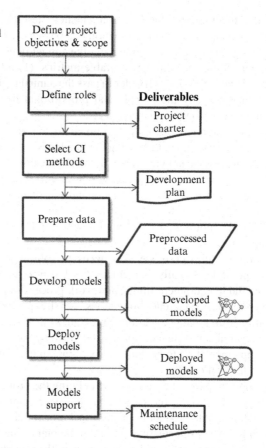

Fig. 12.8 Key steps and deliverables of computational intelligence project management

identified. Examples of adopting the methodology to Six Sigma and a specific project management case for data mining and modeling are also given at the end of the chapter.

The discussed generic project management sequence is valid for any model development technique. The only specific step is selecting the appropriate computational intelligence methods. However, in all the steps we'll use any opportunity to discuss the specific issues related to this technology. It has to be taken into account that computational intelligence methods are often used in combination with other techniques in solving a specific business problem.

12.4.1 Define Project Objectives and Scope

This is the most important and mishandled step in the project management sequence. One of the challenges is finding a performance metric that is measurable, can be tracked, and is appropriate for defining success. An example of an appropriate quantitative objective that satisfies these conditions is to develop an inferential sensor for emission estimation that passes the regulatory tests,[3] increases the operating rate by 2% while reducing the annual noncompliance cases by 30%. This definition includes three measurable metrics: (1) inferential sensor accuracy, which has to be below 7.5% in order to pass the annual regulatory test; (2) operating rate, which is directly measurable and proportional to the economic profit; and (3) the number of noncompliance tickets.

Since the operation rate is proportional to the emissions level, the last two criteria are conflicting, i.e. increasing the rate leads to more emissions and potential noncompliance cases and vice versa. In the absence of on-line emissions estimates, the limits of production rate are very conservative in order to avoid the risk of violating the environmental constraints and the plant is losing money. The economic driving force behind the project is that an inferential sensor will allow simultaneously more aggressive control with higher rates and a reduced risk of noncompliance.

Defining the project scope also needs to follow the quantitative metrics as much as possible. Usually it includes the business geography boundaries, data limits, available equipment limits, and work process requirements. For our example, the project scope includes boundaries like: the specific manufacturing unit, data availability from a data collection campaign with production rate variation range, approved by management, control limits defined by the existing control system, and project implementation to be done in Six Sigma in the existing process information system in the plant.

An important part of this step is also defining the project deliverables in terms of products and impact. In the case of the emissions inferential sensor project, the expected deliverable product is an empirical model that has to be integrated within the existing process information system in the plant. The financial impact from the applied inferential sensor can be calculated by the increased profit from the higher rates and the reduced number of noncompliance cases.

12.4.2 Define Roles

Identification of appropriate stakeholders is another very important step for the success of the computational intelligence project. One possible way to find the necessary stakeholders, such as project sponsors, owners, developers, and model support, is by evaluating the impact of the project on them. In this evaluation, several factors, like stakeholders' needs, career goals, degree of support, and real influence in the organization, have to be assessed.

In the case of our inferential sensor project example, the following stakeholders have been identified: management sponsor (who provides the project funding), process owner (who has the authority to allow changes in the existing

---

[3]Usually the regulatory tests require an average modeling error < 7.5%.

manufacturing system), project leader (who coordinates all project activities), technical experts (who develop, deploy, and maintain the model), process experts (who know the process and the data), and users (who use the new improved control based on estimated emissions).

Based on the defined objectives, scope, and roles, a project charter is delivered to the whole team. The project officially starts after approved funding and entering the project charter into the corresponding project tracking database.

12.4.3   Select Computational Intelligence Methods

A specific task in running a computational intelligence project is mapping the project needs with the most appropriate methods. A very generic mapping is given in Table 12.1. The abbreviations in the table have the following meaning: ANN – Artificial Neural Networks, SVM – Support Vector Machines, EC – Evolutionary Computation, SI – Swarm Intelligence, and ABMS – Agent-Based Modeling System.

As seen from the table, developers can select several methods for the same application area. Different considerations, such as specific performance requirements or software availability, can be taken into account in the final selection.

In our inferential sensor example, the key potential methods for model development were neural networks and symbolic regression via genetic programming. The advantage of the neural network solution was that the entire project could have been developed, deployed, and supported by an established vendor (Pavilion Technologies). However, the limited capability for data collection for model development increases the risk of potential extrapolation mode of the derived model. In contrast to neural networks, symbolic regression models can deliver reliable predictions during minor process changes. The project team also preferred to implement the symbolic regression models directly into the existing control system rather than to use separate run-time neural network software. The selected method was symbolic regression generated by GP.

The deliverable of this step is a model development sequence, which specifies the modeling methods, their tuning parameters, and software tools.

Table 12.1 Most appropriate computational intelligence methods for key application areas

Application area/Method	Fuzzy	ANN	SVM	EC	SI	ABMS
Empirical modeling	√	√	√	√		
Forecasting		√		√		
Optimization				√	√	
Classification	√	√	√			
Scheduling	√			√	√	√
Social systems modeling	√					√
New products invention			√	√		
Games		√		√		√

12.4.4 Prepare Data

Data preparation includes all necessary procedures to explore, clean, and pre-process the available data in order to begin model development with the maximal possible information content in the data.[4] In reality, data preparation is time consuming, nontrivial, and difficult to automate. Very often it is also the most expensive phase of empirical modeling in industrial conditions. A significant part of the cost is data collection, especially when designed experiments are needed.

The key phases and steps of data preparation are shown in the mind-map in Fig. 12.9.

Another more practical way to approach data preparation is by using a checklist of the answers to key questions related to the different phases. An example of a checklist is shown in Table 12.2. We will scan the checklist briefly.

The first phase of data preparation, data definition, includes problem definition, data size definition, data expertise capturing, and data collection. The activities in this phase are aimed at answering the key question "Do we have the data to solve

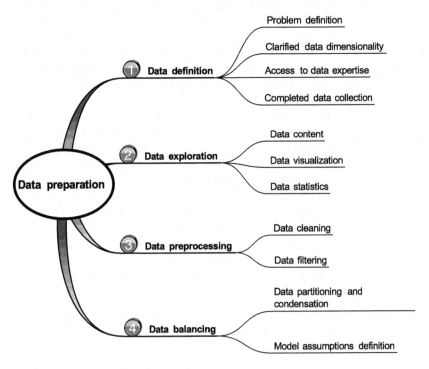

Fig. 12.9 Key data preparation phases and steps

---

[4] A classical book for data preparation is the book of D. Pyle, *Data Preparation for Data Mining*, Morgan Kaufmann, 1999.

Table 12.2 Data checklist preparation

Key questions in data preparation
(1) Do we have the data to solve the problem? (Data definition phase.)
What kind of data do we need to solve the problem?
What is the expected size of the data set?
Number of variables;
Number of data points.
Do we have the necessary knowledge about the available data?
Is the data collection sufficient?
Can we afford a data collection campaign based on designed experiments?
Are there any significant gaps in the collected data?
Are the variables sufficiently disturbed to cover the expected range of change of process parameters?
(2) Can we accept the available data? (Data exploration phase.)
Are the data meaningful?
Are the data visually acceptable?
Are the data statistically acceptable?
(3) How can we improve the information content in the data? (Data preprocessing phase.)
Are the data clean of obvious outliers?
Is it possible to reduce noise in the data?
(4) Are we ready for modeling? (Data balancing phase.)
Are the data prepared according to the specific modeling method requirements?
Can we define realistic modeling assumptions based on the available data quality?

the problem?" This depends on the answers of the questions listed in Table 12.2 and is related to the necessary understanding of the problem, defining the expected size of the data set, allocating the sources of expertise, and assessing the sufficiency of the collected data. The deliverable of the data definition phase is the best possible data collection of raw data that could be used for future data analysis and model building.

The data exploration phase (see Question 2) evaluates the quality of the available data by examining the data content, integrity, and key statistical properties. At the end of this phase we must have a clear view of what can be done with the available data and how we can proceed further. In the case of significant gaps due to measurement or data collection system failures, or statistical problems (like odd distribution or very high variability), it is may be necessary to return to the previous phase and require new data collection.

The objective of the data preprocessing phase (see Question 3) is to increase the information content in the available data using various techniques such as outlier detection, interpolation, smoothing, and filtering. The deliverable of this phase is a clean data set with minimal noise that is ready for the modeling process.

The last phase in data preparation, data balancing (see Question 4), has a dual role. On the one hand, it defines potential clusters and partitions the data according to the requirements of the specific modeling methods (like preparing training and test data sets for neural networks models). On the other hand, it refines the initial assumptions about expected model quality based on the statistical properties of the available data. This leads to more realistic objectives of the forthcoming modeling efforts.

The deliverable of this step of the computational intelligence project sequence is a set of training, validation, and test data files of high-quality data, ready for modeling.

12.4.5   Develop Model

Model development must follow the strategy, based on the selected computational intelligence methods and their specifics. It is good practice, before beginning to derive the more complex computational intelligence solutions, to test the performance of the potential statistical models, based on the available data. It can be used also as a benchmark for the empirical solutions, derived using computational intelligence. In the case of no substantial performance gain, the users may even prefer the simple statistical solutions.

The specific model development sequence for each computational intelligence method, including parameter selection, is given in the corresponding chapters. It is recommended to look at the robustness of the modeling method by generating models with some ranges of the tuning parameters. In the case of population-based methods, such as evolutionary computation, at least 20–30 independent runs of simulated evolution are needed for statistically acceptable reproducibility of the results.

The recommended approach for model development is to use the integrated methodology, described in the previous chapter. In this way the benefits from several computational intelligence methods will be explored and the potential for generating robust empirical models will be increased. Another alternative to deliver robust solutions is by developing an ensemble of models. Of special interest is an ensemble of models derived from different methods, for example neural networks, SVM, and symbolic regression. First, the very fact that models with similar performance are generated using different methods increases the credibility of the derived solution. Second, the average of the individual predictors could be a better metric for prediction. Third, the standard deviation between the models in the ensemble, or the model disagreement, could be used as a measure of confidence of predictions. We have to be careful, though, since model ensemble design is still an open research area and is usually done *ad hoc*.

Model validation is the most critical step in model development. Classical model validation is based on the model performance on three different data sets – training, validation, and test. It is good practice to validate the robustness of the model by varying the percentages of these data sets and different ranges in the data. Another form of model validation is by testing its expected behavior in selected "What-If" scenarios. It is a very good strategy to give the developed model to users and to ask them to assess the expected behavior. User-selected models have higher credibility and better chance for success during deployment and operation. In contrast, model selection without the blessing of or any feedback from the users creates the risk of ignoring the predictions, even competing with the model. The users feel detached and look at the model as a "foreign body".

In our emission estimations example, several symbolic regression models on the Pareto front were selected and explored by developers and users until the final selection was made.

The deliverable of this step is the set of developed models, which are well-validated by the data and accepted by the users.

12.4.6 Deploy Model

The selected solution from the development step may have a different form for the specific computational intelligence methods. For example, the fuzzy system solution can be implemented as "If-Then" rules linked to defined membership functions. Neural networks and support vector machines models are deployed as black-boxes with model structure, defined by the number of layers, neurons, and support vectors, and the corresponding matrices of weights. Symbolic regression models can be represented by mathematical expressions. Usually, the deployed models require a special run-time version of the developed software.

There is a big difference in model deployment for off-line and on-line use, though. Model implementation in a real-time environment requires additional protection in the case of missing data, outliers, or out-of-range operation. Some measure of model performance self-assessment is recommended to increase the reliability of the predictions.[5] This capability is of special importance to inferential sensors, which often operate without output measurements for modeling performance evaluation. In our example for emissions estimation, the model performance is validated with real output measurements only once a year during the regulatory test.

An important procedure of model deployment is the design of the user interface. In most applications, the implemented model is integrated in the existing information or control systems. A clear advantage of this solution is the minimal training of the final user. In our example of an emissions inferential sensor, the selected symbolic regression model was directly coded in the existing process monitoring system as a measurement tag. It made the interface for process operators very convenient since there is no difference in accessing the model and any other hardware sensor in the system.

When a special interface for the deployed model is needed, the dialog must be simplified as much as possible. In addition, training the user is a must. It is also especially important to clarify the meaning and the ranges of the specific tuning parameters, if they are needed. Preparing all the necessary documentation for installing and operating with the model is of critical importance for its use and credibility.

---

[5]Examples of different measures are given in A. Kordon, G. Smits, E. Jordaan, A. Kalos, and L. Chiang, Empirical models with self-assessment capabilities for on-line industrial applications, *Proceedings of CEC 2006*, Vancouver, pp. 10463–10470, 2006.

The deliverable of this step of the computational intelligence project sequence is a model implemented in a user-friendly software environment supported by well-written documentation.

12.4.7   Model Maintenance and Support

Unfortunately, this is the most underestimated and ignored step in the computational intelligence project sequence. The different methods require different maintenance efforts. Usually neural network-based models are more sensitive to process changes and have higher frequency of model adjustment. Part of the neural network maintenance procedure also needs historical data collection that captures the process changes. It is also recommended to add out-of-range indicators for the most statistically significant model inputs to avoid predictions in the case of extrapolation.

Model performance tracking is critical for model maintenance and support. That is why model robustness is so important and becomes a significant economic factor. It is recommended that some built-in model self-assessment criteria are applied to warn the user if the performance is unacceptable. If the performance deterioration becomes a trend, model readjustment or complete redesign is needed.

Unfortunately, there is almost nothing in the literature about the long-term performance of any applied modeling techniques.[6] In our example, there are limited data about the operating conditions of the applied inferential sensors in the chemical industry for a period of three to five years. For different reasons, such as operating regime fluctuations due to different product demand, control system readjustment, or equipment changes, the on-line operating conditions for at least 30% of the applied models exceed the initial off-line model development ranges by > 20% outside the off-line model development range. This extrapolation level is very high and is a challenge for any of the empirical modeling techniques. The high extrapolation level requires model redesign, including derivation of entirely new structures. One potential technical solution that could reduce the maintenance cost is using evolving fuzzy systems which adapt their structure and parameters with the changing operating conditions.[7] An example is given in Chap. 15.

During the introductory and application phases of computational intelligence, the maintenance and support are done either by the vendors or by the developers. In the leveraging phase, however, a viable option is the creation of an internal support service either by a specialized group or by additional training in some of the

---

[6]One exception is the paper: B. Lennox, G. Montague, A. Frith, C. Gent and V. Bevan, Industrial applications of neural networks – An investigation, *Journal of Process Control*, *11*, pp. 497–507, 2001.

[7]P. Angelov, A. Kordon, and X. Zhou, Evolving fuzzy inferential sensors for process industry. *Proceedings of the Genetic and Evolving Fuzzy Systems Conference (GEFS08)*, Witten-Bommerholz, Germany, pp. 41–46, 2008.

existing support functions. In our example of an emissions inferential sensor, the model was supported by the developers.

The expected deliverable from this last step in the computational intelligence project sequence is a well-defined maintenance and support program for at least three years.

12.5 CI for Six Sigma and Design for Six Sigma

A method or set of techniques, Six Sigma,[8] has become a movement and a management religion for business process improvement. It is a quality measurement and improvement program originally developed by Motorola in the 1980s that focuses on the control of a process to the point of $\pm$ Six Sigma (standard deviations) from a centerline, or more specifically 3.4 defects per million items. The specific meaning of reducing the defect level from Three Sigma (where the average quality level typically is before applying the program) to Six Sigma for some well-known cases in the real world is shown in Fig. 12.10.

The Six Sigma systematic quality program provides businesses with the tools to improve the capability of their business processes. Another similar quality program, Design for Six Sigma (DFSS), implements the statistical principles to increase the efficiency of new products and process design.[9] Both programs are based on classical statistics. Computational intelligence methods are a natural extension for

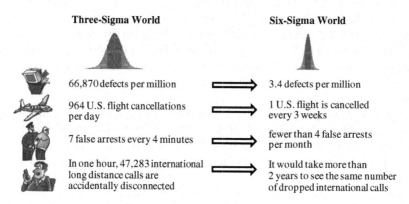

Fig. 12.10 Differences in number of defects for specific cases between Three Sigma and Six Sigma quality

---

[8]One of the best books about Six Sigma is: F. W. Breyfogle III, *Implementing Six Sigma: Smarter Solutions Using Statistical Methods*, 2nd edition, Wiley-Interscience, 2003.

[9]K. Yang and B. El-Haik, *Design for Six Sigma: A Roadmap for Product Development*, McGraw-Hill, 2003.

problems that require solutions beyond the statistical limitations in both Six Sigma and Design for Six Sigma.

12.5.1 Six Sigma and Design for Six Sigma in Industry

At the basis of Six Sigma methodology is the simple observation that customers feel the variance, not the mean, i.e., reducing the variance of product defects is the key to making customers happy. What is important for Six Sigma is that it provides not only technical solutions but a consistent work process for pursuing continuous improvement in profit and customer satisfaction. This is one of the reasons for the enormous popularity of this methodology in industry. According to the *iSixSigma Magazine*,[10] about 53% of Fortune 500 companies are currently using Six Sigma – and that figure rises to 82% for the Fortune 100. Over the past 20 years, use of Six Sigma has saved Fortune 500 companies an estimated $427 billion. The most famous case of Six Sigma embracement is GE where all 300,000+ GE employees must be Six Sigma certified and all new GE products are developed using the Design for Six Sigma (DFSS) approach.

Companies that properly implement Six Sigma have seen profit margins grow 20% year after year for each sigma shift (up to about 4.8 sigma to 5.0 sigma). Since most companies start at about 3 sigma, virtually each employee trained in Six Sigma will return on average $230,000 per project to the bottom line until the company reaches 4.7 sigma. After that, the cost savings are not as dramatic.

One of the key advantages of Six Sigma is the well-defined roles in project development. The typical roles are defined as: Champion, Black Belt, Green Belt, and Master Black Belt. The Champion is responsible for the success of the projects and provides the necessary resources and breaks down organizational barriers. It is typical for a large part of a Champion's bonus to be tied to his/her success in achieving Six Sigma goals.

The project leader is called a Black Belt. Black Belt assignments usually last for two years and have eight to twelve projects. Projects are broken down into quarters and oftentimes the people on the team will change from quarter to quarter.

Project team members are called Green Belts and they do not spend all their time on projects. They receive training similar to that of Black Belts but for less time.

All "Belts" are considered Agents of Change and it is extremely important to get the right people in these roles. They have to like new ideas and generally tend to be the leaders of the organization.

There is also a Master Black Belt level. These are experienced Black Belts who have worked on many projects. They are the ones who typically know about more advanced tools, the business, and have had leadership training and oftentimes have teaching experience. A primary responsibility is mentoring new Black Belts.

---

[10]January/February 2007 Issue at http://www.isixsigma-magazine.com/

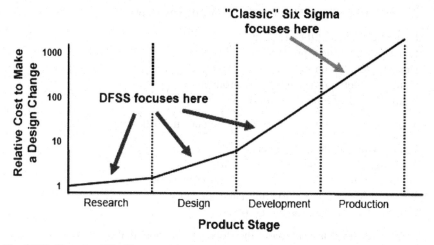

Fig. 12.11 The role of Six Sigma and Design for Six Sigma in different product phases

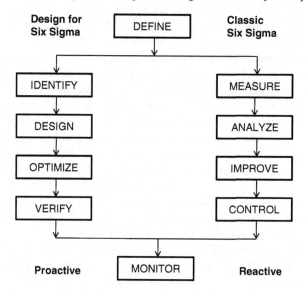

Fig. 12.12 Key steps of Six Sigma and Design for Six Sigma

As illustrated in Fig. 12.11, the two methodologies, classic Six Sigma and Design for Six Sigma (DFSS), are used in different phases of the product or process evolution.

While DFSS is focused on the early phases of identifying new product needs from market research, optimal product design, and prototyping, Six Sigma is focused on improving the production of already designed products. In this case, the relative cost of possible design change is very high and the better alternative to increase profitability is by reducing the variability of existing manufacturing processes. The different focus requires different work processes. The key steps of DFSS and Six Sigma are shown in Fig. 12.12.

Design for Six Sigma includes the following key steps:

Define: Understand the problem
Identify: Define new product
Design: Develop new product
Optimize: Optimal design of new product
Verify: Build new product prototype

The classic Six Sigma methodology includes the following key steps:

Define: Understand the problem
Measure: Collect data on the problem
Analyze: Find root cause as to why the problem occurs
Improve: Make changes to eliminate root causes
Control: Ensure the problem is solved

A more detailed description of the steps and the potential use of computational intelligence are given below.

The key stages of the most popular Six Sigma process, called DMAIC (Define–Measure–Analyze–Improve–Control), as well as the icons of the appropriate computational intelligence technologies are shown in Fig. 12.13. The sequence includes all blocks for computational intelligence project management, as was discussed in the previous section. However, in an organization that has embraced Six Sigma, modeling project management and the related infrastructure are built into the culture. That makes model development, deployment, and maintenance much easier and cheaper. In addition, most of the employees are properly trained and the work processes are well-established and linked to career development, bonuses and promotions. For the majority of the big corporations, the best-case scenario for implementing computational intelligence projects is through Six Sigma.

The biggest benefits from computational intelligence in this case are from integrating its unique features (dealing with nonlinearity, complexity, uncertainty, and social influence) with the statistical methods used in Six Sigma. Some ideas on how to do it in the different DMAIC stages are given below.

In the Define stage, we identify clearly and communicate to all stakeholders the problem we want to solve. We lay down the team members and the timelines. This stage gives the guidelines for what we want to do. Project objectives are based on identifying the needs by collection of the voice of the customer. The opportunities are defined by understanding the flaws of the existing as-is process. The key document in this stage is the project charter, which includes the financial and technical objectives, assessment of the necessary resources, allocation of the stakeholders' roles, and a project plan.

The key computational intelligence method that may contribute in the Define stage is fuzzy logic. The qualitative nature of information in this stage can be quantified with fuzzy logic and used in the next stages. Of special importance is the case when the key defect, such as "bad odor", is not directly measured but can be estimated linguistically by humans. The Six Sigma Define stage includes the following steps from the generic project management methodology: define

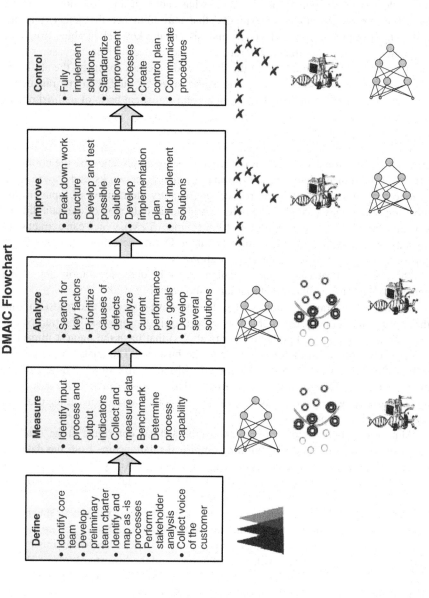

Fig. 12.13 Key blocks of the DMAIC Six Sigma methodology and the corresponding computational intelligence methods

project objectives and scope, define roles, and select computational intelligence methods.

In the Measure stage, we try to understand the problem in more detail by collecting all available data around it. We want to know what really is the problem, where does it occur, when does it occur, what causes it and how does it occur. Identifying the necessary factors (inputs) that may influence the defect (output), collecting and preparing the data for analysis are the key deliverables from this stage. Another very important deliverable is the calculated statistical measures from the data, such as the sigma level of the defect and process capability (the ability of a process to satisfy customer expectations, measured by the Six Sigma range of a process's measured variation). This gives a solid statistical basis of defined objectives and allows us to add another layer of objectivity in decision-making by benchmarking the process with the known standards.

Several computational intelligence methods can complement the statistical analysis that is normally used in this stage. Of special importance are the nonlinear variable selection methods, based on analytic neural networks and Pareto-front genetic programming, which can reduce significantly the number of inputs to the most important ones related to the defect. Additional reduction of the data dimensionality can be done using support vector machines (SVM), which can extract only the data records with high informational content. Outlier detection is another SVM capability that can be used in data preparation.

The Six Sigma Measure stage is equivalent to the data preparation step of the generic project management methodology, shown in Fig. 12.8.

In the Analyze stage, we analyze the data we have collected and try to find out the root causes for why the problem is occurring in the process. The purpose of Analyze is to search for key factors (critical X's) that have the biggest impact in causing the problem to the customer or to the business. The analyses are mostly statistically based and try to identify the critical inputs affecting the defect or the output $Y = f(x)$. The potential cause–effect relationships or models are discussed and prioritized by the experts. As a result, several potential solutions of the problem are identified for deployment.

Almost any computational intelligence approach can contribute in the Analyze stage, but the key methods for building empirical relationships from the data are neural networks, SVM, and symbolic regression via GP. An option very appealing to the Six Sigma community is to deliver the final relationships as linearized statistical models, as was discussed in Chap. 11.

The Six Sigma Analyze stage is equivalent to the model development step of the generic project management methodology, shown in Fig. 12.8.

In the Improve stage, we prioritize the various solutions that we brainstormed and explore the best solution with highest impact and the lowest cost effort. At the end of this stage, a pilot solution is implemented. The purpose of Improve is to remove the impact of the root causes by implementing changes in the process. Before beginning the implementation steps, however, the selected solution is tested with data to validate the predicted improvements. It is also recommended to use a method, called Failure–Mode–Effect–Analysis (FMEA) for recognizing and evaluating potential failures in solution operation. A high-level implementation plan must be reviewed with sponsors and stakeholders. The initial results from pilot implementation are communicated to all stakeholders.

Several computational intelligence methods can be used in the Improve stage, especially in the area of parameter optimization. At the top of the list are Particle Swarm Optimizers (PSO) and Genetic Algorithms (GA). If the selected solutions are nonlinear, they could be implemented using neural networks, symbolic regression, or SVM.

The Six Sigma Improve stage is equivalent to the model deployment step of the generic project management methodology, shown in Fig. 12.8.

In the final, Control, stage, we complete the implementation of the selected solutions and we need to validate that the problem has gone away. This is "the proof of the pudding" stage in Six Sigma. Usually we set up measurement systems to help us figure out if the problem has been solved and the expected performance has been met. One of the key deliverables in the Control stage is the process control and monitor plan. Part of the plan is the transition of ownership from the development team to the final user.

The methods in this stage are the same as in the Improve stage – PSO, evolutionary computation, neural networks, and SVM.

The Six Sigma Control stage is equivalent to the model maintenance step of the generic project management methodology, shown in Fig. 12.8.

An example of applying CI in the Six Sigma work process in the area of data mining[11] at The Dow Chemical Company is given by Kalos and Rey.[12]

12.5.2 How CI Fits in Design for Six Sigma

Design for Six Sigma (DFSS) is a methodology for designing new products and redesigning existing products and/or processes with defects at a Six Sigma level. The driving force behind DFSS is the faster market entry of new products, which gives the benefits of earlier revenue stream and longer patent coverage. DFSS relies on early problem identification through the voice of the customer and market analyses. The key strategy is to implement the Six Sigma methodology as early in the product or service life cycle as possible and to take care of the defects upfront. The key DFSS stages and the corresponding computational intelligence methods icons that may be used are shown in Fig. 12.14.

---

[11]Data mining and modeling is a process of discovering various models, patterns, summaries, and derived values from a given collection of data. Data mining is mostly based on computational intelligence methods.

[12]A. Kalos and T. Rey, Data mining in the chemical industry, *Proceedings of the Eleventh ACM SIGKDD International Conference on Knowledge Discovery and Data Mining*, Chicago, IL, pp. 763–769, 2005.

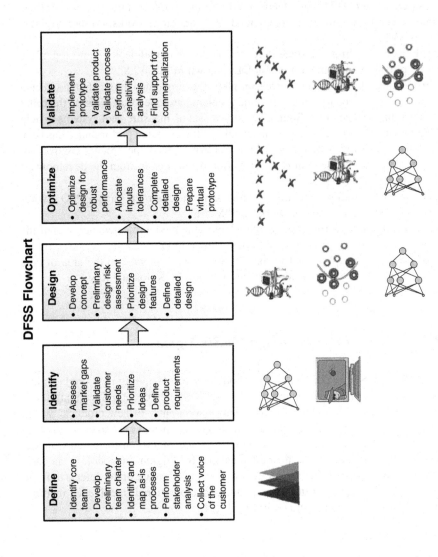

Fig. 12.14 Key blocks of the Design for Six Sigma (DFSS) methodology and the corresponding computational intelligence methods

In contrast to classic Six Sigma, the DFSS activities are focused on innovation and are usually based on small data sets, generated through expensive Design Of Experiments (DOE). In both cases, computational intelligence offers unique capabilities that can increase model development efficiency relative to the existing statistical techniques. The specific methods for each DFSS stage are discussed below.

The Define stage of DFSS is the same as in classic Six Sigma and all steps and recommended computational intelligence methods previously discussed, are still valid.

In the Identify stage, an opportunity for new product development is discovered, based on market analyses and the voice of the customer. The purpose of the market analyses is to assess the existing gaps through future demand forecasting, competitor analysis, and technological trends evaluation. It is usually combined with identifying customer needs through capturing the Voice of the Customer (VOC) and translating it into functional and measurable product requirements or a Critical-to-Customers (CTCs) metric. Several methods are used in DFSS to accomplish this goal. Quality Function Deployment (QFD) is a planning tool for translating the Voice of the Customer (VOC) into explicit design, production, and manufacturing process requirements. The QFD House of Quality (HOQ) transfers customers' needs to requirements based on the strength of the interrelationships.

The ideas about new products, inspired by the marketing gaps and the voice of the customer, are brainstormed and prioritized. The deliverable from this stage is a document that defines the product design requirements.

Computational intelligence can play a significant role in the Identify stage by adding new capabilities in marketing analyses and customer needs identification. For example, neural networks can be used for market pattern recognition and customer clustering. Customer behavior can be simulated by intelligent agents where new needs could emerge.

In the Design stage, the formulated product requirements are translated into a product design specification. Firstly, the conceptual design is performed, followed by a preliminary risk assessment. Secondly, several design alternative solutions are explored and prioritized. Thirdly, the selected solution is developed at a detailed design level.

Computational intelligence adds the following unique capabilities to the statistical methods used in the Design stage. Evolutionary computation can generate solutions with entirely novel design either automatically or in an interactive mode. Empirical models, developed using neural networks, support vector machines, or symbolic regression can be a low-cost alternative to first-principles models for simulating the designed product. Of special importance is the capability of SVM to derive robust models with a small number of data records, as is usually the case in new product development.

In the Optimize stage, the designed product is optimized for robust performance that satisfies the functional requirements at the Six Sigma performance level. The optimal design includes model development, planning and running Design Of Experiments (DOE), inputs sensitivity analysis, setting manufacturing tolerances, and building a virtual prototype.

Computational intelligence can contribute in the Optimize stage by broadening the classical optimization methods with genetic algorithms and swarm intelligence-based optimization. In addition, symbolic regression can reduce the cost of expensive experimental designs, as was discussed in Chap. 11.

In the Validate stage, the virtual prototype is fully implemented and validated on a pilot sale. Its performance is evaluated in a wide range of operating conditions and compared to the specified functionality. If the new product passes all the tests, a plan for full-scale commercial rollout is initiated with a specific sponsor.

Computational intelligence adds unique capabilities in the Validate stage mostly in the area of robust empirical model building. One of the most difficult decisions at this stage with potential for significant savings is the possibility to skip the small-scale manufacturing phase and to jump directly to full-scale production. Having first-principles and robust empirical models with reliable scale-up performance is a prerequisite for exploring the risk of excluding the pilot-scale efforts. Support vector machines and symbolic regression can generate low-cost empirical models with impressive generalization capabilities that can be used to estimate the expected features in scale-up situations.

12.6 Summary

Key messages:

Computational intelligence is the right technology for applications with a high level of complexity and uncertainty or a need for novelty generation.

The key factors that contribute to computational intelligence application success are: management support, defined business problem, competent project team, and communicated incentives for risk-taking.

The key technical requirements for computational intelligence application success are: high-quality data, process knowledge, available infrastructure, and educated stakeholders.

Computational intelligence can be deployed in a business in three key phases: introduction phase (validate technology on pilot projects), application phase (demonstrate value creation on several full-scale applications), and leverage phase (maximize benefits by integrating computational intelligence into existing work processes).

Computational intelligence project management includes project definition, CI method selection, data preparation, model development, deployment, and maintenance.

Integrating computational intelligence applications within the Six Sigma methodology increases efficiency and reduces cost.

The Bottom Line

Successful computational intelligence applications require consistent support of many stakeholders, appropriate infrastructure, and readiness to accept change.

Suggested Reading

The following books and papers cover different application aspects of computational intelligence and Six Sigma:

F. W. Breyfogle III, *Implementing Six Sigma – Smarter Solutions Using Statistical Methods*, 2nd edition, Wiley-Interscience, 2003.

T. Davenport and J. Harris, *Competing on Analytics: The New Science of Winning*, Harvard Business School Press, 2007.

Y. Yang and B. El-Haik, *Design for Six Sigma: A Roadmap for Product Development*, McGraw-Hill, 2003.

A. Kalos and T. Rey, Data mining in the chemical industry, *Proceedings of the Eleventh ACM SIGKDD International Conference on Knowledge Discovery and Data Mining*, Chicago, IL, pp. 763–769, 2005.

D. Pyle, *Data Preparation for Data Mining*, Morgan Kaufmann, 1999

D. Pyle, *Business Modeling and Data Mining*, Morgan Kaufmann, 2003.

Chapter 13
Computational Intelligence Marketing

A model is like a political cartoon. It picks-up the substantial part of a system and exaggerates it.

John Holland

This chapter is about selling research-intensive emerging technologies, such as computational intelligence. The proposed research marketing strategy is based on generic marketing principles, adapted to the specific computational intelligence target markets and research products. Special attention is given to marketing computational intelligence via interaction between universities and industry. Finally, guidelines and examples of marketing the technology to technical and business audiences in industry are presented.

Research marketing relies on capturing the attention of the targeted audience with inspiring messages and attractive presentations. Simple nontechnical language, interesting stories, and informative visualization are a must in this process. Humor is another significant factor for success and is widely used in many forms. One of the most impressive forms is satirical cartoons and we'll begin with one funny example for creating "effective" marketing strategy from the popular Dilbert cartoon strip, as shown in Fig. 13.1.

13.1 Research Marketing Principles

Selling research ideas and methods is not a familiar activity for most of the academic and industrial research communities. The usual attitude is negative and the marketing efforts are treated by many as a "car-dealer-type of activity far below the high academic standards". Common practices for communicating research ideas and methods are via publications, patents, conference presentations, and proposals for funding. In this way the targeted audience is limited to the scientific peers in the corresponding research areas. The average size of this audience is a couple of hundred researchers and barely exceeds several thousand for the most popular

A.K. Kordon, *Applying Computational Intelligence*,
DOI 10.1007/978-3-540-69913-2_13, © Springer-Verlag Berlin Heidelberg 2010

Fig. 13.1 The Dilbert view of "effective" marketing. DILBERT: © Scott Adams/Dist. by United Feature Syndicate, Inc.

specialized scientific communities. If this number is sufficient for spreading the message across the academic world, it is absolutely inadequate for selling the concept to the broad audience of potential users in the real world.

Different research methods require distinct marketing efforts. The well-known and widely used first-principles models or statistics do not need research marketing of their capabilities. However, new technologies, like computational intelligence, need significant and systematic marketing efforts. Of course, the objective is not to transform researchers into advertisement agents or car dealers, but to raise their awareness of the need for research marketing and demonstrate some of the necessary techniques.

13.1.1   Key Elements of Marketing

Firstly, we'll give a condensed overview of the key marketing steps. Marketing is defined as:[1]

> Marketing is the process of planning and executing the conception, pricing, promotion, and distribution of ideas, goods, services, organizations, and events to create and maintain relationships that will satisfy individual and organizational objectives.

It is a common perception that without effective marketing, even a quality product will fail. Quality is determined by a buyer's perception of product performance, not by what company designers and engineers decide to produce. Here lies the critical misconception about marketing, known as Marketing Myopia: to focus on products, not on customer benefits. That very question is the defining moment in understanding the nature of marketing and requires a cultural change by product

---

[1]One of the best references in marketing is the classic book of L. Boone and D. Kurtz, *Contemporary Marketing*, 13th edition, South-Western College Publishers, Orlando, FL, 2007.

developers who are usually obsessed by the technology. The different attitude is best illustrated by the famous quote of Charles Revson, founder of Revlon:

"In our factory we make perfume; in our advertising, we sell hope".

The key elements of a generic marketing strategy are listed below:

- *Define Target Market* – Identify a group of people toward whom a firm markets its goods, services, or ideas with a strategy designed to satisfy their specific needs and preferences.
- *Product Strategy* – Involves more than just deciding what goods or services the firm should offer to the target market. It also includes making decisions about customer service, package design, brand names, trademarks, warranties, etc.
- *Distribution Strategy* – Ensures that the customers find their products and services in the proper quantities at the right times and places. Distribution decisions also involve selection of marketing channels, inventory control, warehousing, and order processing.
- *Promotional Strategy* – Involves appropriate blending of personal selling, advertising, and sales promotion to communicate with and seek to persuade potential customers.
- *Pricing Strategy* – Includes marketing decision-making dealing with methods of setting profitable and justifiable prices.

13.1.2 Research Marketing Strategy

Secondly, we'll adapt the generic marketing steps into the specifics of selling research ideas. From all generic steps, we'll focus on the following three elements, which in our opinion are the most important for developing effective research marketing: (1) target audience identification, (2) product strategy definition, and (3) promotional strategy. The steps are shown in Fig. 13.2 and discussed below.

13.1.2.1 Target Market Identification

The potential users of research ideas and methods can be identified in three broad areas — industry, government, and academic institutions.

- *Industry* – This is the biggest user of research methods, some of them developed internally. The common expectation of this audience is that the proposed methods are ready for real-world applications. With very few exceptions, it makes sense to target the industrial audience after the scientific idea has reached some level of maturity, demonstrated with developed software and impressive simulations. Often, industry accepts only research ideas implemented by established vendors who take responsibility for maintenance of the deployed solutions.

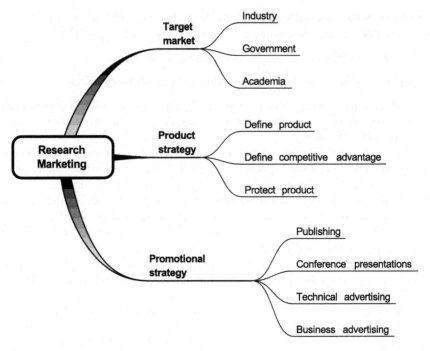

Fig. 13.2 Key steps in research marketing strategy

- *Government* - This plays a dual role as a technology user and a key source of funding. The criteria for satisfying government as a user of research ideas are similar to those already discussed for industry. The defense-related applications, however, require specific marketing efforts due to the mandatory military standards.

    Different governmental agencies offer various ways of funding from purely basic science ideas to applied science methods. This is the audience which is very familiar to the academic community.
- *Academia* - The academic audience is divided into two categories: the specific research community related to the explored technology (for example, the computational intelligence community), and a wide-area scientific community (for example, computer science and engineering societies). While the first category doesn't need intensive technology marketing, the second category requires significant efforts to communicate the advantages of the proposed methods and especially their potential to enhance and integrate with other methods.

13.1.2.2   Product Strategy Definition

One of the key differences between generic and research marketing is the nature of the product. While in generic marketing the advertised product has well-defined properties, the product in research marketing is of a different nature and not directly attached to specific customer needs. That requires additional nontrivial marketing efforts to clarify the most appropriate form of the research method and to link it to the desired customer needs. The steps in product strategy development are as follows:

- *Define Product* – The following products could be defined as possible outcomes from research efforts: published paper, filed patent, internally defined trade secret, software package, and designed process, device or material. Each of these products requires different marketing techniques, which will be discussed later in the chapter. The very nature of these products could be different as well. One category is related to products, based on specific research methods. An example is a toolbox that includes several genetic algorithm implementations. Another category is related to products that cover a given application area. An example is an optimization toolbox that includes genetic algorithms as an option. The third category is related to products, generated by the research methods. An example is an electric circuit or inferential sensor, derived automatically using genetic programming.
- *Define Competitive Advantage* – Proving technical superiority of the proposed research method by comparative analysis with its key competitors is the first step in this process. The more difficult step is linking the technical advantages with the business needs of the potential customers and defining the business competitive advantage. The key is clarifying and communicating the value gain from applying the solution based on the proposed research method relative to the available solutions in the market. The generic methods for defining the competitive advantages of computational intelligence are given in Chap. 9, and specific examples are given in this chapter.
- *Protect Product* – It is strongly recommended that the defined product is protected either by publication, patent or trade secret at the very early phases of the research marketing campaign.

13.1.2.3 Promotional Strategy Implementation

Relative to generic marketing with its broad advertisement efforts, research marketing is limited to direct interactions with potential customers either by publications or by different types of presentations. On the one hand, the marketing cost is significantly lower. On the other hand, the size of the audience is very limited. The situation could be changed, however, with the growing influence of Web-based marketing.

- *Publishing* – This step of the promotional strategy is well-known to the academic and industrial research communities and no additional recommendations are needed. The only recommendation is to go beyond the narrow technical field and to address the broad technical and nontechnical audiences using survey papers and popular articles in well-known scientific, technical, and trade journals. Of course, the method description and the language in these types of publications must be less technical and more vivid to capture the attention and the interest of the potential customers. In addition, sharing some ideas about potential application areas is always a plus.

- *Conference Presentations* – This is also a well-known step for the academic and industrial research communities. The key strategy again is to target a broad audience beyond the specific scientific community. One possible way is by presenting introductory tutorials. Another effective way is to attend the application sessions and to begin a dialog with industrial participants and vendors.
- *Technical Advertising* – This includes the necessary tools to capture the attention of technical experts that may open the door for the proposed research method in practice. Examples are given at the end of this chapter.
- *Business Advertising* – This includes the necessary tools to map the technical advantages of the proposed research method with customer business needs and to convince the management to support potential projects. Examples are also given at the end of this chapter.

13.2   Techniques – Delivery, Visualization, Humor

One of the key specifics of research marketing is that the predominant marketing efforts are of direct promotion of the product to the potential user. In this case, the quality of the presentation and the attractiveness of the delivery are critical. We'll focus on three techniques that are of high importance for marketing success – attractive message delivery, effective visualization, and, of course, humor.

13.2.1   Message Delivery

Sending a clear message and combining it with a touching story are critical in the promotion process. Successful marketers are just the providers of stories that consumers choose to believe. With the almost universal access to the Internet, there is a fundamental shift in the way ideas are spread. Either you're going to tell stories that spread, or you will become irrelevant. The key suggestions for effective message delivery, recommended by marketers, are given below:[2]

- *Define Attractive Headlines* – As we know, company and product icons and headlines are a substantial part of any form of advertisement. They try to capture

---

[2]S. Godin, *All Marketers Are Liars: The Power of Telling Authentic Stories in a Low-Trust World*, Portfolio Hardcover, 2005.

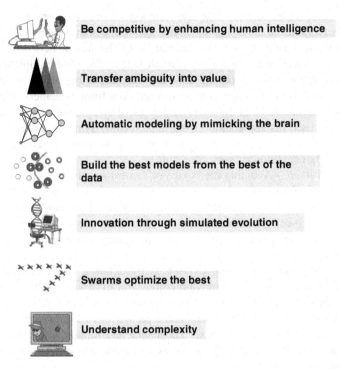

Fig. 13.3 Ideas for icons and headlines of the discussed computer intelligence approaches

the essence of the message and to succeed in the tough competition for customer's attention. The same idea is valid for research marketing, and some suggestions for computational intelligence methods, presented in this book, are given in Fig. 13.3. Ideally, the headline must represent the substance of the method and the key source of value creation.

• *Tell a True Story* – In direct marketing, the best way to capture immediate attention is by telling interesting stories that stay in the audience memory longer than product features. The stories must be short, shocking, and true. Here is an example from the author's experience.

In a presentation promoting symbolic regression-based inferential sensors, the author begins with a real story of customer dissatisfaction with the existing neural-network-based solutions, applied in a chemical plant. The plant engineer was asked how the neural network model predictions had been used recently. She looked ironically and asked: "Do you really want to know the truth?" She then walks over to the trashcan and retrieves from the garbage the last printout of model estimates. The predictions for a week were a constant negative number, which made no physical sense. The credibility of the model was at the level of the place it was "used" – a piece of junk in a trashcan. No more explanations were needed.

- *Deliver, and Then Promise* – The most important part of the message is to define clear deliverables to the customer. In the case of the above-mentioned example, the advantages of the proposed alternative to the neural network fiasco — symbolic regression — were well emphasized. They were illustrated with delivered inferential sensors in different manufacturing plants, which were heavily used. Good practice is to give references from customers. Plant engineers and operators liked the simplicity of the derived solutions. One operator even advertised his appreciation from implemented symbolic-regression inferential sensors to his peers with the highest possible form of admiration in this straight-talk community: "very cool stuff".
- *Do Not Oversell* – Avoiding the "snake oil" salesmen image is critical in direct marketing. Unfortunately, the relative ignorance of the generic audience towards the capabilities of computational intelligence gives opportunities for the "irrational exuberance" type of marketing. A typical example was the initial oversell of neural networks in the 1990s, often with disappointing results, like the one given above.

The best strategy to avoid oversell is by defining clear deliverables and discussing openly the technology limitations. For the given symbolic regression inferential sensor example, the well-known issues with this type of model, such as dependence on data quality, unreliable predictions 20% outside the range of training data, and more complex maintenance and support, were identified. In addition, examples of how these issues had been addressed in the previous applications were given.

13.2.2   Effective Visualization

Images are critical for marketing. Without attractive visual representation, the defined product has low chances for sale. In the case of research marketing, finding a simple and memorable image is not trivial. Exploring successful ads, imagination and using different visualization techniques are needed. Some examples are given below.

13.2.2.1   Combining Mind-maps and Clip Art

One recommended visualization approach, which is broadly used in the book, is combining the power of mind-mapping[3] with the attractiveness of clip art.[4] An example is given in Fig. 13.4.

---

[3]T. Buzan, *The Mind-Map Book*, 3rd edition, BBC Active, 2003.

[4]One possible source, which was used in this book is: www.clipart.com

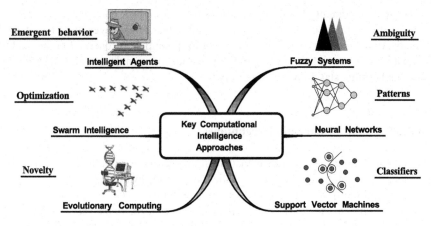

Fig. 13.4 An example of combining mind-maps and clip art

Key Computational Intelligence Techniques

- **Fuzzy Systems**
  – Handle ambiguity in knowledge and data

- **Neural Networks**
  – Discover patterns and dependencies in data

- **Support Vector Machines**
  – Classify data and build robust models

- **Evolutionary Computation**
  – Generate novelty by simulated evolution

- **Swarm Intelligence**
  – Optimize by using social interactions

- **Intelligent Agents**
  – Capture emergent macrobehavior from microinteractions

Fig. 13.5 A bullet-type PowerPoint presentation on the same topic, represented by the mind-map and clip art in Fig. 13.4

The figure clearly illustrates the advantage of exploring the synergy between both techniques. While mind-maps represent structures and links very effectively, they are not visually attractive and the chance that a mind-map will stick in our memory is very low. Let's not forget that, in the end, remembering the presentation is one of the main marketing objectives. Adding vivid clip-art icons and images brings "life" into the presentation and increases the chance of attracting potential customers' attention.

Compare the impact of the combined mind-map and clip-art presentation in Fig. 13.4 with the boring alternative of a text and bullets PowerPoint (PP) slide in Fig. 13.5.

The textual content of both presentations is almost equal; moreover the explanations of the methods of the PP slide are more detailed. However, the lack of any visualization significantly reduces the effect for potential customers. Let the reader estimate the impact of the message by answering the simple question: How long will this PP slide stay in her/his memory?

13.2.2.2   Some Visualization Techniques

Among the multitude of known visualization techniques, we'll focus on some suggestions, given by the famous visualization guru Edward Tufte.[5]

His definition of effective visualization is as follows: "*Graphical excellence is that which gives to the viewer the greatest number of ideas in the shortest time with the least ink in the smallest space.*" Some of the key techniques he recommends to achieve this goal are:

- Show comparison
- Show causality
- Show multivariate data
- Integrate information
- Use small multiples
- Show information adjacent in space

An example of a presentation based on some of these techniques is shown in Fig. 13.6.

The presentation represents the results of a comparative study between classic first-principles modeling and accelerated fundamental modeling, using symbolic regression (described in detail in Chap. 14). The competitive nature of the study is emphasized by horizontal parallel comparison in the similar steps of the approaches. Each step is additionally visualized either with icons or graphs. The key synergetic message is sharpened by a clear verbal equation and supported by the yin-yang icon, a well-known symbol of synergy. The final results from the competition – the striking differences in model development time – are directly represented by a calendar and a clock.

13.2.2.3   To PP or Not to PP?

One of the critical factors in research marketing is mastering PP (PowerPoint) presentations. According to Tufte, between 10 and 100 billion PP slides are

---

[5]E. Tufte, *Visual Explanations: Images and Quantities, Evidence and Narrative*, Graphics Press, 2nd edition, Cheshire, CT, 1997.

Case Study with Structure-Property Relationships

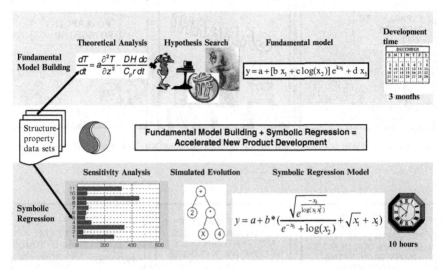

Fig. 13.6 An example of applying horizontal visual comparison between first-principles modeling and symbolic regression

produced yearly.[6] Unfortunately, the average information content of bullet-rich PP slides is extremely low, not to mention the sleeping pill effect. The Dilbert comic strip cartoon creator Scott Adams even defined the ultimate negative impact of PP boredom – a new disease, called PP poisoning, shown in the cartoon in Fig. 13.7.

There is a lot of criticism of the prevailing cognitive style of PP, which, rather than providing information, allows speakers to pretend that they are giving a real talk, and audiences to pretend that they are listening. This prevailing form of the bureaucratic-style of communication may lead to complete information abuse, satirically demonstrated in another Dilbert cartoon in Fig. 13.8, and must be avoided.

One alternative, recommended by Tufte, is preparing paper handouts of text, graphics, and images. The big advantage of handouts is that they leave permanent traces with the potential customers. One suggested paper size for presentation handouts is A3, folded in half to make four pages. According to Tufte, that one piece of paper can contain the content-equivalent of 50–250 typical PP slides! From our experience, even using a double-sided A4 format paper is very effective for capturing the attention of the audience. All marketing slides for computational intelligence methods given in the corresponding chapters (Chaps. 2–7) can be used as handouts on this format.

---

[6]E. Tufte, *Beautiful Evidence*, Graphics Press, Cheshire, CT, 2006.

Fig. 13.7 PowerPoint poisoning. DILBERT: © Scott Adams/Dist. by United Feature Syndicate, Inc.

Fig. 13.8 The new version of the old proverb "Lies, Damn Lies, and PP slides." DILBERT: © Scott Admas/Dist. by Unoted Feature syndicate, Inc.

Another more realistic alternative, having in mind the ubiquitous role of PP, is increasing the information content and attractiveness of the slides and reducing their number as much as possible. It is our experience that even the most complex topics can be condensed into fewer than 20–25 information-rich slides. Several examples in the book, like the slide shown in Fig. 13.6, demonstrate the recommended design of such slides.

13.2.3 Humor

Successful marketing must be funny. The paradox is that very often the short funny part of a presentation that includes cartoons, jokes, quotations, and Murphy's laws stays longer in the memory than any technical information. That is why it is very important to include humor in the marketing process. Several examples with Dilbert cartoons (see Fig. 13.9), useful quotations, and Murphy's laws, related to computational intelligence, are given below. They are a good starting point to inspire future presenters to develop their own humor libraries.

13.2.3.1 Dilbert Cartoons

Fig. 13.9 Examples of Dilbert cartoons related to data mining and Six Sigma. DILBERT: © Scott Adams/Dist. by United Feature Syndicate, Inc.

13.2.3.2 Useful Quotations

All models are wrong; some models are useful.

George Box

Don't fall in love with a model.

George Box

Complex theories do not work, simple algorithms do.[7]

Vladimir Vapnik

---

[7]The caricatures in this section are from the Bulgarian cartoonist Stelian Sarev.

It's not a question of how well each process works; the question is how well they all work together.

Lloyd Dobens

If you are limited to a restricted amount of information, do not solve the particular problem you need by solving a more general problem.

Vladimir Vapnik

Nothing is more practical than a good theory.

Vladimir Vapnik

A silly theory means a silly graphic.

Edward Tufte

The machine does not isolate man from the great problems of nature but plunges him more deeply into them.

Antoine de Saint-Exupery

The complexity of a model is not required to exceed the needs required of it.

D. Silver

It is better to solve the right problem approximately than to solve the wrong problem exactly.

John Tukey

13.2.3.3 Murphy's Laws Related to Computational Intelligence

Murphy's law
If anything can go wrong, it will.

Box's law
When Murphy speaks, listen.
   Corollary:
   Murphy is the only guru who is never wrong.

Law of ubiquitous role of data mining
In God we trust. All others bring data.

Murphy's law on machine learning
The number one cause of computer problems is computer solutions.

Kordon's law on computational intelligence
Computers amplify human intelligence.

Positive Corollary:
Smart guys get smarter.
Negative Corollary:
Dumb guys get dumber.

Murphy's law on rule-based systems
Exceptions always outnumber rules.

Murphy's law on sources of fuzzy logic
Clearly stated instructions will consistently produce multiple interpretations.

Menger's law on modeling success
If you torture data sufficiently, it will confess to almost anything.

Murphy's law effects on data
If you need it, it's not recommended.
If it's recommended, it's not collected.
If it's collected, it's missing.
If it's available, it's junk.

Knuth's law on evil optimization
Premature optimization is the root of all evil.

Murphy's law on evolutionary computation[8]
Simulated evolution maximizes the number of defects which survive a selection process.

13.3   Interactions Between Academia and Industry

Interaction between universities, where the scientific basis of computational intelligence is developed, and industry, where it delivers value through successful applications, is the key factor driving the implementations and moving the technology. Promoting research ideas to industrial people is almost as important as convincing the research community of the technical brilliance of the proposed methods. While the latter paves the way to scientific recognition, the former builds the bridge to future value creation, without which the research approach will gradually become irrelevant.

Some marketing recommendations on how to sell academic ideas in the key ways of interacting with industry, such as intellectual property protection, conference advertising, technology development, and approaching vendors, are given below.

13.3.1   Protecting Intellectual Property

Since value creation is the key issue of applied computational intelligence, protecting intellectual property has top priority. There are three available options of protection: (1) filing a patent; (2) filing a trade secret; and (3) defensive publication.[9] With patent protection, an inventor gives all details of the invention to the

---

[8]R. Brady, R. Anderson, R. Ball, *Murphy's Law, the Fitness of Evolving Species, and the Limits of Software Reliability*, University of Cambridge, Computer Laboratory, Technical Report 471, 1999.

[9]A good reference on intellectual property protection is the book of H. Rockman, *Intellectual Property Law for Engineers and Scientists*, Wiley, 2004.

government for a granted period of time of exclusive advantage, i.e. nobody else can create value with the invention without payment.

A trade secret can protect any form of business-related information for an unlimited time. However, if somebody else uncovers the secret independently or by reverse engineering, it is permitted under the law to use it and to make a profit.

A defensive publication establishes prior art[10] against competitors. It disables potential patent filing and prevents a competitor's claim on specific intellectual property.

A simplified flow chart navigating the decision-making process of proper intellectual property protection selection is shown in Fig. 13.10.

The decision-making process is divided into six steps:[11]

1. *Is public disclosure necessary?* – In some cases, like Federal Drug Administration (FDA) regulations in the U.S.A., there exist legal regulations necessitating disclosure of the invention. In this case, the idea must be either patented or published, since trade secret protection is not a viable option.

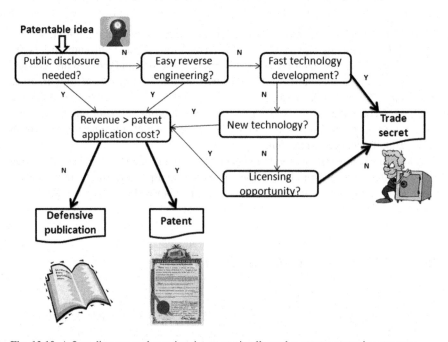

Fig. 13.10 A flow diagram to determine the proper intellectual property protection strategy

---

[10]Criteria that the invention is sufficiently different from those found in the known literature, in public use, or on sale.

[11]I. Daizadeh, D. Miller, E. Glowalla, M. Leamer, R. Nandi, and C. Numark, A general approach for determining when to patent, publish, or protect information as a trade secret, *Nature Biology*, *20*, pp.1053–1054, 2002.

2. *Is the idea easy to reverse engineer or discover independently?* – In this step, the risk of potential independent rediscovery or reverse engineering from a competitive group is evaluated. If the risk is too high, filing a patent is recommended.
3. *Is the technology area evolving quickly?* – If the pace of technology development is too high (as is the case with computational intelligence), then the speed of market penetration of the invention becomes a critical factor. Often the speed of technology growth exceeds the average time of patent release (2–3) years) and that can make the invention obsolete from the beginning and the filing cost worthless.
4. *Is it a new area of technology?* – A unique opportunity for patenting is the birth of a brand new technology since there will be few records of prior art. This golden chance allows one to define unusually broad claims and establish a virtual monopoly by limiting potential competition.
5. *Are there benefits from licensing?* – If licensing to third parties may provide additional revenue to the inventor, a patent is the preferable solution, because licensing trade secrets creates the risk of potentially unauthorized disclosure to third parties.
6. *Does the potential revenue outweigh the patent filing and associated costs?* – It has to be taken into account that patent and associated costs can be quite expensive, since legal expertise is needed. The average cost for patent filing in American industry is $20,000. In addition, associated enforcement and litigation costs can easily reach seven figures.

If the expected revenue cannot cover the patent filing and associated costs, a defensive strategy with journal publication is recommended. This strategy has several additional advantages, including securing a market brand and preventing others from claiming the invention.

13.3.2 Publishing

From the point of view of protecting intellectual property, computational intelligence inventions are mostly related to the last category of defensive publications in respected technical journals. This form of validating the scientific claims and promoting the research ideas into the specific technical communities is familiar to any active researcher. A good scientific publication record is the critical measure in academic evaluation and in establishing a researcher's credibility. It is also the required basis to be recognized by industrial partners and to begin the dialog. However, it is a necessary but not a sufficient condition for establishing fruitful academic-industrial interaction.

The key to success in academic-industrial interaction mostly depends on the initiative of the academic researcher. Marketing skills in approaching the broad industrial audience are a must. One recommended option is publishing application-related papers or technology surveys in popular technical and nontechnical journals. The purpose of these papers is to explain the proposed methods in plain English and to demonstrate the value creation potential. It is good to know the specific journals of the targeted business, which are read by industrial technical gurus and especially by managers.

13.3.3 Conference Advertising

For most academic researchers conferences are the best places to begin a research marketing campaign. However, the process must go beyond technical presentations only. Several other efforts have to be pursued.

The first option is obvious – grabbing the attention with attractive technical presentations of the research methods and using any opportunity to demonstrate the potential for applicability and value creation.

The second option is attending the application sessions and trying to understand industrial needs. Another benefit from these sessions is looking at the limitations of the approaches, applied in real-world problems. An example of sessions with established industrial participation is the track Evolutionary Computation in Practice at the annual Genetic and Evolutionary Computation Conference (GECCO).

The third option for conference advertising is by participation in technical competitions, which are part of almost any significant technical meeting. Most of these competitions are based on industrial case studies and give a good opportunity for a reality check of the proposed methods. The publicity of the winners is also very high and goes beyond the narrow technical community. In addition, they gain credibility as effective real-world solvers, which is critical for grabbing the attention of industrial researchers.

The key objective of conference research marketing, however, is building and maintaining the social network of application-oriented academics, industrial researchers, potential customers from industry and government, software vendors, and consultants. It is impossible to operate in the area of any technology, including computational intelligence, without it.

13.3.4 Technology Development

In most cases, the definition of research marketing success is inclusion of the proposed scientific idea in projects or being funded by industry or the government. There are various forms of academic-industrial collaboration in technology development, such as direct or joint industrial-governmental project funding, support of graduate students or post-docs, industrial internships, consortia support, and consultant services.[12] It has to be taken into account that handling intellectual property,

---

[12]A good example of academic-industrial collaboration is given in the chapter by R. Roy and J. Mehnen, Technology transfer: Academia to industry, in the book of T. Yu, L. Davis, C. Baydar, R. Roy (Eds), *Evolutionary Computation in Practice*, Springer, pp. 263–281, 2008.

generated during technology development, may become an issue. Very often the financial support from industry is linked to intellectual property ownership or restrictions on publishing some of the results.

The normal practice of building bridges between academia and industry is by individual interactions, which vary depending on the strength of the researcher's social networks. The efficiency of this approach, however, is not very high and random factors play a significant role. An alternative way to link leading academic researchers with key practitioners in the area of GP in a more systematic way has been explored since 2003 at the Center for Complex Systems, University of Michigan, Ann Arbor. In a three-day workshop, called Genetic Programming Theory and Practice (GPTP), both sides present ideas, discuss issues, and, above all, listen carefully to the needs of the other side. Gradually, the differences in the mindsets have been reduced and theoreticians and practitioners have begun a fruitful exchange of ideas.[13] However, it took 2–3 years of face-to-face discussions until this level of understanding had been reached.

13.3.5 Interaction with Vendors

Often direct marketing of a research idea to industry fails for the simple reason that it may require nonstandard software. With very few exceptions, such as specific R&D needs, industrial users prefer to apply computational intelligence on software tools developed by established vendors. Often, the applied solutions have to be leveraged globally and global support becomes a critical issue, which can be handled only by established vendors. From that perspective, the right strategy to promote a research idea in industry is by contacting vendors first. Of special interest are vendors, such as the SAS institute, Aspen Technologies, and Pavilion Technologies,[14] that continuously upgrade their products with new computational intelligence algorithms and are open to fresh ideas on working algorithms.

13.4 Marketing CI to a Technical Audience

The success of research marketing depends on two key audiences – technical and business. The technical audience includes the potential developers of computational intelligence-based applications. Usually this audience has an engineering and mathematical background and is interested in the technology principles and functionality. The business audience includes the potential users of the applied systems,

---

[13]The results from the workshops have been published since 2003 in several books under the title *Genetic Programming Theory and Practice* by Kluwer Academic and recently by Springer.

[14]Recently, a division of Rockwell Automation, Inc.

who are more interested in the value creation capability of the technology and how easily it can be integrated into the existing work processes. The marketing approach for the technical audience is discussed in this section, and for the nontechnical audience it is presented in the next section. We'll concentrate on two topics: (1) how to prepare an effective presentation for an audience with a technical background; and (2) how to approach technical gurus.

13.4.1 Guidelines for Preparing Technical Presentations for Applied Computational Intelligence

The main objective of a good technical presentation is to demonstrate the technical competitive advantage of the proposed approaches. The key expectations from a technical introduction of a new technology are: well-explained main principles and features, defined competitive advantage, potential application areas within users' interest, and assessment of implementation efforts. One of the challenges in the case of computational intelligence is that the technology is virtually unknown to the technical community at large. The assumption is that computational intelligence must be introduced from scratch with minimal technical details. In order to satisfy the different levels of knowledge in the audience, a second, more detailed presentation can be offered off-line to those who are interested.

One of the key topics in the presentation is the differences between the proposed methods and the most popular approaches, especially those used frequently by the targeted audience, such as first-principles models and statistics. The examples, given in Chaps. 2 and 9 are a good reference for preparing of this type of presentation. The audience must be convinced about the technical competitive advantage of a given approach relative to another technology with arguments based on their scientific principles, simulation examples, and application capabilities. A realistic assessment with a balance between the strengths and weaknesses is strongly recommended. An example of an introductory slide for neural networks is shown in Fig. 13.11.

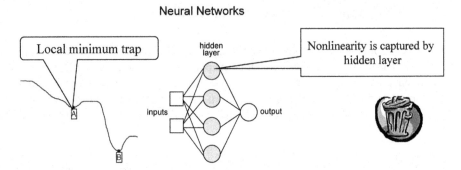

Fig. 13.11 An example of introducing neural networks to a technical audience

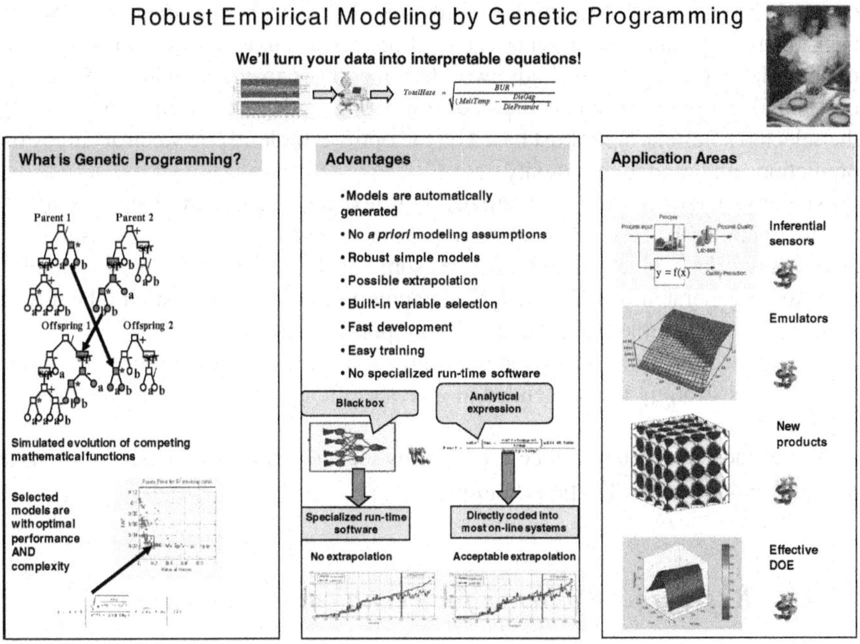

Fig. 13.12 An example of introducing genetic programming to technical audience

Two central topics of neural networks are addressed: (1) the capability of a neural network to generate nonlinear models; and (2) the danger that these solutions are inefficient due to local optimization. The first topic is demonstrated by showing how nonlinearity is captured by the hidden layer of a three-layer neural network. The second topic, generating nonoptimal models due to the potential of the learning algorithm to be entrapped in local minima, is represented visually in the left part of the slide. The trashcan icon can be used during real presentations as a reminder to tell the real story of lost credibility due to inefficient design of neural networks (the trashcan story in Sect. 13.2.1).

Another, more complicated slide for representing the key technical features of genetic programming (GP) is shown in Fig. 13.12.

This is a typical slide for introducing a new method to technical audience. We call it the "kitchen slide" since it includes information about the technology kitchen in generating the final solution. The slide template is organized in the following way. The title section includes a condensed message that represents the essence of the method, supported by an appropriate visualization. In the case of GP, the condensed message is "We'll turn your data into interpretable equations!" visualized by the icons showing the transformation of the data into equations by GP (in this case, the standard evolutionary computation icon is used).

The slide is divided into three sections: (1) method description; (2) advantages; and (3) application areas. The method description section presents a very high-view visual description of the approach. In the case of GP, it includes the images of competing mathematical functions, represented by trees and the image of a Pareto front figure with an equation on the zone of optimal accuracy and complexity. The method description section sends two clear messages to the audience: (1) at the basis of GP is the simulated evolution of competing mathematical functions, and (2) the selected solutions are robust based on the optimal trade-off between accuracy of predictions and model complexity.

The advantages section summarizes with text and images the key benefits of the approach. In the case of the GP example in Fig. 13.12, a visual comparison with neural networks is given. It demonstrates some of the key advantages of symbolic regression, generated by GP – analytical expressions, directly coded into the user's system with acceptable performance in extrapolation mode.

The application area section focuses on the key implementation successes of the method. It is strongly recommended to give specific numbers of the created value, if possible.

The proposed template can be used for any technology. It can also be given as a printed handout before the presentation.

13.4.2 Key Target Audience for Technical Presentations

A typical technical audience includes industrial researchers, technical subject matter experts, practitioners, system integrators, and software developers. Their incentives to break the cozy status quo and introduce a new technology depend on the benefits they'll get as a result of increased productivity, ease of use, and rewarding their risk-taking with potential career development and bonuses. In principle, it is very difficult to expect that introducing computational intelligence will satisfy all of these incentives. Thus, management support is required to start the efforts. The critical factor in gaining the needed support is the opinion of the technical leaders (gurus) in the organization or the research community. It is practically impossible to open the door for the new technology without their blessing. That is why in the rest of the section we'll focus our attention on approaching effectively this key category of the technical audience.[15]

In order to specify the marketing efforts as much as possible, we divide technical gurus into six categories based on their support of a new idea. The bell curve of the different types is shown in Fig. 13.13, where the level of support increases to the right and the level of rejection of the new idea increases to the left.

Two types of gurus – The Visionary and The Open Mind – can fully support the new technology. On the opposite side are The 1D Mind and The Retiring Scientist gurus, who will try by all means to kill or postpone the introduction of the new idea. Most technical gurus are neutral to the new technology and their support will

---

[15]A very good source for analyzing targeted audiences for high-tech products is the classic book of G. Moore, *Crossing the Chasm: Marketing and Selling Disruptive Products to Mainstream Customers*, HarperCollins, 2002.

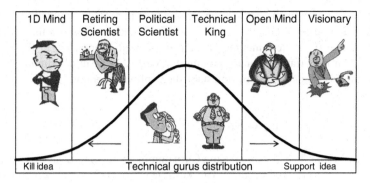

1D Mind	Retiring Scientist	Political Scientist	Technical King	Open Mind	Visionary
Kill idea		Technical gurus distribution		Support idea	

Fig. 13.13 Technical guru distribution based on their support of a new idea

depend on increasing their personal technical influence (in the case of The Technical King) or gaining political benefits (in the case of The Political Scientist). The typical behavior of the different gurus and the recommended approaches for presenting the new idea to them are given below.

13.4.2.1 Visionary Guru

Typical Behavior: Having the blessing of a Visionary Guru is the best-case scenario of full enthusiastic support of the new idea. The Visionary Guru is the driving force of innovation in an organization. She/he is technically sharp and has an abundance of ideas, i.e. shares her/his own ideas and enjoys exploring new ones. Other important characteristics of the Visionary Guru are: outside focus with respect to external expertise, risk taking, and political skills to convince management. The Visionary Guru is well informed and her/his office is full with bookshelves of technical journals and books from diverse scientific areas.

Recommended Approach: Prepare an inspirational talk with a very solid technical presentation. Be ready to answer detailed technical questions and to demonstrate competence about the scientific basis of the idea. Convincing the Visionary Guru is not easy, but if successful, the chance of a positive decision to apply the proposed approach is almost 100%.

13.4.2.2 Open Mind Guru

Typical Behavior: The Open Mind Guru accepts new ideas but will not initiate change without serious technical arguments. She/he is more willing to support methods that require gradual changes or have an established application record. The risk-taking level and enthusiasm is moderate. Before supporting the new idea, the Open Mind Guru will carefully seek the opinion of key technical experts and especially managers. Her/his bookshelf is half the size of the Visionary Guru.

Recommended Approach: Focus on detailed technical analysis that demonstrates competitive advantages over the methods closer to the Open Mind Guru's personal experience. Show specific application examples or impressive simulations in areas similar to the targeted business. It is also recommended to discuss the synergetic options between the new method and the most used approaches in the targeted business. An important factor for gaining the Open Mind Guru's support is if the new method is implemented by an established software vendor.

13.4.2.3 Technical King Guru

Typical Behavior: The Technical King Guru dominates an organization with her/his own ideas, projects, and cronies. The key factor in her/his behavior is gaining power, i.e. the best way to have the blessing for a new idea is if it will increase her/his technical influence. In this case, the Technical King Guru will fully support the idea and will use all of her/his influence to apply it. Otherwise, the chances for success depend only on top management push. The Technical King Guru shines in her/his area of expertise and has a good internal network and business support. On her/his bookshelf one can see only books related to her/his favorite research topics.

Recommended Approach: Understand key areas of expertise of the Technical King Guru and prepare a presentation that links the proposed new approach with the identified areas. Recognize the importance of the Technical King and her/his contribution and try to demonstrate how the new idea will fit in with her/his technical kingdom and will increase her/his glory and power. Independently, top management support could be pursued to counteract eventual idea rejection.

13.4.2.4 Political Scientist Guru

Typical Behavior: The Political Scientist Guru is on the negative side of "new idea support" distribution, shown in Fig. 13.13. By default, she/he rejects new approaches, since they increase the risk of potential technical failure with corresponding negative administrative consequences. Technically, the Political Scientist Guru is not in the list of "the best and the brightest" and this is another reason for looking suspiciously at any new research idea. The real power of the Political Scientist is in using effectively political means to achieve technical objectives. From that perspective, new idea support depends on purely political factors, such as top management opinion, current corporate initiatives, and the balance of interests between the different parts of the organization related to the new technology. On her/his bookshelves one can see a blend of books on technical and social sciences with favorite author Machiavelli.

Recommended Approach: The marketing efforts must include a broader audience than technical experts. It is critical to have a separate presentation to top management first, which emphasizes the competitive advantages of the proposed approach. The ideal argument will be if some of the competitors are using the new methods. The presentations must have minimal technical details, be very visual, and application-oriented.

13.4.2.5 Retiring Scientist Guru

Typical Behavior: The Retiring Scientist Guru is counting the remaining time to retirement and trying to operate in safe mode with maximum political loyalty and minimum technical effort. In order to mimic activity, she/he uses the sophisticated rejection technique, known as "killing the new idea by embracing it". The Retired Scientist is a master of freezing the time by combining bureaucratic gridlock with departmental feuds. As a result, the new idea is buried in an infinite interdepartmental decision-making loop. Often the possible solution is in her/his potential career plans as a consultant after retirement. Don't expect bookshelves in her/his office. Only outdated technical manuals occupy the desk.

Recommended Approach: Broaden marketing efforts to several departments. Sell the new idea directly to top management first. Be careful to avoid the "embrace and kill" strategy. Try to understand if the new technology may fit in with Retirement Scientist's consultant plans after leaving office.

13.4.2.6 1D Mind Guru

Typical Behavior: The 1D Mind Guru is the worst-case scenario with almost guaranteed new idea rejection. She/he has built her/his career on one approach only, which has created value. This mode of ideas deficiency creates fear of novelty and aggressive job protection. Any new idea is treated as a threat that must be eliminated. The 1D Mind Guru is well informed on the weaknesses of any new approach and actively spreads negative information, especially to top management. Bookshelves are also absent in 1D Mind Guru's office. However, the walls are often filled with certificates and company awards.

Recommended Approach: Don't waste your time. Try other options.

13.5 Marketing to a Nontechnical Audience

The marketing approach for nontechnical audience will be discussed in this section. We'll concentrate on two topics: (1) how to prepare an effective presentation for an audience with a nontechnical background; and (2) how to approach managers.

13.5.1 Guidelines for Preparing Nontechnical Presentations for Applied Computational Intelligence

The main objective of a good nontechnical presentation is demonstrating the value creation capabilities of computational intelligence. The key expectations from a

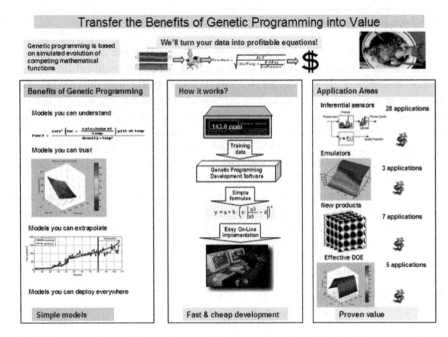

Fig. 13.14 An example of introducing genetic programming to a business audience

nontechnical introduction of a new technology, such as computational intelligence, are: well-explained benefits of the technology, defined business competitive advantage, suggested potential application areas within users' interest, and assessment of implementation efforts.

An example of a nontechnical introduction of GP is given in Fig. 13.14. This is a typical slide for introducing a new method to nontechnical audiences. We call it the "dish slide".

In contrast to the "kitchen slide", shown in Fig. 13.12, it does not include information about the technology principles. The focus is on the benefits of the final product from the technology – "the dish". The slide template is organized in the following way. The title section includes a condensed message that represents the source of value creation, supported by an appropriate visualization. In the case of GP, the source of value creation is defined as "We'll turn your data into profitable equations!", visualized by the icons showing the transformation of the data into equations that create value (represented by the dollar sign).

The slide is divided into three sections: (1) benefits; (2) deployment; and (3) application areas. The benefits section focuses on the key advantages of symbolic regression models, generated by GP – interpretable models, trusted by users, increased robustness, and almost universal deployment. The interpretability benefit is illustrated by an equation that can be easily understood by the experts. The advantage of having a trustable model is shown by a "What-If" scenario response surface. Increased robustness is demonstrated by model performance outside the training range.

The deployment section is illustrated by a simple sequence on how the technology is applied. The application area section focuses on the key implementation successes of the method. It is strongly recommended to give specific numbers of the applications and the created value, if possible. As in the case of the technical presentation, the proposed template can be used for any technology and can be given as a printed handout before the presentation.

13.5.2 Key Target Audience for Nontechnical Presentations

A typical nontechnical audience includes professionals in different nontechnical areas, such as accounting, planning, sales, human resources, etc. Their incentives to introduce a new technology are the same as for the technical audience. In principle, this audience is more resistant to new and complex technologies. An important factor for support is the ease of use of the technology and the minimal required training.

The final fate of the new technology, however, will be decided by the management. That is why in the rest of the section we'll focus our attention on approaching effectively this key category of nontechnical audience. We have to make an important clarification, though. The critical factor in the management decision-making process is the opinion of technical gurus. Very often, it automatically becomes the official management position. Smart managers even go outside the organization for the advice of external technical gurus before making the decision.

The hypothetical bell curve distribution for managers, supporting or rejecting a new technology, is similar to the one for technical gurus, shown in Fig. 13.13. On both edges of the distribution we have the two extremes: (1) on the positive side – visionary managers who initiate and embrace change and (2) on the negative side – low-risk "party line soldiers" who shine in following orders and hate any novelty. In the same way as the size and the content of the bookshelves is a good indicator of the research guru's type, the frequency of using the word *value* can be used for manager type categorization. Visionary managers who embrace novelty are using it appropriately. The other extreme - innovation-phobic managers who are resistant to supporting new technologies use the word *value* two-three times in a sentence.

The majority of managers, however, are willing to accept innovation if it fits in the business strategy and their career development. We'll focus on some recommendations on how to be prepared for approaching this type of manager.

- *Understand Business Priorities* – Do your homework and collect sufficient information about targeted business priorities. Proposing technologies and research ideas that are outside the scope of the business strategy and the manager's goals is marketing suicide.

- *Understand Management Style* – Information about the style of the manager is very important. Is it a hands-on, detail-oriented, need-to-be-involved-in-every-decision approach, or a hands-off, bring-me-a-solution-with-results style? Is she/he a quick learner? How rigid or flexible is she/he?
- *Understand the Paramount Measure of Success in the Targeted Business* – The proposed new technology must contribute to the key business metric of success. If on-time delivery is the paramount measure of success in the business, any solutions in this direction will be fully supported. If low-cost development is the primary goal, any suggestion that is seen as adding cost will not be accepted. Promises of future savings that offset a current addition may fall on deaf ears.
- *Avoid the Problem Saturation Index (PSI)* – The meaning of PSI is that management is saturated with today's problems. Any attempt to promise results that will occur three years from now is a clear recipe for rejection.
- *Offer Solutions, not Technologies* – managers are more open to solutions that address current problems. For example, if one of the known problems in the targeted business is improved quality, better offer on-line quality estimators using robust inferential sensors (the solution) rather than symbolic regression models generated by genetic programming (the technology).

13.6   Summary

Key messages:

Research marketing includes three main steps: (1) target audience identification, (2) product strategy definition, and (3) promotional strategy implementation.

The techniques critical for research marketing success are: attractive message delivery, effective visualization, and, of course, humor.

Research marketing through interaction between academia and industry can be accomplished in different ways, such as: intellectual property protection, conference advertising, technology development, and approaching vendors.

A differentiated approach for research marketing to technical and nontechnical audience is recommended.

The main objective of research marketing to a technical audience is to demonstrate the technical competitive advantage of the proposed approaches.

The main objective of research marketing to a nontechnical audience is to demonstrate the value creation capabilities of the proposed approaches.

The Bottom Line

Applying computational intelligence requires significant marketing effort to sell the research methods to technical and business audience.

Suggested Reading

The following books give generic knowledge on marketing and visualization techniques:

L. Boone and D. Kurtz, *Contemporary Marketing*, 13th edition, South-Western College Publishers, Orlando, FL, 2007.

T. Buzan, The Mind-Map Book, 3rd edition, BBC Active, 2003.

S. Godin, *All Marketers Are Liars: The Power of Telling Authentic Stories in a Low-Trust World, Portfolio* Hardcover, 2005.

G. Moore, Crossing the Chasm: *Marketing and Selling Disruptive Products to Mainstream Customers*, HarperCollins, 2002.

E. Tufte, *Visual Explanations: Images and Quantities, Evidence and Narrative*, Graphics Press, 2nd edition, Cheshire, CT, 1997.

E. Tufte, *Beautiful Evidence*, Graphics Press, Cheshire, CT, 2006.

Chapter 14
Industrial Applications of Computational Intelligence

Theory is knowledge that doesn't work. Practice is when everything works and you don't know why.

Hermann Hesse

If we follow the proverb that "the proof is in the pudding", the presented computational intelligence industrial applications in this chapter are the "pudding" of this book. The described examples reflect the personal experience of the author in applying computational intelligence in The Dow Chemical Company. However, they are typical of the chemical industry in particular, and for manufacturing and new product development in general.

For each example, first the business case will be described briefly, followed by the technical solution. Some of the technical details require more specialized knowledge than the average level and may be skipped by nontechnical readers. However, technically savvy readers may find the specific implementation data very helpful.

The chapter includes three computational intelligence applications in manufacturing, two in new product development, and two examples of failed projects. The details for the discussed examples are publicly available in the given references.

14.1 Applications in Manufacturing

The three examples in the chemical manufacturing area are related to process monitoring and optimization. The first application covers the broad area of inferential sensors and focuses on robust solutions, based on symbolic regression. The application uses the integrated methodology discussed in Chap. 11. Since the reader is already familiar with the nature of the derived symbolic regression models through the example in Chap. 11, the focus in this chapter is on their broad applicability in manufacturing. The second manufacturing application is in one critical

A.K. Kordon, *Applying Computational Intelligence,*
DOI 10.1007/978-3-540-69913-2_14, © Springer-Verlag Berlin Heidelberg 2010

area for successful plant operation – an automated operating discipline. The following technologies have been integrated in the application: expert systems, genetic programming, and fuzzy logic. The third computational intelligence application is in on-line process optimization by using emulators of complex first-principles models. Both neural networks and symbolic regression have been used in this application.

14.1.1 Robust Inferential Sensors

Some critical parameters in chemical processes are not measured on-line (composition, molecular distribution, density, viscosity, etc.) and their values are captured either by lab samples or off-line analysis. However, for process monitoring and quality supervision, the response time of these relatively low-frequency (several hours or even days) measurements is very slow and may cause loss of production due to poor quality control. When critical parameters are not available on-line in situations with potential for alarm "showers", the negative impact could be significant and eventually could lead to shutdowns. One of the approaches to address this issue is through development and installation of expensive hardware on-line analyzers. Another solution is by using soft or inferential sensors that infer the critical parameters from other easy-to-measure variables like temperatures, pressures, and flows.[1]

There are several empirical modeling methods, such as statistics, neural networks, support vector machines, and symbolic regression that can extract relevant information out of historical data and be used for inferential sensor development. Most of the currently applied inferential sensors, however, are based on neural networks.[2] The most popular product name used by the key vendors and associated with neural networks is "soft sensor". However, neural networks have some limitations such as:

- Low performance outside the ranges of process inputs used for model development (i.e. soft sensor predictions are unreliable in new operating conditions);
- Selection of model structure is not parsimonious (i.e. the developed models are not simple and are sensitive to even minor process changes);
- Model development and maintenance requires specialized training;
- Model deployment requires specialized run-time licenses.

As a result, there is a necessity for frequent retraining which increases significantly the maintenance cost. In many applications, decreased performance and credibility have been observed even after several months of on-line operation.

---

[1] The state of the art of soft sensors is given in the book of L. Fortuna, S. Graziani, A. Rizzo, and M. Xibilia, *Soft Sensors for Monitoring and Control of Industrial Processes,* Springer, 2007.

[2] http://www.pavtech.com

An alternative technology, called robust inferential sensors, has been under development in The Dow Chemical Company since 1997. It is based on symbolic regression, generated using genetic programming, and resolves most of the issues of the neural network-based soft sensors currently available on the market. The detailed description of the technology is given by Kordon *et al.*[3] and is based on the integrated methodology described in Chap. 11. Robust inferential sensors have been successfully leveraged in different businesses in the company and some typical examples are given below.

14.1.1.1 Robust Inferential Sensor for Alarm Detection

This is the first robust inferential sensor implementation that initiated the development of symbolic regression-based inferential sensing. The objective of this robust inferential sensor was an early detection of complex alarms in a chemical reactor.[4] In most cases, late detection of the alarms causes unit shutdown with significant losses. The alternative solution of developing hardware sensors was very costly and required several months of experimental work. An attempt to build a neural network-based inferential sensor was unsuccessful, due to the frequent changes in operating condition.

Twenty-five potential inputs (hourly averaged reactor temperatures, flows, and pressures) were selected for model development. The output was a critical parameter in the chemical reactor, measured by lab analysis of a grab sample every 8 hours. The selected model was an analytical function of the type:

$$y = a + b * \left[e * \left(\frac{x_3}{x_5} - d \right) \right]^c \tag{14.1}$$

where x_3 and x_5 are two temperatures in the reactor, y is the predicted output, and a, b, c, d, and e are adjustment parameters.

The simplicity of the selected function for critical reactor parameter prediction allows its implementation directly in the Distributed Control System. In addition, the GP-generated robust inferential sensor was implemented in Gensym G2. This was done because the predictor is a critical alarm indicator in the Expert System for Automated Operating Discipline, described in the next section. The system has been in operation since November 1997 and initially it included a robust inferential sensor for one reactor. An example of successful alarm detection several hours before the lab sample is shown on Fig. 14.1.

---

[3]A. Kordon, G. Smits, A. Kalos, and E. Jordaan, Robust soft sensor development using genetic programming, in *Nature-Inspired Methods in Chemometrics*, R. Leardi (Editor), Elsevier, pp. 69–108, 2003.

[4]A. Kordon and G. Smits, Soft sensor development using genetic programming, *Proceedings of GECCO 2001*, San Francisco, pp. 1346–1351, 2001.

Fig. 14.1 Successful alarm detection in real time by the inferential sensor. The sensor (with the continuous line) triggered the alarm (the alarm level is at 30) several hours in advance of the laboratory measurements (with the dashed line)

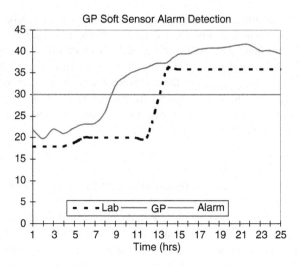

The robust performance for the first six months gave the process operation the confidence to ask for leveraging the solution to all three similar chemical reactors in the unit. No new modeling efforts were necessary for model scale-up and the only procedure to fulfill this task was to fit the parameters a, b, c, d, and e of the GP-generated function (14.1) to the data set from the other two reactors. Since the fall of 1998 the robust inferential sensors have been in operation without need for retraining. The prediction quality is with standard deviation close to that of the lab measurement (between 2.9% and 4.1% vs. 2% for the lab measurement). The robust long-term performance of the GP-generated soft sensor convinced the process operation to reduce the lab sampling frequency from once a shift to once a day since July 1999. The maintenance cost after the implementation is minimal and covers only the efforts to monitor the performance of the three soft sensors.

14.1.1.2   Robust Inferential Sensor for Product Transition Monitoring

This robust inferential sensor application is for interface level estimation in a chemical manufacturing process with multiple product types.[5] The engineers in the plant were very interested in keeping the level under control during the product-type transition. The loss of control usually leads to a process upset. Initially, attempts were made to estimate the interface level with several hardware sensors (e.g. level device and level density meter). However, these measurements were not reliable or consistent and proved of little use in preventing process upsets. The only credible estimate to process operators was a manual reading of the interface level which was taken every two hours in a five-level scale.

---

[5]A. Kalos, A. Kordon, G. Smits, and S. Werkmeister, Hybrid model development methodology for industrial soft sensors, *Proc. of ACC 2003*, Denver, CO, pp. 5417–5422, 2003.

The original data set included 28 inputs and 2900 data points from 14 different product types. After evaluating the different solutions, the following function was selected by process engineers and proposed for on-line implementation of the interface level robust inferential sensor:

$$y = 7.93 - 0.13 * \frac{x_{24}\sqrt{x_{27}}}{x_{22}}, \tag{14.2}$$

where the selected inputs have the following physical meaning:

Input	Description
x_{22}	brine flow
x_{24}	organic/acid ratio setpoint
x_{27}	caustic flow

A unique feature of this particular robust inferential sensor is that the output data used for development are not quantitative measurements but qualitative estimates of the level determined manually by process operators using a scale from one to five. Since this is a subjective procedure it has limited accuracy of 20% (1/5 of the original scale) in a 0–100% scale. The robust inferential sensor has been on-line since October 2000. Although the robust inferential sensor was trained in a broad range of operating conditions, it still had to operate 11.7% of the time outside its training range.

14.1.1.3 Robust Inferential Sensor for Biomass Estimation

Biomass monitoring is fundamental to tracking cell growth and performance in bacterial fermentation processes. During the growth phase, biomass determination over time allows for calculation of growth rates. Slow growth rates can indicate nonoptimal fermentation conditions which can then be a basis for optimization of growth medium or conditions. In fed-batch fermentations, biomass data can also be used to determine feed rates of growth substrates when yield coefficients are known.

Usually the biomass concentrations are determined off-line by lab analysis every 2–4 hours. This low measurement frequency, however, can lead to poor control, low quality, and production losses, i.e. on-line estimates are needed. Several neural network-based soft sensors have been implemented since the early 1990s. Unfortunately, due to the batch-to-batch variations, it is difficult for a single neural network-based inferential sensor to guarantee robust predictions in the whole spectrum of potential operating conditions. As an alternative, an ensemble of GP-generated predictors was developed and tested in a real fermentation process.[6]

---

[6]A. Kordon, E. Jordaan, L. Chew, G. Smits, T. Bruck, K. Haney, and A. Jenings, Biomass inferential sensor based on ensemble of models generated by genetic programming, *Proceedings of GECCO 2004*, Seattle, WA, pp. 1078–1089, 2004.

Data from eight batches was used for model development (training data) and the test data included three batches. Seven process parameters like pressure, agitation, oxygen uptake rate (OUR), carbon dioxide evolution rate (CER), etc. were used as inputs to the model. The output was the measured Optical Density (OD) which is proportional to the biomass. Several thousand candidate models were generated by 20 runs of GP with different numbers of generations. An ensemble of five models with an average R^2 performance above 0.94 was selected (a detailed description of the design is given by Jordaan et al.).[7]

The prediction of the ensemble is defined as the average of the predictions of the five individual models. The accuracy requirements for the ensemble were to predict OD within 15% of the observed OD level at the end of the growth phase. The performance of the ensemble on the training and test data can be seen in Fig. 14.2.

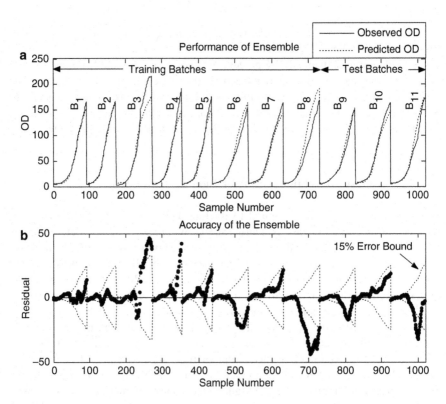

Fig. 14.2 Performance of the ensemble of biomass predictors on the training and test data

---

[7]E. Jordaan, A. Kordon, G. Smits, and L. Chiang, Robust inferential sensors based on ensemble of predictors generated by genetic programming, *Proceedings of PPSN 2004*, Birmingham, UK, pp. 522–531, 2004

In Fig. 14.2(a) the OD level is plotted against the sample number, which corresponds to the time from the beginning of the fermentation. The solid line indicates the observed OD-level. The dashed line corresponds to the prediction of the ensemble. In Fig. 14.2(b) the residuals of the ensemble's prediction with respect to observed OD is shown. The 15% error bound is also shown. For the training data one sees that for three batches (B_3, B_4 and B_8), the ensemble predicts outside the required accuracy. For batch B_3 it was known that the run was not consistent with the other batches. However, this batch was added in order to increase the range of operating conditions captured in the training set. The performance of the ensemble at the end of the run for all the batches of the test data (batches B_9, B_{10} and B_{11}) is within the required error bound.

14.1.2 Automated Operating Discipline

Operating discipline is a key factor for competitive manufacturing. Its main goal is to provide a consistent process for handling all possible situations in the plant. It is the biggest knowledge repository for plant operation. However, this documentation is static and is detached from the real-time data of the process. The missing link between the dynamic nature of process operation and the static nature of operating discipline documents is traditionally carried out by the operating personnel. It is the responsibility of process operators and engineers to assess the state of the plant, to detect the problem, to find and follow the recommended actions. The critical part in this process is problem recognition. The capability to detect a complex problem strongly depends on the experience of process operators. This makes the existing operating discipline process very sensitive to human errors, competence, inattention, or lack of time.

One approach to solving the problems associated with operating discipline and making it adaptive to the changing operating environment is to use real-time intelligent systems. Such a system runs in parallel with the chemical manufacturing process, independently monitoring events in the process, analyzing the state of the process, and performing fault detection. If a known problem is detected, which is not accommodated by the existing control system and which requires human intervention, the system can automatically suggest the appropriate corrective actions.

The advantages of the proposed approach are illustrated with an industrial application for automating operating discipline in a large-scale chemical plant.[8]

---

[8] A. Kordon, A. Kalos, and G. Smits, Real-time hybrid intelligent systems for automating operating discipline in manufacturing, *Artificial Intelligence in Manufacturing Workshop Proceedings of the 17th International Joint Conference on Artificial Intelligence IJCAI 2001*, pp. 81–87, 2001.

The specific application is based on using alarm troubleshooting as the key factor in establishing a consistent operating discipline in handling difficult operating conditions. The objectives of the system are as follows:

- Automatically identify root causes of complex alarms;
- Reduce the number of unplanned unit shutdowns;
- Provide proper advice to the operator for the root cause of the alarms in real time;
- Streamline the operating discipline through automatically linking the corrective actions to the identified problems;
- Assist the training of new operators to detect problems and carry out corrective actions.

The system was implemented in Gensym Corporation's G2.[9] The integration of the various intelligent systems techniques into the hybrid intelligent system was done in the following sequence:

14.1.2.1 Knowledge Acquisition from the Experts

The knowledge acquisition was carried out in two main steps: (1) Definition of alarm cases by the experts and (2) validation of the alarm cases by the knowledge engineer through analysis of the potential causes of chemical process upsets. After discussions between the experts, the knowledge for detecting and handling process upsets was organized into building blocks called "alarm cases". An alarm case defines a complex root cause and usually contains several low-level alarms linked with logical operators. The description of the alarm case is given in natural language. An example of an alarm case is given in Fig. 14.3:

Part of knowledge acquisition is linked to fuzzy logic in order to quantify with membership functions expressions like "fast temperature change" and "not enough liquid in the reactor". Of special importance is the validation process where the defined alarm cases have been evaluated with the operating history of the plant.

14.1.2.2 Organization of the Knowledge Base

The alarm logic that was defined during the knowledge acquisition phase was implemented using Gensym G2/GDA (G2 Diagnostic Assistant) logic diagrams. The diagram is organized in the following manner (see Fig. 14.4).

First, the raw data from the chemical process are linked to the diagram, followed by trending. The processed data are then supplied to feature detection, by testing against a threshold or a range. Every threshold contains a membership function as defined by the fuzzy logic. The values for the parameters of the membership function are based either on experts' assessment or on statistical analysis of the data.

---

[9]http://www.gensym.com

Alarm Case "Excess feed pre-cooling"

IF

    High value of critical variable

AND

    Rapid temperature decrease in sections 3-4

OR

    Fast change of reactor pressure

AND

    Decreased flow to unit 12

THEN

    Raise alarm and follow corrective actions for
    "Excess feed pre-cooling"

Fig. 14.3 An example of an alarm case

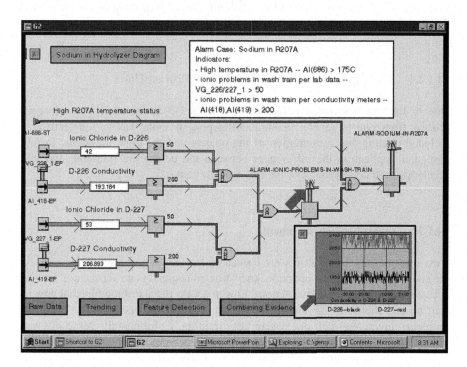

Fig. 14.4 An example of an alarm logic diagram

The combining evidence phase of the diagram includes the necessary logic to implement the alarm logic that triggers the alarm block. When this block is active, it has red stress lines around its icon. The logic path can be traced from the alarm-triggering block backward to the data sources. The state of the logic can be estimated by its color (red or green) and is very convenient for fast interpretation from process operators.

14.1.2.3　Implementation of Prototype for One Process Unit

This phase included the full integration of all components into the Gensym G2 software environment. A critical task for validation of the knowledge base was the alarm threshold adjustment. This process is essential for establishing credibility of the alarm expert system because incorrect thresholds have negative effect in alarm triggering. They either trigger the alarm too often (nuisance alarms) or respond to the alarm condition too late. In order to improve this process it is necessary to verify the thresholds defined by the experts with statistical information. If significant differences are detected, they need to be resolved before putting the alarm system into operation. Otherwise, it will lose credibility at the beginning of on-line operation and this may have a negative psychological effect on process operators.

14.1.2.4　Scaling up to the Full System for All Process Units

Due to the object-oriented nature of Gensym's G2 environment, the scale-up process from one to several manufacturing process units was very effective. It included cloning and instantiation of the objects and the logic diagrams. In this particular application the differences between the units were minimal, and this accelerated the development of the full-scale system.

14.1.2.5　Operators' Involvement

Two operators from each shift were trained on how to interact with the system. At the end of the training, the operators knew how to respond to an alarm case, how to access and print the corrective actions, how to access the alarm log file, and how to start up and shutdown the intelligent system. Most of the operators did not have problems interacting with the system after training. Unfortunately, some of them did not use the system regularly and their experience in its use gradually decreased. However, other operators used the system regularly and even suggested important improvements in the logic diagrams and in the corrective actions.

14.1.2.6　Value Evaluation

After an evaluation period of six months, the operations personnel identified the following items that they believed provided value to the plant:

- Detecting problems the operators may miss;
- Eliminating upsets or minimizing the number of upsets to just a few each year;
- Catching operator mistakes, specifically on complex alarm cases that include parameters from different process units;
- Training new operators on problem detection and corrective actions;
- Catching operator mistakes, specifically on complex alarm cases that include parameters from different process units;
- Training new operators on problem detection and corrective action;
- Fast problem recognition due to reliable soft sensors.

The main issue in value assessment of this class of systems is the time span for performance evaluation. The key source of value is prevention of major upsets. These are rare events in principle and usually are not regular. That is why it is very difficult to assess the cost/benefits of the intelligent system in a short time-scale (less than one year). A reasonable time period for cost evaluation is 3–5 years.

14.1.3 Empirical Emulators for On-line Optimization

Empirical emulators mimic the performance of first-principles models by using various data-driven modeling techniques. The driving force to develop empirical emulators is the push for reducing the time and cost for new product or process development. Empirical emulators are especially effective when hard real-time optimization of a variety of complex fundamental models is needed.

14.1.3.1 Motivation for Developing Empirical Emulators

The primary motivation for developing an empirical emulator of a first-principles model is to facilitate the on-line implementation of a model for process monitoring and control. Oftentimes it may prove difficult or impractical to incorporate a first-principles model directly within an optimization framework. For example, the complexity of the model may preclude wrapping an optimization layer around it. Or, the model may be implemented in a different software/hardware platform than the Distributed Control System (DCS) of the process, again preventing its on-line use. In other occasions, the source code of the model may not even be available. In such circumstances, an empirical emulator of the fundamental model can be an attractive alternative. An additional benefit is the significant acceleration of the execution speed of the on-line model (10^3–10^4 times faster).

14.1.4 Empirical Emulators Structures

The most obvious scheme for utilization of empirical emulators is for complete "replacement" of a fundamental model. The key feature of this scheme, shown in Fig. 11.10 in Chap. 11, is that the emulator represents the fundamental model entirely and is used as a stand-alone on-line application. This scheme is appropriate when the fundamental model does not include too many input variables and a robust and parsimonious empirical model can be built from the available data generated by Design Of Experiments (DOE).

In the case of higher dimensionality and model complexity, a hybrid scheme of fundamental model and emulator integration is recommended (see Fig. 14.5). Emulators based only on submodels with high computational load are developed off-line using different training data sets. These emulators substitute the related sub-models in on-line operation and enhance the speed of execution of the original fundamental model. This scheme is of particular interest when process dynamics has to be taken into account in the modeling process.

Finally, an item of special importance to on-line optimization is the scheme (shown in Fig. 14.6) where the empirical emulator is used as an integrator of different types of fundamental models (steady-state, dynamic, fluid, kinetic, and thermal).

In this structure, data from several fundamental models can be merged and a single empirical model can be developed on the combined data. The empirical emulator, as an integrator of different fundamental models, offers two main advantages for on-line implementation. The first advantage is that it is simpler to interface only the inputs and outputs from the models than the models themselves. More importantly, when constructing the data sets using DOE, the developer selects only those inputs/outputs that are significant to the optimization. Hence, the emulator is a compact empirical representation of only the information that is pertinent to the optimization.

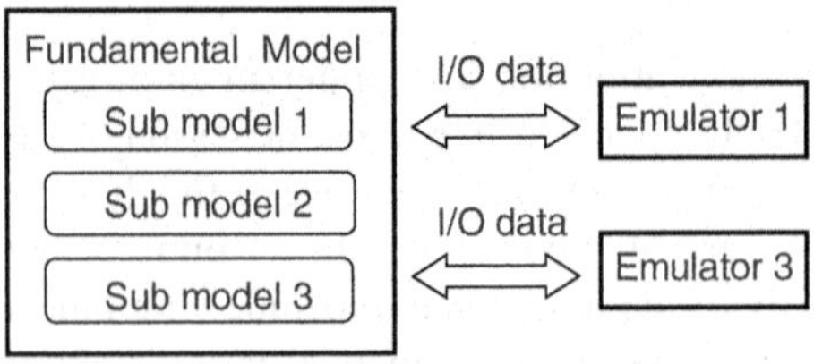

Fig. 14.5 Hybrid scheme of empirical emulators and fundamental models

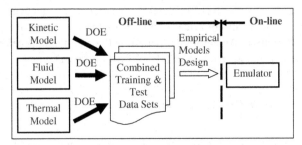

Fig. 14.6 Empirical emulator as integrator and accelerator of fundamental models

The second advantage is that one optimizer can address the whole problem, rather than trying to interface several separate optimizers. The optimization objectives, costs, constraints, algorithm, and parameters are more consistent, allowing the multimodel problem to be solved more efficiently.

14.1.5 A Case Study: an Empirical Emulator for Optimization of an Industrial Chemical Process[10]

14.1.5.1 Problem Definition

A significant problem in the chemical industry is the optimal handling of intermediate products. Of special interest are cases where intermediate products from one process can be used as raw materials for another process in different geographical locations. The case study is based on a real industrial application of intermediate products optimization between two plants in The Dow Chemical Company, one in Freeport, Texas and the other in Plaquemine, Louisiana. The objective is to maximize the intermediate product flow from the plant in Texas and to use it effectively as a feed in the plant in Louisiana. The experience of using a huge fundamental model for "What-If" scenarios in planning the production schedule was not favorable because of the specialized knowledge required and the slow execution speed ($\sim$20–25 min/prediction). Empirical emulators are a viable alternative to solve this problem. The objective is to develop an empirical model which emulates the existing fundamental model with acceptable accuracy (with $R^2 \sim 0.9$) and calculation time (< 1 sec).

14.1.5.2 Data Preparation

Ten input variables (different product flows) were selected by the experts from several hundred parameters in the fundamental model. There are 12 output variables (Y_1 to Y_{12}) that need to be predicted and used in process optimization. The assumption was that the behavior of the process can be captured with these most significant variables and that a representative empirical model could be built for each output. A specialized 32-run Plackett&2011;Burman experimental design with 10 factors at four levels was used as the DOE strategy. The training data set consisted of 320 data points. For 15 of these cases the fundamental model did not

---

[10]A. Kordon, A. Kalos, and B. Adams, Empirical emulators for process monitoring and optimization, *Proceedings of the IEEE 11th Conference on Control and Automation MED 2003*, Rhodes, Greece, p.111, 2003.

converge for three of the outputs. The test data set included 275 data points where the inputs were randomly generated within the training range.

14.1.5.3 Empirical Emulator Based on Analytic Neural Networks

Several runs with different numbers of hidden nodes were done and the results for all 12 neural networks-based emulators are summarized in Table 14.1

The structure of the neural network for each emulator includes 10 inputs and one output. The same set of inputs is used for all emulators. Since single hidden-layer analytic neural networks are based on direct optimization, the only design parameter to be adjusted is the number of neurons in the hidden layer. A number of different structures (with between 1 and 50 hidden nodes) were constructed and each neural net was optimized based on the test data set. The optimal number of hidden nodes was then determined by applying each neural network to the test data set and selecting the structure with the minimal R^2 value. This procedure was repeated for each emulator.

One special feature of the analytical neural network is the method of initializing the random weights between the input and the hidden layer. In order to minimize the effect of randomization, it is possible to use a stack of many neural networks with the same complexity (i.e. with the same number of hidden nodes) for each emulator. An advantage of this approach is that the final prediction is based on the average of all models in the ensemble. Of even greater practical importance is that the standard deviation between the individual predictors can be used to develop a model disagreement measure which is a type of a confidence indicator for the stacked neural net models and also adds some self-assessment capability to the emulator.

As shown in Table 14.1, all emulators have acceptable accuracy on the training and test data. An example of emulator performance for emulator Y_5 is shown in Fig. 14.7 for the training data set, and in Fig. 14.8 for the test data set.

Table 14.1 Performance of all emulators on training and test data

Output	R^2 NN training	R^2 NN test	# Hidden nodes
Y_1	0.910	0.890	30
Y_2	0.994	0.989	20
Y_3	0.984	0.979	20
Y_4	0.987	0.981	20
Y_5	0.991	0.967	30
Y_6	0.999	0.999	1
Y_7	0.995	0.999	1
Y_8	0.995	0.993	10
Y_9	0.994	0.992	10
Y_{10}	0.992	0.993	1
Y_{11}	1.000	1.000	1
Y_{12}	0.997	0.989	20

In both cases, the stacked ensemble model disagreement is shown (at the bottom of the figures) on the same scale as the actual data and thus the range of its magnitude is relatively small. As expected, it is somewhat larger for the testing data.

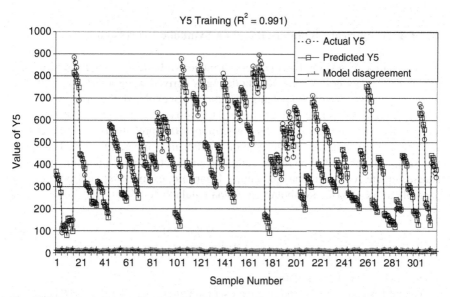

Fig. 14.7 Emulator Y_5 actual and predicted values and model disagreement indicator based on training data

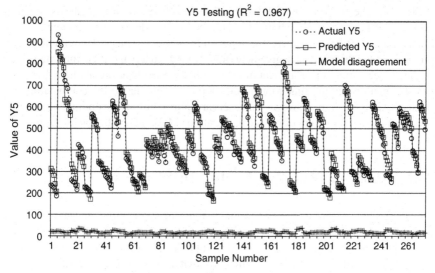

Fig. 14.8 Emulator Y_5 actual and predicted values and model disagreement indicator based on test data

The emulators' performance varies between $R^2 = 0.89$ for Y_1 and the perfect fit for Y_{11}. The neural network complexity also varies – from an almost linear structure of one hidden node for Y_6, Y_7, Y_{10}, and Y_{11} to a structure with 30 hidden nodes for Y_1 and Y_5. The prediction quality is good in all ranges.

14.1.5.4 Empirical Emulators Based on Symbolic Regression

The alternative of developing emulators based on symbolic regression has also been explored. The symbolic regression modeling is derived using a GP-run simulated with a population size of 200 potential functions that evolve during 300 generations with 0.5 probability for random crossover, 0.3 probability for mutation of functions and terminals, 4 reproductions per generation, 0.6 probability for selecting a function as the next node, and correlation coefficient as fitness function. The initial functional set for the GP includes: {addition, subtraction, multiplication, division, square, square root, sign change, natural logarithm, exponential, and power}. An example of a GP-based symbolic regression emulator for Y_5 is:

$$Y_5 = \frac{6x_3 + 2x_4x_6 - x_2x_9 - \sqrt{x_6e^{-x_{10}}}}{x_1 \log(x_2^3)} \qquad (14.3)$$

where, x_1—x_{10} are the emulator inputs.

The performance on the training data set (shown in Fig. 14.9, $R^2 = 0.94$) and on the test data set are comparable.

In summary, the performance of stacked analytic neural nets is generally better than that of GP-generated emulators (training $R^2 = 0.99$ vs. 0.94; test $R^2 = 0.97$ vs. 0.94). Still, the performance of GP-generated functions is within the acceptable accuracy of prediction. Deciding which method to use depends on the application: Stacked analytic neural nets offer the potential for self-assessment of unreliable model predictions and the model disagreement indicator, which are of critical importance for on-line process monitoring and optimization. On the other hand, symbolic regression-based emulators require much longer development time due to the computationally intensive GP algorithm. However, end users are more comfortable optimizing the process with an analytical function, such as equation (14.3), than with black-box models.

14.2 Applications in New Product Development

One of the key differences between manufacturing applications and new product development is the amount of available data. While manufacturing applications deal with tons of data, new product development efforts rely on very few data records. Thus, of key interest are methods that can build credible models from small data sets. One of these approaches is symbolic regression. Two examples of using symbolic regression in new product development are given in this section.

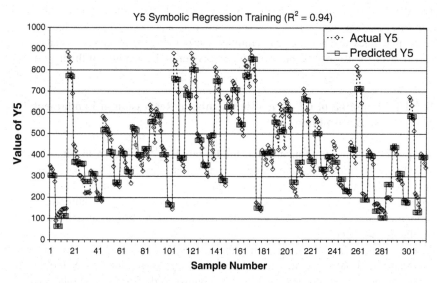

Fig. 14.9 Emulator Y_5 actual and predicted values from symbolic regression on training data

The first example, which could revolutionize this application area, is related to the key problem of structure—property relationships. A methodology for accelerating first-principles model building by using symbolic regression-based proto-models is described and illustrated in a case study. The second application demonstrates the capability of symbolic regression to deliver robust empirical models with small data sets in the area of modeling blown film process effects.

14.2.1 Accelerated Fundamental Model Building[11]

With the increasing dynamics of a global market economy there is an increasing need to accelerate the decision-making processes leading to new products and manufacturing. Thus, the speed of development and use of effective modeling of phenomena becomes a key factor for success. Those models based on solid fundamentals, or first principles, are the most robust. The development of these modeling "crown jewels" often requires significant research and development time. One of the key factors to shorten this time-consuming process and to enrich creativity is to "accelerate" and "navigate" hypothesis search and validation. An inevitable step in

---

[11]The material in this section was originally published in: A. Kordon, H. Pham, C. Bosnyak, M. Kotanchek, and G. Smits, Accelerating industrial fundamental model building with symbolic regression, *Proc. of GECCO 2002, Volume: Evolutionary Computation in Industry*, pp. 111–116, 2002. With kind permission of the International Society for Genetic and Evolutionary Computation.

hypothesis search and validation is some form of experimental work and data analysis. With the availability of several new data analysis approaches like support vector machines, and genetic programming, it is possible to obtain results that can not only deliver robust models but contribute to better understanding of the hypothesis.

14.2.1.1 Potential of GP-Generated Symbolic Regression in Fundamental Model Building

A typical sequence for fundamental model building[12] is shown in Fig. 14.10. It includes seven model building steps, covering problem definition, key variables identification, analyzing the problem data, constructing the model based on the proper physical laws, solving the model either analytically or numerically by simulation, and iteratively verifying the model solution until an acceptable model is reliably validated.

The key creative process, behind the scenes, is hypothesis search. For each hypothesis it usually requires the definition of an assumption space, looking for various hypothetical physical/chemical mechanisms, developing hypothetical fundamental models, and testing the hypothesis on selected data. Unfortunately, the

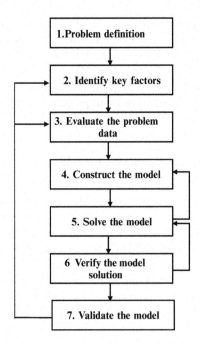

Fig. 14.10 Classical accelerated fundamental model building steps

---

[12]K. Hangos and I. Cameron, *Process Modeling and Model Analysis*, Academic Press, San Diego, 2001.

effectiveness of hypothesis search depends very strongly on the creativity, experience, and imagination of model developers. The broader the assumption space is (i.e. the higher the complexity and dimensionality of the problem), the larger the differences in modeler's performance will be and the higher the probability for ineffective fundamental model building.

The problem can be further amplified in team modeling efforts when various types of fundamental models (fluid dynamics, kinetic models, thermodynamic models, etc.) have to be integrated by different experts. An additional obstacle to effective hypothesis search is the limited data for hypothesis testing. Very often the initial available data set covers a fraction of the process variability and is insufficient for model validation. In order to improve the efficiency of hypothesis search and to make the fundamental model discovery process more consistent, a new "accelerated" fundamental model building sequence is proposed. The key idea is to reduce the fundamental hypothesis search space by using proto-models, generated by symbolic regression. The main steps in the proposed methodology are shown in Fig. 14.11.

The key difference from the classical modeling sequence is in running simulated evolution before beginning the fundamental model building. As a result of the GP-generated symbolic regression, the modeler can identify the key variables and assess the physical meaning of their presence/absence. Another significant side-effect from the simulated evolution is the analysis of the key transforms with high fitness that persist during the GP run.

Very often some of the transforms have a direct physical interpretation that can lead to better process understanding at the very early phases of fundamental model development. The main result from the GP-run, however, is the list of potential empirical models in the form of symbolic regression, called proto-models. The expert may select and interpret several empirical solutions or repeat the GP-generated symbolic regression until an acceptable proto-model is found. It is also possible to simulate the behavior of the selected empirical models in expected operating conditions and to assess its physical relevance in cases of difficult physical interpretation. The fundamental model building step 5 is based either on a direct use of empirical models or on independently derived first-principles models induced by the results from the proto-models. In both cases, the effectiveness of the whole modeling sequence could be significantly improved.

The proposed methodology differs from the "classical" AI approaches that try to mimic the expert by incorporation of a variety of reasoning techniques (qualitative reasoning, qualitative simulation, geometric reasoning, etc.) to build ordinary differential equation models of nonlinear dynamic systems. It also differs from the other extreme of using GP as an automated invention machine that replaces the human being entirely through the simulated evolutionary process of model discovery.[13] The objective of the "accelerated" methodology is not to eliminate

---

[13]J. Koza, F. Bennett III, D. Andre, and M. Keane, *Genetic Programming III: Darwinian Invention and Problem Solving*, Morgan Kaufmann, 1999.

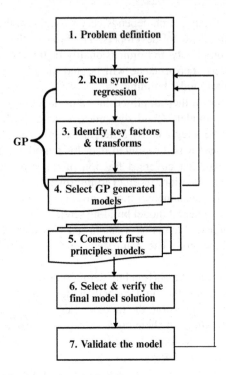

Fig. 14.11 Accelerated fundamental model building steps

the expert but to improve her/his efficiency by reducing the fundamental model hypothesis search space. It is our belief that there is no substitute for human creativity and imagination in fundamental model building, i.e. it is preferable to improve these qualities instead of replacing the experts.

The potential of the proposed accelerated model building approach based on GP-generated symbolic regression is illustrated with a case study from a real model development process in The Dow Chemical Company.

14.2.2 Symbolic Regression in Fundamental Modeling of Structure–Properties

14.2.2.1 Case Study Description

The case study is based on a typical industrial problem of modeling a set of structure–property relationships for specified materials. A large number of variables like molecular weight, molecular weight distribution, particle size, the level and type of crystallinity, etc. affect these properties. These factors all interact with each other in different ways and magnitude to give the specific performance attributes. By traditional experimental design there would be a huge number of experiments to perform. Of even more difficulty, many of these variables cannot be controlled systematically and independently of each other. As a result, validating the first-principles models is time-consuming and developing "black-box" models is unreliable due to insufficient data. The objective of the case study is to validate to what extent the proposed "accelerated" fundamental model building methodology will reduce the hypothesis search space and the development time in comparison to the classical first-principles model building process. The specific structure–property relationship includes five input variables and one output.

14.2.2.2 Fundamental Model Building Approach

In the traditional fundamental model building approach a careful review of the literature was made and hypotheses for structure–property relationships were developed. A set of experimental results from 33 experiments was selected in such a way that the variables were covered in a systematic manner. Different types of assumptions and physical mechanisms were discussed and explored. As a result, the following simplified model was derived, based on four input properties only:

$$y = a + [bx_1 + c\log(x_2)] \, e^{kx_3} + dx_5, \qquad (14.4)$$

where a, b, c, d, and k are constants, x_1, x_2, x_3, and x_5 are the input properties, and y is the output property. The developed fundamental model has a clear physical interpretation and was validated on the selected data set. It took about three months for one of the best experts to build and validate the model.

14.2.2.3 Symbolic Regression Approach

The symbolic regression model was derived from a population size of 200 potential functions that evolve during 50 generations with probability for random crossover 0.5, probability for mutation of functions and terminals 0.3, number of reproductions per generation of 4, probability for a function as next node as 0.6, and correlation coefficient as optimization criterion. The same data set of 33 experimental data records was used for both first-principles and symbolic regression modeling. The initial functional set for the GP includes: {addition, subtraction, multiplication, division, square, change sign, square root, natural logarithm, exponential, and power}. In the process of deriving the symbolic regression-based models, the following relevant transforms were obtained (shown in the order of increasing complexity):

Transform 1: $\frac{x_3}{x_1}$ with R^2 of 0.74;

Transform 2: $\frac{x_3}{\sqrt{x_5}}$ with R^2 of 0.81;

Transform 3: $\frac{x_1 x_3}{\sqrt{x_5}}$ with R^2 of 0.84.

These three transforms are a side result during the evolutionary process of generation of the final solution. However, they can be used in the fundamental modeling process as indicators for variable influence or potential physical or chemical relationships.

One of the key differences between "classical" empirical modeling and GP-generated empirical regression is the multiplicity of solutions as a final result. Usually there are several functional expressions with similar quality and structure. The selected function in the case study is in the form:

$$y = a + b * \left[\frac{\sqrt{e^{\frac{-x_3}{\log(x_1 x_5^2)}}}}{e^{-x_3} + \log(x_2)} + \sqrt{x_1} + x_5 \right], \tag{14.5}$$

where a and b are constants, x_1, x_2, x_3, and x_5 are the input properties, and Y is the output property. A plot of the predicted and measured output property on the selected data set is shown in Fig. 14.12. The performance of the model measured by its R^2 of 0.9 is very good and within the acceptable limits.

All of the functional forms for the output property that were identified by the symbolic regression agreed with the model generated as above from human deduction (see, for reference, equation 14.4). Only the independent variable x_1 was not taken as square root. However, this discrepancy can be explained from the physics of the process and it led to the development of an improved first-principles model. An additional result from the simulated evolution is that the irrelevant input property x_4 was automatically eliminated during the evolutionary process.

It took about 10 hours of an expert's time to define the problem, select the symbolic solutions, analyze, and interpret the results. Most of this time was spent on meetings to define and understand the problem, to prepare the final results from the GP-generated reports and to find physical interpretations.

In summary, the case study shows that this specific case clearly demonstrates the large potential of GP-generated symbolic regression for accelerated fundamental model building. The generated symbolic regression solution was similar to the fundamental model and was delivered in significantly less time (10 hours vs. three months).

14.2.3   Fast Robust Empirical Model Building

Another big opportunity of computational intelligence in new product development is generating low-cost solutions as robust empirical models. In many cases

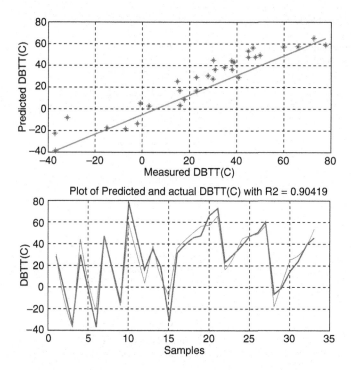

Fig. 14.12 Performance of the symbolic regression solution

development of first-principles models is too expensive and the only available option is empirical modeling. Usually it is based on classical statistics. However, the increasing complexity of chemical processes, due to the high dimensionality, process interactions, and the nonlinear type of physical relationships, requires more sophisticated methods. Another factor that contributes to the problem is the reduced volume of available data. Usually the data sets are small and in many cases are not based on Design Of Experiments (DOE). This typical feature of new product modeling limits the implementation of neural networks since they require a sub-stantial amount of training and test data. Another limitation of the neural networks is the black-box nature of their models which makes the physical interpretation of the nonlinear relationship practically impossible.

Symbolic regression, generated by GP, has the necessary capabilities to resolve the discussed issues. Especially attractive to practical applications is the feature of automatic empirical model generation from small data sets. These advantages are the basis of a methodology for GP-based empirical model development of robust empirical models, shown in Fig. 14.13.

The first two blocks (experimental data collection and linear regression model-ing) reflect the state of the art of new product modeling. The final result is a linear empirical model based on the statistically significant process inputs. It can be used

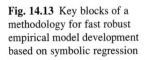

Fig. 14.13 Key blocks of a
methodology for fast robust
empirical model development
based on symbolic regression

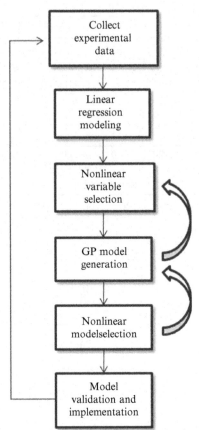

for model prediction and organizing the next phase of designed experiments that
will improve process knowledge or lead to optimal process conditions.

For many practical occasions linear models can predict within specific process
ranges and are an adequate solution. However, in many cases a nonlinear model is
required. One possible option is to use linearized transforms of the inputs and the
output. Unfortunately, this is time-consuming and some key nonlinear transforms
could be missed. A second option is to develop a fundamental model, which leads to
increased model development cost. A third option, based on symbolic regression, is
explored in the methodology. The principles of the key blocks in the methodology
that are related to symbolic regression have already been described in this book.
The practical application of the methodology will be illustrated with a case study
for empirical modeling of blown film process effects.[14]

---

[14]A. Kordon and C.T. Lue, Symbolic regression modeling of blown film process effects,
Proceedings of CEC 2004, Portland, OR, pp. 561–568, 2004.

14.2.4 Symbolic Regression Models of Blown Film Process Effects

14.2.4.1 Modeling Scope

The lack of a well-established theory of molecular parameter rheology processing property relationships of blown film requires development of reliable empirical models that relate critical processing variables to key film properties. The scope of the case study includes nine processing parameters (inputs) and 21 film properties for a specific product. The selected inputs are given in Table 14.2.

The number of data points is 20, derived from two set of experiments (not based on DOE). A linear multivariate analysis has shown that acceptable linear models with $R^2 > 0.9$ can be built for 13 film properties. Eight film properties, however, need nonlinear models. The proposed methodology has been successfully used to automatically generate the models for all of them. We will illustrate the method with the results for one of these key film properties – DART impact.

14.2.4.2 Symbolic Regression Model for DART Impact

The R^2 of the linear model of DART impact is 0.64. Thus, there is a need for non-linear model development and this is a good candidate to illustrate the symbolic regression approach.

In developing the symbolic regression model we have followed the methodology shown in Fig. 14.13. The results from the nonlinear sensitivity analysis and variable selection are shown in Fig. 14.14 with the accumulated nonlinear sensitivities of the nine inputs towards DART impact during 2000 generations of the simulated evolution. The conclusion from the nonlinear sensitivity analysis is that there are four inputs (x_3, Blow Up Ratio(BUR); x_4, Melt Temp; x_5, Output; and x_9, MD Stretch Ratio) that have much stronger influence on DART impact than the other input variables.

Table 14.2 Process inputs	Input	Description
	x_1	Die Gap (mil)
	x_2	Film Gauge (mil)
	x_3	Blow Up Ratio
	x_4	Melt Temp (F)
	x_5	Output (lb/hr)
	x_6	Die Pressure (psi)
	x_7	Frost Line Height (in)
	x_8	Take-up Speed (fpm)
	x_9	MD Stretch Ratio

Fig. 14.14 Input sensitivity towards DART impact

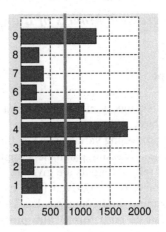

The next step of the methodology is GP-model generation by running a second set of 20 simulated evolutions on the selected four inputs:

x_1 = Blow Up Ratio (BUR)
x_2 = Melt Temp
x_3 = Output
x_4 = MD Stretch Ratio

The selected model on the Pareto front has the following nonlinear expression:

$$y = A + Be^{\left[e^{-x_1} + \frac{x_4 - x_2}{x_3 - x_1}\right]^2}, \tag{14.6}$$

where $A = 151.15$, $B = 0.358$, y is DART impact, and x_1, x_2, x_3, and x_4 are the four selected inputs.

The performance of the model is illustrated in Fig. 14.15. The performance is better than the linear multivariable model (R^2 is 0.71 in comparison to 0.64).

14.2.4.3 Blown Film Process Effects Model Implementation

All linear and nonlinear models for the key properties were implemented in Visual Basic for Applications (VBA) and integrated in an Excel spreadsheet. This allowed easy model distribution to final users without any training. An example of a "What-If" scenario when increasing Film Gauge from 1 mil to 4 mil for the key film properties is shown in Fig. 14.16. The radar chart captures the necessary changes in the properties.

The radar chart in Fig. 14.16 is constructed on a scale normalized to the reference case and in such a way that when a point is "moving outward" on an axis, the corresponding property is improved (in contrast to increase of value). For

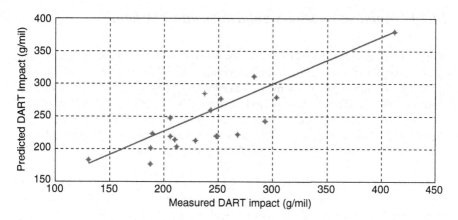

Fig. 14.15 Performance of the selected nonlinear model for DART impact

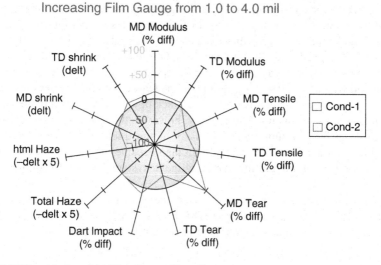

Fig. 14.16 Empirical model - Film Gauge effects diagram

example, increasing the Film Gauge from 1 to 4 mil will improve MD Tear and decrease internal Haze, TD Tear, and MD Tensile.

14.3 Unsuccessful Computational Intelligence Applications

It is unrealistic to expect that every computational intelligence application will be successful. We'll illustrate the negative experience of applied computational intelligence with two typical cases of failed implementation efforts: (1) introduction of significant cultural changes and (2) low-quality data.

14.3.1 Application with Significant Cultural Change

A critical factor for project success is the level of organizational and cultural changes triggered by the applied computational intelligence system. The challenge is very high when many stakeholders need to modify substantially their working habits in a short period of time. It is especially difficult in manufacturing when 40–50 process operators spread in different shifts are involved in this process. Often, the resistance towards the administratively imposed new work process overcomes the technical benefits from using the more advanced system.

This was the case with the intelligent alarm system described in Sect. 14.1.2. On the one hand, the system fully accomplished its technical objectives and was recognized by all stakeholders as a technical success. On the other hand, it introduced a significant change in the existing culture of plant operation, which affected everybody, from the unit manager to process operators. The support of the new work process varied significantly. The biggest variation was among process operators. Some saw an opportunity to train new, less experienced operators, others thought they were capable of detecting the problems that the intelligent system was designed to identify without help. One operator even said that he would not use the system and was smarter than computers and did not want a computer telling him what to do. It was expected that the least experienced operators would use the system more frequently due to their lack of knowledge for reliable pattern recognition of the problems. Surprisingly, however, the most experienced operators were the biggest supporters and users of the intelligent system. They not only enriched the knowledge content, but also became the natural leaders in the difficult process of integrating the intelligent system in the decision-making process.

Unfortunately, the support of experienced operators and management was insufficient to promote the cultural change. The resistance and the inertia were too high and the system was gradually ignored after several years in operation.

14.3.2 Applications with Low-Quality Data

Poor data quality or insufficient data are some of the most frequent causes for failed computational intelligence applications. Two unsuccessful attempts to develop robust inferential sensors were in this category. The first case was related to very noisy data from a wastewater treatment plant. In the second case, data from a faulty gas chromatograph was inappropriately used in the initial design of a chemical property inferential sensor. However, after correcting the problem and using high-quality data from a well-calibrated source, a solution with very good performance was developed and applied in a similar plant.

14.4 Acknowledgements

Almost any industrial application requires teamwork. The author participated in different roles in all presented application projects but the real credit for their success goes to the corresponding project teams. The author would like to acknowledge the contribution in the development and deployment of the systems described in this chapter of the following colleagues from The Dow Chemical Company: Elsa Jordaan, Sofka Werkmeister, Guido Smits, Alex Kalos, Brian Adams, Hoang Pham, Ed Rightor, David Hazen, Lawrence Chew, and Torben Bruck; Mark Kotanchek from Evolved Analytics; Ching-Tai Lue from Univation Technologies; and the former Dow colleagues Clive Bosnyak, Ioa Gavrila, and Judy Hacskaylo.

14.5 Summary

Key messages:

Examples of successful computational intelligence applications in manufacturing are inferential sensors, automated operating discipline, and empirical emulators for process optimization.

Examples of successful computational intelligence applications in new product development are accelerated first-principles model development and fast robust empirical model building from small data sets.

Examples of unsuccessful computational intelligence applications in industry are implementations triggering substantial cultural change and projects based on low-quality data.

The Bottom Line

Computational intelligence delivers valuable solutions to industrial problems.

Suggested Reading

The following references give detailed technical description of the applications, described in this chapter:

E. Jordaan, A. Kordon, G. Smits, and L. Chiang, Robust inferential sensor based on ensemble of predictors generated by genetic programming, *Proceedings of PPSN 2004*, Birmingham, UK, pp. 522–531, 2004.

A. Kalos, A. Kordon, G. Smits, and S. Werkmeister, Hybrid model development methodology for industrial soft sensor, *Proc. of ACC 2003*, Denver, CO, pp. 5417–5422, 2003.

A. Kordon and G. Smits, Soft sensor development using genetic programming, *Proceedings of GECCO 2001*, San Francisco, pp. 1346–1351, 2001.

A. Kordon, A. Kalos, and G. Smits, Real-time hybrid intelligent systems for automating operating discipline in manufacturing, *Artificial Intelligence in Manufacturing Workshop Proceedings of the 17ᵗʰ International Joint Conference on Artificial Intelligence IJCAI 2001*, pp. 81–87, 2001.

A. Kordon, G. Smits, E. Jordaan and E. Rightor, Robust soft sensor based on integration of genetic programming, analytical neural networks, and support vector machines, *Proceedings of WCCI 2002*, Honolulu, pp. 896–901, 2002.

A. Kordon, H. Pham, C. Bosnyak, M. Kotanchek, and G. Smits, Accelerating industrial fundamental model building with symbolic regression, *Proc. of GECCO 2002, Volume: Evolutionary Computation in Industry*, pp. 111–116, 2002.

A. Kordon, G. Smits, A. Kalos, and E. Jordaan, Robust soft sensor development using genetic programming, in *Nature-Inspired Methods in Chemometrics*, pp. 69–108, (R. Leardi, editor), Elsevier, 2003.

A. Kordon, A. Kalos, and B. Adams, Empirical emulators for process monitoring and optimization, *Proceedings of the IEEE 11ᵗʰConference on Control and Automation MED 2003*, Rhodes, Greece, p.111, 2003.

A. Kordon A. and C.T. Lue, Symbolic regression modeling of blown film process effects, *Proceedings of CEC 2004*, Portland, OR, pp. 561–568, 2004.

A. Kordon, E. Jordaan, L. Chew, G. Smits, T. Bruck, K. Haney, and A. Jenings, Biomass inferential sensor based on ensemble of models generated by genetic programming, *Proceedings of GECCO 2004*, Seattle, WA, pp. 1078–1089, 2004.

A. Kordon, G. Smits, E. Jordaan, A. Kalos, and L. Chiang, Empirical model with self-assessment capabilities for on-line industrial applications, *Proceedings of CEC 2006*, Vancouver, pp. 10463–10470, 2006.

Part IV
The Future of Computational Intelligence

Chapter 15
Future Directions of Applied Computational Intelligence

> *The practical applications of a science often precede the birth of the science itself.*
>
> N. K. Jerne
>
> *All things happen by virtue of necessity.*
>
> Democritus

The objective of the last chapter of the book is to estimate the future trends and areas of applied computational intelligence. It includes analysis of two different trends. The first trend focuses on the next generation of computational intelligence technologies, which are still in the research domain, but demonstrate potential for future industrial applications. Examples are technologies such as computing with words, evolving intelligent systems, co-evolution, and artificial immune systems. The second trend explores the projected needs in industry for the next 10–15 years, such as predictive marketing, accelerated new products diffusion, high-throughput innovation, etc. The key assumption is that the expected industrial needs will drive the development and deployment of emerging technologies, like computational intelligence. A method that describes the principles and mechanisms of applied research driven by industrial demand is discussed in this chapter. Fortunately, almost any existing and new computational intelligence methods may contribute to satisfying current and future industrial demand. However, the long-term sustainability of computational intelligence depends on broadening the application audience from large corporations to small businesses and individual users.

15.1 Supply-Demand-Driven Applied Research

In contrast to Nostradamus or astrology, technology forecasts are based on solid scientific methods and are supported by available data. Usually long-term

A.K. Kordon, *Applying Computational Intelligence*,
DOI 10.1007/978-3-540-69913-2_15, © Springer-Verlag Berlin Heidelberg 2010

predictions rely on generic trends,[1] such as computational power growth, economic forecasts for different geographic areas, and increased energy efficiency. An obvious limitation of this approach is its inability to predict unexpected innovations, such as the Internet or most computational intelligence methods. A potential solution is "shaping the future" by funding some promising scientific fields, such as nanotechnology or quantum computing. However, in several cases like, the superconductivity euphoria in the early 1990s, the huge directed funding didn't change the future in the desired direction and didn't lead to the expected technology breakthrough.

In contrast to predicting scientific trends, applied research forecasts are based on balancing expected industrial needs with available technologies and projected innovations. By analogy from economics, we call this way of estimating future trends, supply-demand-driven applied research. In the proposed framework, the supply side represents the flow of research methods and ideas, such as the known computational intelligence approaches with different levels of maturity. The demand side includes the current and expected industrial needs, such as an optimal global supply-chain in diverse local infrastructures, dynamic planning in large price swings, high-throughput invention process, etc. The hypothesis is that in the same way as the dynamics of supply-demand balance of goods and services moves the economy, a similar parity of academic ideas and industrial needs drives applied research.

15.1.1 Limitations of Supply-Driven Research

The supply-driven research model, visualized in Fig. 15.1, represents the classical academic *modus operandi*. The key driving force, according to this model, is the perpetuation of scientific ideas in society. The process begins with a published idea from a Founding Father, which inspires several enthusiastic supporters. Eventually, the enriched idea perpetuates through the growing number of papers by increased number of academics. At some level of growth the idea is pronounced as a new approach and a scientific community of university professors, researchers, graduate students, and post-docs is formed with its corresponding conferences and journals. For further perpetuation of the approach, however, increased funding is needed.

Sustainable funding becomes the critical issue of supply-driven research. As is well known to the academic world, the funding process is unpredictable, highly competitive, and slow. Both potential sources for basic research funding – governmental

---

[1]An interesting reference on using generic trends in long-term forecasts is the visionary book of R. Kurzweil, *The Singularity is Near: When Humans Transcend Biology*, Viking, 2005. An opposite approach of predictions based on different microtrends is presented in the book of M. Penn and K. Zalesne, *Microtrends: The Small Forces Behind Tomorrow's Big Changes*, Twelve, New York, NY, 2007.

Fig. 15.1 A clip art view of the nature of supply-driven research

Government   Industry

agencies and industry have limited resources, which need to be distributed between growing numbers of new approaches. Unfortunately, the funding deficiency may replace creativity with academic schisms, politics and bureaucracy. As a result, the research productivity could be significantly reduced since the lion's share of an academic's time is occupied with writing proposals and hobnobbing with the decision-makers. The inefficiency of the existing funding process is one of the limitations of supply-driven research.

Another limitation is the slow dynamics of some of the classical forms of research relative to the hectic industrial tempo. A typical example is the most established form of academic research – graduate studies. Three to five years of graduate research is an eternity in the business world of the 21st century. In this period of time, the economic situation changes significantly, which drives organizational changes and corresponding business priority shifts. Having in mind the annual cycle of industrial budgets, it is almost impossible to guarantee external funding for such a long period of time. This is one of the reasons for declining industrial support of Ph.D. students.

The most important limitation of supply-driven research, however, is the lack of value assessment mechanisms. Estimating the applicability of the proposed research idea is still not a significant factor in the fitness criteria of an academic's performance evaluation. The focus is mostly on the quality and the number of publications. This leads to expanding the paper universe and delaying the reality check as long as possible. As a result, the critical real-world feedback is often replaced by "proofs" based on simulations of well-designed toy problems.

15.1.2   What Is Supply-Demand Research?

The nature of supply-demand research is illustrated in Fig. 15.2 and discussed below.

On the supply side we have the available research methods with different levels of maturity for industrial applications. Some of these methods, like classical statistics, have a high level of maturity (i.e. developed scientific basis, available professional software, and low training efforts for potential users) and can be applied with low total cost of ownership and minimum marketing efforts. Other methods, like the discussed computational intelligence methods, are still in the research development phase but have demonstrated their application potential in many industries. There are also research methods at a very early stage of development with an incomplete scientific basis but with promising application potential.

On the demand side we have the industrial needs, such as customer loyalty analysis, dynamic optimization, optimal supply-chain, etc. prioritized by potential value creation. These needs must be satisfied with some available tools or methods for some critical time period, which may vary from several months to a maximum of two-three years. In a number of cases, the complex nature of the industrial need will require using sophisticated research methods. For example, dynamic optimization of complex plants may benefit from the new approaches offered by swarm intelligence or evolutionary computation. When a fit between the identified need and appropriate research methods is found, some form of industrial funding can be justified and offered. If the particular industrial problem has been resolved successfully, it could open the door for long-term collaboration.

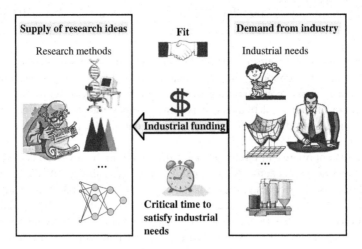

Fig. 15.2 A clip art view of the nature of supply–demand research

15.1.3 Advantages of Supply–Demand Research

Supply-demand-driven research is not a universal mechanism and is limited to applied science with a short-term focus. However, it has several advantages over the classical idea-driven research.

Firstly, the research efforts are naturally driven by expected value creation. The funding is direct and specific and the bureaucracy involved is minimal.

Secondly, the dynamic of research is high and aligned to the tempo of the industrial user. The imposed high-speed research mode changes the working habits of academics and prepares the students for their future work in industry.

Thirdly, supply-demand research has a built-in feedback mechanism for evaluating the capabilities of the applied methods. The strengths and the limitations of the explored approaches are clearly defined after the reality check on industrial applications. Very often, academics enriched their research ideas as a result of solving practical problems. In addition, publications based on industrial applications are much more credible than demonstrating method capabilities on simulations.

Fourthly, industrial demands are more predictable than the supply of new academic ideas, which are unpredictable in principle.

15.1.4 Mechanisms of Supply-Demand Research

The benefits of supply-demand research are well known to the academic community and several established forms of academic-industrial collaboration have been explored. One widespread form is the consortium of a number of industrial participants who shared the research cost of prioritized scientific efforts. Usually the supported activities are in one broad application area (for example, process control) and the funding priorities are defined by the consensus of the industrial participants and are definitely demand-driven. Another form of governmental funding is based on proper industrial-academic collaboration in selected research areas. For example, the Industrial Technologies Program of the Department of Energy supports strategic applied research. Most of the established mechanisms, however, are slow, bureaucratic, and detached from immediate industrial needs. It is our opinion that more flexible and dynamic mechanisms are needed to make the link between the supply of academic ideas and industrial demands more effective. The definition of success is creating an open market of ideas and practical needs with low regulations and large numbers of participants. A necessary condition for functioning of such mechanism is that both sides, the academics and industrial users, will actively participate in this process and will communicate either directly or with minimal intermediate brokers.

Communicating the supply side of new emerging technologies with high application potential needs effective marketing and the techniques described in Chap. 13

can be used. Communicating the demand side to a research audience can be done in various forms.

The first form with growing popularity is defining a specific industrial need and then opening a competition. The most famous case in this category is the competition announced in October 2006 by Netflix. The DVD rental company offered an award of one million dollars for improving the efficiency of their movie recommendation algorithm by 10%. The response was enormous. Within a month, more than a thousand solutions had been submitted, including research groups from AT&T, Princeton, and the University of Toronto. The best algorithm improved the performance by 8.43%, close to the dreamed 10%.[2]

Another popular competition is the recent Grand Urban Challenge, announced by DARPA in 2007 with an award of two million dollars for the wining solution.[3] The competition required teams to build an autonomous vehicle capable of driving in traffic, performing complex maneuvers such as merging, passing, parking and negotiating intersections. Eleven teams competed for the award in a complex environment where for the first time autonomous vehicles interacted with both manned and unmanned vehicle traffic in an urban environment.

The competition form has several advantages. It is closer to the classic market mechanism. It attracts many participants since it eliminates bureaucracy and favoritism (i.e. gives equal rights to everybody). On the industrial side, it saves tremendous time and effort to evaluate potential academic partners and to coordinate joint projects.

The second form of communicating industrial needs to the academic community and encouraging the dialog is by preparing industrial benchmarks. The benchmarks capture typical real-world cases and give the academic community a solid reality check for their ideas. They also introduce credibility in the academic claims for algorithms performance. A typical case of an industrial benchmark is the Tennessee-Eastman problem,[4] widely used for validating new process monitoring and control algorithms.

The third promising form of communicating industrial needs to the academic community and establishing long-term relationships is an industrial sabbatical. Until recently, it was a one-way street of academic sabbaticals in industry. Some companies, like The Dow Chemical Company, have begun to explore the benefits of the other direction of industrial sabbatical by offering to leading industrial researchers opportunities to spend at least three months in world-class academic institutions.

---

[2]J. Ellenberg, The Netflix challenge, *Wired,* March 2008, pp. 114–122, 2008

[3]http://www.darpa.mil/grandchallenge/

[4]A description of the Tennessee-Eastman problem and an archive of models can be found in http://depts.washington.edu/control/LARRY/TE/download.html

15.2 Next-Generation Applied Computational Intelligence

We'll use the supply-demand research approach to estimate future trends in applied computational intelligence. First, we'll focus on the supply side of academic ideas in this booming research area. From the multitude of promising new approaches, the following four methods, shown in Fig. 15.3, have been selected by estimating their value creation potential: computing with words, evolving intelligent systems, co-evolving systems, and artificial immune systems.

15.2.1 Computing with Words

It is a well-known fact that computers are based on numerical calculations and symbol manipulation. Humans, however, base their thought process on words from natural language premises. The key idea behind computing with words is to close the gap between human and machine reasoning by using mathematical models for natural language semantics. Computing with words is based on the concept of fuzzy set and fuzzy logic, described in Chap. 3. This new technology allows us, by extending the capabilities of fuzzy logic, to define machine process statements expressed in natural language and draw computer-generated conclusions in verbal statements that are meaningful and acceptable for humans.[5]

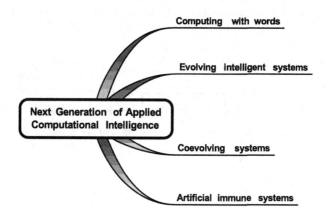

Fig. 15.3 Selected approaches from the next generation of applied computational intelligence

---

[5]The state of the art of this technology is given in the paper: L. Zadeh, Toward human level machine intelligence – Is it achievable? The need for a paradigm shift, *IEEE Computational Intelligence Magazine*, *3*, 3, pp. 11–22, 2008.

15.2.1.1 Basic Principles of Computing with Words

Computing with words and perceptions is a mode of computing in which, as in humans, the objects of computation are words, propositions and perceptions described in a natural language. The generic scheme of this revolutionary technology is shown in Fig. 15.4.

Computing with words is a methodology in which words are used in place of numbers for computing and reasoning. According to Lotfi Zadeh, the father of fuzzy logic, there are two main reasons to use this technology.[6] First, computing with words is a necessity when the available information is too imprecise to justify the use of numbers, and, second, when there is a tolerance for imprecision which can be exploited to achieve tractability, robustness, low solution cost, and better rapport with reality. Exploitation of the tolerance for imprecision is an issue of central importance in computing with words.

Computing with words can be developed on different computational engines, such as Precisiated Natural Language,[7] granular computing,[8] and type 2 fuzzy sets.[9] We'll focus only on the first engine, Precisiated Natural Language, where computing and reasoning with perceptions is reduced to operating on propositions expressed in a natural language.

The basic structure of the Precisiated Natural Language is shown in Fig. 15.5.

The central concept of Precisiated Natural Language, proposed by Lotfi Zadeh, is that an effective way of representing the meaning of natural language in a computer is by using the so-called generalized constraint. A typical example of a generalized constraint is a fuzzy If-Then rule.

The Precisiated Natural Language includes two key operations over the propositions in a natural language, as shown in Fig. 15.5. The first operation is called

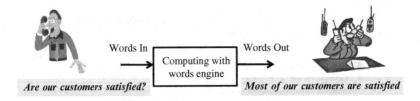

Fig. 15.4 Generic scheme of computing with words

---

[6]L. Zadeh, Fuzzy logic = Computing with words, *IEEE Trans. Fuzzy Systems, 90*, pp. 103–111, 1996.

[7]L. Zadeh, Precisiated natural language (PNL), *AI Magazine, 25*, pp. 74–91, 2004.

[8]S. Aja-Fernandez and C. Alberola-Lopez, Fuzzy granules as a basic word representation for computing with words, *SPECOM 2004*, St. Petersburg, Russia, 2004.

[9]J. Mendel, An architecture for making judgment using computing with words, *Int. J. Appl. Math. Comput. Sci, 12*, pp. 325–335, 2002.

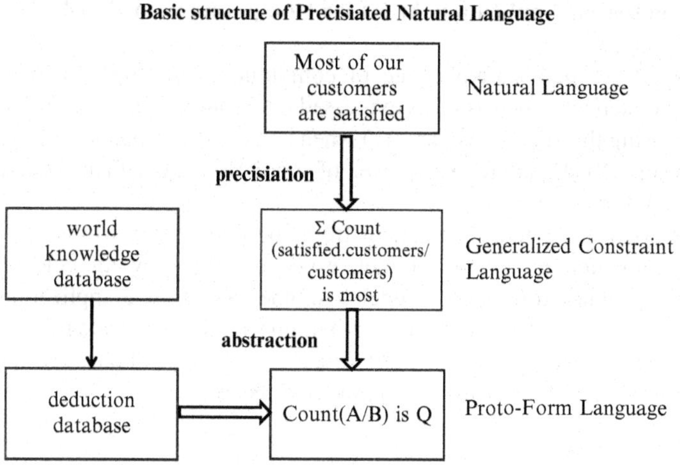

Basic structure of Precisiated Natural Language

Fig. 15.5 Basic structure of Precisiated Natural Language (PNL)

precisiation and translates the verbal proposition into some form of computerized relationship defined as generalized constraint. However, not every proposition is precisable even with the relatively broad range of potential generalized constraints. It has to be taken into account that precisiation may include a significant number of numerical calculations, such as weighted summations. An example of precisiation is shown in Fig. 15.5, where the proposition "Most of our customers are satisfied" is decomposed into precisiation of the corresponding words. One generalized constraint for the word *most* can be the weighted sum of satisfied customers over all customers.

The second operation in Precisiated Natural Language is called abstraction and translates the precisiated proposition into a generic expression, called a proto-form. This is an abstract representation of the semantic meaning of the proposition. An example for the proto-form of the proposition "Most of our customers are satisfied" is the generic expression *Count (A/B) is Q*. The abstraction process includes also a world knowledge database and a deduction database.

Precisiated Natural Language can be used as a high-level concept definition language. It allows us to describe the perception of a concept, such as *most*, in a natural language, precisiate it in numerical terms via generalized constraints, and generalize the description via proto-forms.[10]

---

[10]A good example of using computing with words in time series analysis is the chapter of J. Kacprzyk, A. Wilbik, and S. Zadrozny, Towards human consistent linguistic summarization of time series via computing with words and perceptions, in *Forging New Frontiers: Fuzzy Pioneers 1*, M. Nikravesh, J. Kacprzyk, and L. Zadeh (Eds), Springer, pp. 17–35, 2007.

15.2.1.2 Potential Application Areas for Computing with Words

The expected competitive advantage of computing with words is in modeling imprecise phenomena, such as ideas expressed in a natural language. The economic drivers for using this technology are: (1) when acquisition of precise information is too costly and (2) the expressive power of words is greater than the expressive power of numbers.

Several application areas can benefit from computing with words. An obvious implementation domain is the Internet, especially the growing field of search engines with improved semantic models. Another prospective application area is simulating new product perceptions and defining marketing strategies. However, computing with words is still in an early research development phase with minimal available software and no industrial application record.

15.2.2 Evolving Intelligent Systems

Handling changing operating conditions is one of the biggest challenges in real-world applications. The potential solutions are based on three key methods - adaptive systems, evolutionary algorithms, and evolving intelligent systems. Adaptive systems are limited to slow minor changes within the 10–15% range of the operating conditions. The process variation is captured by update of the parameters of a model (usually linear) with a fixed structure. As we discussed in Chap. 5, evolutionary algorithms respond to change by applying simulated evolution of a population of candidate solutions with entirely different structures and parameters. This approach, however, is not very appropriate for on-line applications due to the unpredictable swings of simulated evolution and the high computational requirements. For this class of problems, one of the best solutions is the new emerging technology, called evolving intelligent systems.

Evolving intelligent systems are self-adaptive structures with learning and summarization capabilities.[11] In contrast to evolutionary algorithms, they respond to the unknown environment by appropriately changing their structure and adapting their parameters based on incoming data. Their response to significant changes is by structural transformations of the model while gradual changes are handled with parameter updates. By handling process variations in this way, the evolving intelligent systems deliver smooth on-line performance during the operating condition changes at very low computational cost.

---

[11]P. Angelov, http://www.scholarpedia.org/article/Evolving_fuzzy_systems

Fig. 15.6 Visualization of a
Takagi-Sugeno fuzzy rule-
based system

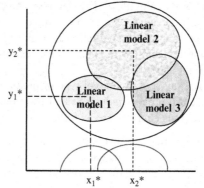

IF x is around x_1^* THEN y is y_1^* (Linear model 1)
IF x is around x_2^* THEN y is y_2^* (Linear model 2)

15.2.2.1   Basic Principles of Evolving Intelligent Systems

The most developed form of evolving intelligent systems is evolving fuzzy systems. They are based on a popular type of model, called Takagi-Sugeno, which is very convenient for structural and parametric identification. A visual interpretation of a very simple Takagi-Sugeno fuzzy model is shown in Fig. 15.6.

Takagi-Sugeno models are based on the assumption that the data can be decomposed into fuzzy clusters. Within each cluster, a linear model is defined and its parameters can be easily updated. An example of partitioning the data into three clusters with corresponding linear models is shown in Fig. 15.6. The key advantage of Takagi-Sugeno models is that they represent nonlinear systems in a very flexible and effective way for practical applications. The nonlinearity is captured by fuzzy If-Then rules (the clusters) and the functional input-output relationships are represented by simple locally linear models (see the examples of simple If-Then rules in Fig. 15.6).

The rules are not fixed and the structure building can begin from scratch with no *a priori* knowledge. The structure of an evolving fuzzy system is a flexible and open set of Takagi-Sugeno fuzzy rules, which can grow, shrink, and update in real time according to the information content of the incoming data. It is based on simple algorithms, requires low computational resources in real time, and the models are linguistically interpretable.

15.2.2.2   Potential Application Areas for Evolving Intelligent Systems

The expected competitive advantage of evolving intelligent systems is their potential to create low-cost self-healing systems with minimum maintenance. An example of such a system is the evolving fuzzy inferential sensor *eSensor* with embedded evolving Takagi-Sugeno models that brings the ability to self-develop, self-calibrate, and self-maintain.[12] The learning method of evolving Takagi-Sugeno is based on two stages that are performed during a single time interval: (i) automatically separating the available data through evolving clustering and defining the fuzzy rules; and (ii) recursive linear estimation of the parameters of the consequents of the defined rules.

Some of the advantages of this novel type of predictive on-line model over the existing sensors used in process industries and from the robust inferential sensors discussed in the previous chapters are as follows:

- It requires minimal maintenance efforts due to its *evolving* structure which automatically tracks process changes;
- It requires minimal maintenance efforts due to its *evolving* structure which automatically tracks process changes;
- It requires very low model development efforts since the *evolving* structure process can start from scratch with minimal process knowledge and historical data;
- It has a Multiple-Input Multiple-Output (MIMO) structure, which allows a very compact representation of processes with multiple outputs;
- It can automatically detect *shifts* in the data pattern by online monitoring the quality of the clusters and fuzzy rules.

The proposed novel approach has been tested on four problems from the chemical industry (prediction of the properties of three compositions and propylene in a simulated online mode). The four test cases include a number of challenges, such as operating regime change, noise in the data, and a large number of initial variables. These problems cover a wide range of real issues in industry when an inferential sensor is to be developed and applied. In all of these tests, the proposed new *eSensor* proved to be capable of being an advanced replacement for the existing less-flexible solutions.

The operation sequence of the *eSensor* is as follows. First, it starts to learn and generate its fuzzy rule-base from the first data sample it reads. Second, *eSensor* evolves the fuzzy rule-base structure on a sample-by-sample basis (see an example of the evolution of the number of rules for Composition 3 in Fig. 15.7). Third, it adjusts the parameters of each rule in the rule-base online. In this way, the evolving fuzzy inferential sensor continuously adapts its structure and self-calibrates.

Another advantage of *eSensor* is the simplicity and interpretability of the evolving models. An example of a Takagi-Sugeno model at the end of the simulated period for the test case of Composition 2 is shown in Fig. 15.8. It could be easily interpreted by plant engineers and process operators.

Evolving intelligent systems are one of the new computational intelligence methods with the fastest-growing application record. Some of the application

---

[12]P. Angelov, A. Kordon, and X. Zhou, Evolving fuzzy inferential sensors for process industry, *3rd Workshop on Genetic and Evolving Fuzzy Systems*, Witten-Bommerholz, Germany, pp. 41–46, 2008.

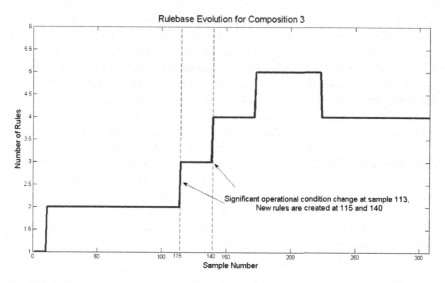

Fig. 15.7 Evolution of the fuzzy rule base of *eSensor* for Composition 3

FINAL RULE-BASE for COMPOSITION 2:

R_1: **IF**$(x_1$ *is around 183.85) **AND** $(x_2$ is around 170.31)***THEN***

$(\overline{y} = 84.0 - 0.9\overline{6}x_1 + 0.6\overline{1}x_2)$

R_2: **IF**$(x_1$ *is around 178.09) **AND** $(x_2$ is around 166.84)***THEN***

$(\overline{y} = 0.87 - 0.9\overline{8}x_1 + 0.5\overline{4}x_2)$

R_3: **IF**$(x_1$ *is around 172.70) **AND** $(x_2$ is around 166.01)***THEN***

$(\overline{y} = 0.87 - 1.0\overline{2}x_1 + 0.6\overline{4}x_2)$

Fig. 15.8 Final rule base for Composition 2

areas are mobile robots, mobile communications, process modeling and control, and recently in machine health prognosis at Ford Motor Company.[13]

15.2.3 Co-evolving Systems

Co-evolution is a form of evolutionary computation in which the fitness evaluation is based on interactions between multiple individuals.[14] It is inspired by biology, where *all* evolution is co-evolution because individual fitness is a function of other

[13]D. Filev and F. Tseng, Novelty detection-based machine health prognostics, *In Proc. 2006 International Symposium on Evolving Fuzzy Systems,* IEEE Press, pp. 193–199, 2006.

[14]E. de Jong, K. Stanley, R. P. Wiegand, Introductory tutorial on co-evolution, *Proceedings of GECCO 2007,* London, UK, pp. 3133–3157, 2007.

individuals. An example is predator-prey co-evolution where we have inverse fitness interaction between the two species. This mode is called an *arms race* where a win for the one species is a failure for the other and vice versa. To survive, the losing species changes its behavior to counter the winning species in order to change the status quo and to become the new winner. As a result, the complexity of both species increases.

Co-evolutionary algorithms are very similar to traditional evolutionary algorithm methods, discussed in Chap. 5. For example, in co-evolution the individuals are encoded according to the specific problem, they are altered during search with genetic operators, and the search is directed by selection based on fitness. However, both algorithms differ substantially on several features. For example, the evaluation requires interaction between multiple individuals which may belong to the same population or in different populations; the fitness of an individual depends on the relationship between that individual and other individuals; and new modes of cooperation and competition are possible.

15.2.3.1   Basic Principles of Co-evolving Systems

In simulated co-evolution, two populations of individuals (called *parasites* and *hosts*) evolve concurrently, with the fitness of individuals in each population depending on their interactions with individuals in the other population.[15] The success of simulated co-evolution depends on three factors: (1) the constantly changing environment produced for each population by the other, (2) the maintenance of diversity in the *host* and *parasite* populations, and (3) the *arms race* between *hosts* and *parasites*. In such an *arms race*, the *hosts* evolve to successfully solve the problems encoded as *parasites*, forcing them to evolve to be more effective in fighting the *hosts*. As a result, the evolution of the *hosts* is directed to solve these new problems, and so on. The hypothesis is that co-evolution has a built-in mechanism for continuous improvement of both populations and delivering very robust solutions.

There are two key modes of operation in simulated co-evolution, depending on whether the two populations benefit from each other or if they are in conflict. Those two modes of interaction are called *cooperative* and *competitive*, respectively. The first mode is appropriate for applications in complex optimization and the second mode has great potential to be used in computer games.

In *cooperative* co-evolution the interacting individuals succeed or fail together. In this mode of operation, improvement on one side results in positive feedback on the other side and vice versa. As a result, there is a reinforcement of the relationship between the two populations. An application area where *cooperative* co-evolution

---

[15]M. Mitchell, M. Thomure, and N. Williams, The role of space in the success of co-evolutionary learning. In *Proceedings of Artificial Life X: Tenth Annual Conference on the Simulation and Synthesis of Living Systems*, MIT Press, 2006.

is an appropriate solution is complex optimization. The large problem is decomposed into a collection of easier subproblems, the solutions to which work together to solve the original, larger problem. More specifically, the objective of *cooperative* co-evolution is to solve a difficult problem by co-evolving an effective set of solutions to a decomposition of the problem into *n* subproblems. Then these *n* isolated populations are co-evolved to "cooperatively" solve the difficult problem. The less the sub-problems interact with each other, the more effective the *cooperative* co-evolution.

In *competitive* co-evolution some individuals succeed at the expense of other individuals, i.e. the two populations are in conflict. This mode can lead to two possible results – populations with stable and unstable behavior. In the stable case, equilibrium is reached when one of the populations consistently outperforms the other. This result could be a proof of a very strong strategy discovered by the individuals of the winning population.

In the case of unstable behavior, individuals in one population outperform members of the other population until those develop a strategy to defeat the first one and vice versa. This leads to a cyclic pattern where species appear then disappear and reappear in later generations.

Unstable behavior is one of the cases, called pathologies, when co-evolution cannot deliver practical solutions. Unfortunately, a generic theoretical analysis of co-evolutionary pathologies is nonexistent. As a result, there is always a risk that co-evolution will not deliver the expected results and will be stuck in pathology.

15.2.3.2 Potential Application Areas for Co-evolving Systems

The expected competitive advantage of co-evolutionary systems is based on the benefits of the *arms race* hypothesis. It is assumed that during the *arms race* increased performance is generated by each population making incremental improvements over the others in such a way that steady progress is produced. Simulated co-evolution lowers the fitness bar and allows the system to evolve faster toward better solutions. During the evolutionary process, the coevolved structures are getting better and better by raising the fitness bar and increasing the selection pressure.

Co-evolution could be a good solution when the search space is very large and when the objective function is difficult to formalize or is unknown. Most of the computer games discussed in Chap. 8 are in this category. A co-evolutionary approach has been successfully applied for analysis of electricity market equilibrium for a three-firm market.[16] Another broad potential application area of co-evolution is simulation of economic and social systems.[17]

---

[16]H. Chen, K. Wong, D. Nguyen, and C. Chung. Analyzing oligopolistic electricity market using co-evolutionary computation. *IEEE Transactions on Power Systems, 21*, pp. 143–152, 2006.

[17]B. LeBaron. Financial market efficiency in a co-evolutionary environment. In *Proceedings of the Workshop on Simulation of Social Agents: Architectures and Institutions*, pp. 33–51. Argonne National Laboratory and the University of Chicago, 2001.

15.2.4 Artificial Immune Systems

Artificial immune systems are another clear example of computational intelligence algorithms inspired by biology. The immune system exists to protect organisms from dangerous agents such as bacteria, viruses, and other foreign life forms. These harmful non-self agents are called pathogens. The immune system accomplishes the protective functions by carefully distinguishing the self (parts of the organism protected by this immune system) from non-self (anything else). This is done using special detectors called lymphocytes. Although they are created randomly, they are trained and remember infections so that the organism is protected from future intrusions as well as past ones.

The natural immune system offers two lines of defense, the innate and adaptive immune system,[18] shown in Fig. 15.9.

The first line of defense, based on the innate immune system, includes cells that can counteract a known set of attackers, or "antigens". The important feature is that the innate immune system does not require previous exposure to pathogens. The antigen can be an intruder or part of the cells of the organism itself.

The second line of defense is based on an adaptive immune system that can learn to recognize, eliminate and remember specific new antigens. The key pathogen warriors are bone marrow and thymus, which continuously produce lymphocytes.

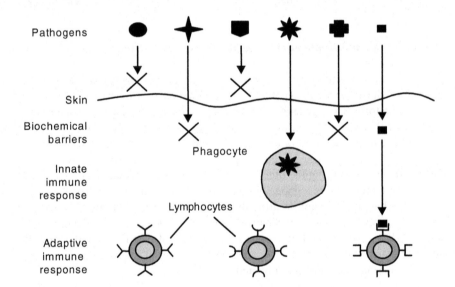

Fig. 15.9 The two lines of defense for organism protection, innate and adaptive immune systems

[18]L. de Castro and J. Timmis, *Artificial Immune Systems: A New Computational Intelligence Approach*, Springer, 2002.

Each of these lymphocytes can neutralize a specific type of antigen and produce a large number of cloned cells, called clonal selection.

The immediate reaction of the innate and adaptive immune system cells toward intruders is called the primary immune response. A selection of the activated lymphocytes during the immediate reaction are turned into sleeper memory cells. The secondary immune response happens when the cells are activated again if a new intrusion of the same antigen occurs, resulting in a much quicker and more effective neutralization.

15.2.4.1 Basic Principles of Artificial Immune Systems

Artificial immune systems emerged in the 1990s driven by the growing need to improve computer security. Artificial immune systems are adaptive systems inspired by biological immune system, which are applied to complex problem domains. It is based on three key bio-inspired algorithms – for positive, negative, and clonal selection, shown in Fig. 15.10.

The biological basis of the positive selection algorithm is those cells that don't recognize self. They become mature, i.e. anything a mature cell recognizes is defined as non-self or an antigen. As a result, the mature cell is activated when it recognizes an antigen. The computer analogy of cell maturity is defining non-self detectors that represent chunks of typical patterns of abnormal activity. The positive selection algorithm generates a set of detectors, each of which must not fail on at least one member of the defined set of normal activity patterns; otherwise that detector must be rejected. The algorithm continuously monitors activity patterns and if any pattern triggers at least one of the detectors then non-self has been detected and action must be taken.

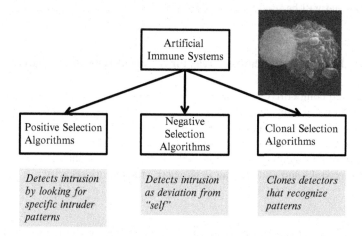

Fig. 15.10 Key artificial immune systems algorithms

The biological basis of the negative selection algorithm is the T-cells, produced in the thymus, and the B-cells produced in bone marrow. Both of them operate on the principle that the cells recognized as self are destroyed. The first step in the negative selection algorithm is to define self as a normal pattern of activity. Then the algorithm generates a set of detectors, each of which must fail to match any normal pattern of activity. The negative selection algorithm continuously monitors new observations for changes by continually testing the detectors matching against representatives of the defined self. If any detector ever matches, a change must have occurred in system behavior.

Both positive and negative selection algorithms are basically similar. However, the positive selection algorithm is based on some *a priori* knowledge of abnormal behavior. In contrast, the negative selection algorithm makes assumptions about the normal state of the system, which from a practical consideration is a very weak assumption.

The biological basis of clonal selection algorithms is the mechanism of proliferation of recognizing cells to guarantee sufficient B-cells to trigger the immune response. This process is similar to natural selection and is known as immune microevolution. The promoted B-cells have high affinity to the recognized antigen and possess long-term memory.

The most popular clonal selection algorithm, called CLONALG,[19] is very similar to the evolutionary algorithm, discussed in Chap. 5. However, there are two main differences: (1) the reproduction of immune cells is proportional to its affinity with the detected antigen and (2) the mutation rate of the cells is inversely proportional to the affinity. The clonal selection algorithm has some advantages over evolutionary algorithm, such as dynamically adjusted population size, capability of maintaining local optima solutions, and a defined stopping criterion.

The artificial immune system development process requires a non-trivial step of mapping the problem into the artificial immune systems framework, i.e. defining the type of immune cells, their mathematical representation, and the immune principles that will be used for the solution.

15.2.4.2   Potential Application Areas for Artificial Immune Systems

The expected competitive advantage of artificial immune system is the hypothesis that they do not require exhaustive training with negative (non-self) examples to distinguish between self and non-self and can reliably identify items as foreign bodies or intruders which have never been encountered. This unique capability may open the door for potential applications in fault detection in technical systems, fraud detection in banks and auditing, and computer and network security.

---

[19]L. de Castro and F. J. von Zuben, The clonal selection algorithm with engineering applications, *Proc. GECCO Workshop on Artif. Immune Syst. Their Appl.*, pp. 36–37, 2000.

The application opportunities of artificial immune systems in the above-mentioned areas have already been explored.[20] We'll illustrate the capabilities of this new computational intelligence technology with an application involving aircraft fault detection. It is based on experiments of simulated failure conditions using the NASA Ames C-17 flight simulator and five different simulated faults were targeted for detection: one for the engine, two for the tails and two for the wings. A special type of real-valued negative selection algorithm, called MILD,[21] was utilized to detect a broad spectrum of known as well as unforeseen faults. Once the fault was detected and identified, a direct adaptive control system used the detection information to stabilize the aircraft by utilizing available resources.

15.3 Projected Industrial Needs

The projected industrial needs are based on the key trends of increased globalization, outsourcing, progressively growing computational power, and expanding wireless communication. It is believed that these industrial requirements could be fulfilled with solutions generated by applied computational intelligence. The selected needs are summarized in the mind-map in Fig. 15.11 and are discussed shortly below:[22]

15.3.1 Predictive Marketing

Successful acceptance of any new product or service by the market is critical for industry. Until now most modeling efforts have been focused on product discovery and manufacturing. Usually the product discovery process is based on new composition or technical features. Often, however, products with expected attractive new features are not accepted by potential customers. As a result, big losses are recorded and the rationale of the product discovery and the credibility of the related technical modeling efforts are questioned.

One possible solution to this generic issue in industry is to improve the accuracy of market predictions with better modeling. Recently, there has been a lot of activities in using the enormous amount of data on the Internet for customer

---

[20]A representative survey of the state of the art and the selected example are from the paper: D. Dasgupta. Advances in artificial immune systems. *IEEE Computational Intelligence Magazine, 1,* 5, pp. 40–49, 2006.

[21]http://www.nasa.gov/vision/earth/technologies/mildsoftware_jb.html

[22]The material in this section was originally published in: A. Kordon, Soft computing in the chemical industry: Current state of the art and future trends, In: *Forging the New Frontiers: Fuzzy Pioneers I: Studies in Fuzziness and Soft Computing,* M. Nikravesh, J. Kacprzyk, and L. A. Zadeh (eds), pp. 397–414, Springer, 2007. With kind permission of Springer-Verlag.

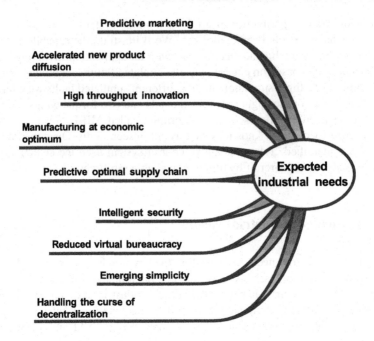

Predictive marketing

Accelerated new product diffusion

High throughput innovation

Manufacturing at economic optimum

Predictive optimal supply chain

Expected industrial needs

Intelligent security

Reduced virtual bureaucracy

Emerging simplicity

Handling the curse of decentralization

Fig. 15.11 Expected industrial needs related to computational intelligence

characterization and modeling by intelligent agents.[23] Computational intelligence may play a significant role in this growing new type of modeling. The key breakthrough can be achieved by modeling customer perceptions. The subject of marketing is the customer. In predicting her/his response to a new product, the perception of the product is at the center point of the decision-making process. Unfortunately, perception modeling is still an open area of research. However, this gap may create a good opportunity to be filled by combining the capabilities of agent-based systems with the new computational intelligence approach – computing with words.

With the growing speed of globalization, of special importance is predicting the perception of a new product in different cultures. If successful, the impact of predictive marketing to any type of industry will be enormous.

15.3.2   Accelerated New Products Diffusion

The next big gap to be filled with improved modeling is optimal new product diffusion, assuming a favorable market acceptance. The existing modeling efforts in this area vary from analytical methods to agent-based simulations. The objective

---

[23]D. Schwartz, Concurrent marketing analysis: A multi-agent model for product, price, place, and promotion, *Marketing Intelligence & Planning*, *18*(1), pp. 24–29, 2000.

is to define an optimal sequence of actions to promote a new product into research and development, manufacturing, supply-chain, and different markets. There are different solutions for some of these sectors.[24] What is missing, however, is an integrated approach across all sectors, since it will optimize the value in the most effective way.

15.3.3 High-Throughput Innovation

Recently, high-throughput combinatorial chemistry is one of the leading approaches for generating innovative products in the chemical and pharmaceutical industries.[25] It is based on intensive design of experiments through a combinatorial sequence of potential chemical compositions and catalysts on small-scale reactors. The idea is fast experiment combined with data analysis to discover and patent new materials. The bottleneck, however, is the speed and quality of model generation. It is much slower than the speed of experimental data generation. Applied computational intelligence could increase the efficiency of high-throughput innovation by adding additional modeling methods that could generate robust empirical dependencies from small data sets and capture the accumulated knowledge during the experimental work.

By combining the capabilities of statistics, evolutionary computation and swarm intelligence, it is possible to develop high-throughput systems with self-directing experiments, which could minimize the new product discovery time. The biggest benefit, however, will be in integrating the high-throughput system with a system for predictive marketing. In this way, a critical feedback of potential market response to the newly discovered material will be given from agent-based simulations of targeted customers. As a result, the risk involved with investment in the new product could be significantly reduced.

15.3.4 Manufacturing at Economic Optimum

Some of the best manufacturing processes operate under model predictive control systems or well-tuned PID controllers. It is assumed that the setpoints of the controllers are calculated based on either optimal or sub-optimal conditions. However, in the majority of cases the objective functions include technical criteria and are not directly related to economic profit. Even when economics is explicitly used in optimization, the resulting optimal control is not adequate to fast changes and swings in raw material prices. The high dynamics and sensitivity to local events of

---

[24]B. Carlsson, S. Jacobsson, M. Holmén, and A. Rickne, Innovation systems: Analytical and methodological issues, *Research Policy, 31*, pp. 233–245, 2002.

[25]J. Cawse (Editor), *Experimental Design for Combinatorial and High-Throughput Materials Development*, John Wiley, 2003.

the global economy require process control methods that continuously fly over the moving economic optimum. It is also desirable to include a predictive component based on economic forecasts.

One potential approach that could address the need for continuously tracking the economic optimum and could revolutionize process control is swarm-based control.[26] It combines the design and controller development functions into a single coherent step through the use of evolutionary reinforcement learning. In several case studies, the designed swarm of neuro-controllers finds and continuously tracks the economic optimum, while avoiding the unstable regions. The profit gain in case of a bioreactor control is $> 30\%$ relative to classical optimal control.[27]

15.3.5   Predictive Optimal Supply-Chain

As a consequence of increased outsourcing of manufacturing and the growing tendency of purchasing over the Internet, the share of supply-chain operations in total cost have significantly increased.[28] One of the challenges for the future supply-chain systems is the exploding amount of data they need to handle in real time because of the application of Radio-Frequency Identification (RFID). It is expected that this new technology will require online fast pattern recognition technology, trend detection, and rule definition on very large data sets. Of special importance are the extensions of supply-chain models to incorporate demand management decisions and corporate financial decisions and constraints. It is obviously suboptimal to evaluate strategic and tactical supply-chain decisions without considering marketing decisions that will shift future sales to those products and geographical areas where profit margins will be highest. Without including the future demand forecasts the supply-chain models lack predictive capability. Current efforts in supply chain are mostly focused on improved data processing, analysis, and exploring different optimization methods. What is missing in the supply-chain modeling process is including the growing experience of all participants in the process and refining the decisions with the predictive methods of computational intelligence.

15.3.6   Intelligent Security

It is expected that with technological advancements in distributed computing and communication the demand for protecting security of information will grow. Even at some point the security challenges will prevent the mass-scale applications of

---

[26]A. Conradie, R. Miikkulainen, and C. Aldrich, Adaptive control utilising neural swarming, *Proceedings of GECCO 2002*, New York, NY, pp. 60–67, 2002.

[27]A. Conradie and C. Aldrich, Development of neurocontrollers with evolutionary reinforcement learning, *Computers and Chemical Engineering, 30* (1), pp. 1–17, 2006.

[28]J. Shapiro, *Modeling the Supply Chain*, Duxbury, Pacific Grove, CA, 2001.

some technologies in industry because of the high risk of intellectual property theft, manufacturing disturbances, even process incidents. One technology in this category is distributed wireless control systems that may include many communicating smart sensors and controllers. Without secure protection of the communication, however, it is doubtful that this technology will be accepted in manufacturing even if it has clear technical advantages.

Computational intelligence can play a significant role in developing sophisticated systems with built-in intelligent security features. Of special importance are the co-evolutionary approaches based on intelligent agents.

15.3.7   Reduced Virtual Bureaucracy

Contrary to the initial expectations of improved efficiency by using global virtual offices, an exponentially growing role of electronic communication not related to creative work activities is observed. Virtual bureaucracy enhances the classical bureaucratic pressure with new features like bombardment of management emails from all levels in the hierarchy; transferring the whole communication flow to any employee in the organization; obsession with numerous virtual training programs, filling electronic feedback forms, or exhaustive surveys; continuous tracking of bureaucratic tasks and pushing the individual to complete them at any cost, etc. As a result, the efficiency of creative work is significantly reduced. Computational intelligence cannot eliminate the root cause of this event, which is based on business culture, management policies, and human nature. However, by modeling management decision-making and analyzing efficiency in business communication, it is possible to identify criteria for bureaucratic content of messages. It could be used in protecting the individual from virtual bureaucracy in a similar way as spam filters operate.

Another option to demonstrate the consequences of management decisions and the potential for creating virtual bureaucracy is by using agent-based simulations.

15.3.8   Emerging Simplicity

An analysis of survivability of different approaches and modeling techniques in industry shows that the simpler the solution the longer it is used and the lower the need to be replaced with anything else. A typical case is the longevity of the PID controllers which for more than 60 years have been the backbone of manufacturing control systems. According to a recent survey, PID is used in more than 90% of practical control systems, ranging from consumer electronics such as cameras to industrial processes such as chemical processes.[29] One of the reasons is their simple

---

[29]Y. Li, K. Ang, and G. Chong, Patents, software, and hardware for PID control, *IEEE Control Systems Magazine*, 26 (1), pp. 42–54, 2006.

structure that appeals to the generic knowledge of process engineers and operators. The defined tuning rules are also simple and easy to explain.

Computational intelligence technologies can deliver simple solutions when multiobjective optimization with complexity as an explicit criterion is used. An example of this approach with many successful applications in the chemical industry is using symbolic regression generated by Pareto front genetic programming. Most of the implemented empirical models are very simple, as was demonstrated in Chaps. 5 and 14, and were easily accepted by process engineers. Their maintenance efforts were low and the performance over changing process conditions was acceptable.

15.3.9   Handling the Curse of Decentralization

The expected growth of wireless communication technology will allow mass-scale introduction of "smart" components of many industrial entities, such as sensors, controllers, packages, parts, products, etc. at relatively low cost. This tendency will give a technical opportunity for the design of totally decentralized systems with built-in self-organization capabilities. On the one hand, this could lead to the design of entirely new flexible industrial intelligent systems capable of continuous structural adaptation and fast response to the changing business environment. On the other hand, the design principles and reliable operation of a totally distributed system of thousands, even millions, of communicating entities of a diverse nature is still a technical dream. Most existing industrial systems avoid the curse of decentralization through their hierarchical organization. However, this imposes significant structural constraints and assumes that the designed hierarchical structure is at least rational, if not optimal. The problem is the static nature of the designed structure, which is in striking contrast to the expected increased dynamics in the industry of the future. Once set, the hierarchical structure of industrial systems is changed either very slowly or not changed at all, since significant capital investment is needed. As a result, the system does not operate effectively or optimally in dynamic conditions.

The appearance of totally decentralized intelligent systems of free communicating sensors, controllers, manufacturing units, equipment, etc., can lead to the emergence of optimal solutions in real time, and effective response to changes in the business environment is needed. Computational intelligence technologies could be the basis in the design of such systems and intelligent agents could be the key carrier of the intelligent component of each distributed entity.

A supply-demand mind-map, based on the selected computational intelligence methods and the expected industrial needs, is shown in Fig. 15.12. The future trends in applied computational intelligence depend on the best fit between the industrial demand and the supply of the corresponding research methods. It has to be taken into account that this is a competitive relationship with all available approaches and it is possible that some of the expected industrial needs will be satisfied by non-computational intelligence methods.

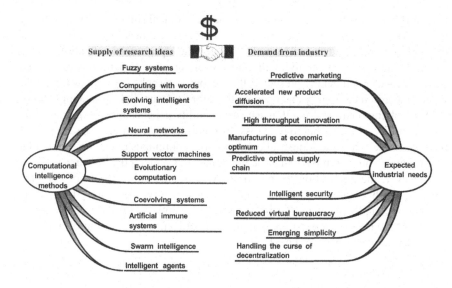

Fig. 15.12 Supply–demand mind-map of established and new computational intelligence methods and expected industrial needs

An analysis of the expected industrial needs shows that a shift of demand from manufacturing-type process-related models to business-related models, such as marketing, product diffusion, supply-chain, and security, is expected. As was discussed in Chap. 8, there is saturation in manufacturing with different models, and some existing plants operate close enough to the optimum. Using computational intelligence for modeling the human-related activities in the business process could provide benefits similar to or larger than plant optimization. By their very nature human-related activities are based on fuzzy information, perceptions, and imprecise communication. There is no doubt that computational intelligence techniques are the appropriate methods to address these modeling challenges.

15.4 Sustainability of Applied Computational Intelligence

The objective answer about applied computational intelligence sustainability depends on how successfully it will satisfy the expected industrial needs. There are several factors for optimism, though. The academic growth of the field is obvious and is supported by the expanding number of publications and filed patents. It is one of the most dynamic areas of computer science which generates not only new algorithms but entirely new paradigms, like computing with words or artificial immune systems.

Another important factor in favor of computational intelligence sustainability is that this academic growth is combined with very fast transfer into credible real-world applications. We observe even the paradoxical situations when industry did not have the patience to wait for complete theoretical development of some of the presented approaches but took the risk for their practical applications. Such cases of relatively fast proofs of the great value creation capabilities of applied computational intelligence, presented throughout the book, are critical for establishing the initial credibility and opening the door for future industrial applications.

Applied computational intelligence sustainability depends mostly on the growing credibility of the technology across different industries. It assumes a trend of more effective academic-industrial collaboration, continuous reduction of the total cost of ownership and broadening the potential users and application areas.

However, the application lessons from AI show the vulnerability of complex emerging technologies, i.e. the door can be easily closed for a long time if the methods are implemented inappropriately. Some potential roadblocks that can slow down the progress of applied computational intelligence and the fun of applying this technology are discussed next.

15.4.1   Potential Roadblocks

The key factors that may slow down, even damage the long-term sustainability of applied computational intelligence are as follows:

- *Growing Gap Between Theory and Practice* – Breaking the link between new theoretical ideas and their practical use is always a recipe for failure. In some research areas, such as advanced process control, academia produces mostly theoretical results and the gap between theory and practice is growing. For example, hundreds of new theoretical control algorithms generated annually are of no practical use.
- *Expensive Infrastructure* – The high development cost limits the potential users of computational intelligence to big prosperous businesses. If the situation does not change, this market will be saturated in the near future. One of the reasons for the expensive infrastructure is the limited number of software vendors. The cost could be reduced by higher competition and involvement of more vendors in computational intelligence systems development. A more active role by leading academic developers in approaching software vendors is recommended.
- Another way to reduce the cost of infrastructure is to integrate computational intelligence applications into existing work processes, such as Six Sigma, as was discussed in Chap. 12.
- *Elitism* – At this early phase of technology introduction, applying computational intelligence is treated as an extraordinary activity. Usually it involves leading companies, the best brains in R&D and highly respected application area experts. For further development of the technology, however, such an elitist approach could limit progress and must be transferred into mass-scale applications. The key criterion for success is the gradual transition of computational intelligence users from big corporations to small businesses. Initially, small businesses can play the

important role of brokers between leading researchers, vendors, and big businesses in defining key applications from the current supply-demand market of research methods and industrial needs. Gradually, they could find their own market niches with specialized applications or software.

- *Insufficient Efforts at Integration* – As was emphasized in Chap. 11, different forms of integration of computational intelligence methods, either among themselves or with other available approaches, is critical for real-world applications. Unfortunately, the research efforts in this direction are minimal since new method discovery is more prestigious than integration of the different modeling approaches. In addition, the software for integrating various computational intelligence methods is virtually nonexistent. This may create a big obstacle for applied computational intelligence and may increase the dangerous gap between theory and practice.

15.4.2 The Fun of Computational Intelligence

The last topic in the book is devoted to a theme which formally is too far away from value creation – the fun of exploring, applying and using computational intelligence. In reality, however, it is and it will be one of the key drivers for implementing this amazing technology. Let's look at some opportunities to create fun with computational intelligence, which could spawn new application areas.

One example is funny education, based on amusing intellectual games on almost any subject. It can revolutionize learning and attract students with the same success as the current PlayStation obsession.

Another opportunity is evolutionary art, which can add new creative dimensions and fun in various art forms and open the door for many new talents, most of them nonprofessionals.

The biggest opportunity, however, is in the growing army of highly educated computer-savvy baby-boomers. Their retirement will be significantly different and they will be hungry to spend their unlimited free time on challenging technologies, such as computational intelligence. It is expected that baby-boomers will open the market of intelligent health analysis and monitoring systems where computational intelligence can play a substantial role.

15.5 Summary

Key messages:

Supply–demand-driven applied research is an effective mechanism for dynamic academic support by solving immediate industrial needs.

The supply side of academic ideas in the area of computational intelligence technologies includes new technologies, such as computing with words, evolving intelligent systems, co-evolution, and artificial immune systems.

The demand side of projected industrial needs for the next 10-15 years includes: predictive marketing, accelerated new products diffusion, high-throughput innovation, manufacturing at economic optimum, predictive optimal supply chain, intelligent security, reduced virtual bureaucracy, emerging simplicity, and handling the curse of decentralization.

The future trends in applied computational intelligence depend on the best fit between industrial demand and the supply of the corresponding research methods.

Long-term sustainability of computational intelligence depends on broadening the application audience from large corporations to small businesses and individual users.

The Bottom Line

The future directions of applied computational intelligence depend on effective solutions of projected industrial needs.

Suggested Reading

The following references describe some of the new computational intelligence techniques:

P. Angelov, D. Filev, and N. Kasabov (Eds), *Evolving Intelligent Systems: Methodology and Applications*, Wiley, in press.

D. Dasgupta, *Advances in Artificial Immune Systems: IEEE Computational Intelligence Magazine*, *1*, 5, pp. 40–49, 2006.

L. de Castro and J. Timmis, *Artificial Immune Systems: A New Computational Intelligence Approach*, Springer, 2002.

L. Zadeh and J. Kacprzyk (Eds), *Computing with Words in Information/Intelligent Systems 1 (Foundations)*, Springer, 1999.

L. Zadeh, Toward human level machine intelligence – Is it achievable? The need for a paradigm shift, *IEEE Computational Intelligence Magazine*, *3*, 3, pp. 11–22, 2008.

A good reference on using generic trends in long-term forecasts is the visionary book of R. Kurzweil, *The Singularity is Near: When Humans Transcend Biology*, Viking, 2005.

An opposite approach to predictions based on different micro-trends is presented in the book of M. Penn and K. Zalesne, *Microtrends: The Small Forces Behind Tomorrow's Big Changes*, Twelve, New York, NY, 2007.

Glossary

Accuracy When applied to models, accuracy refers to the degree of fit between the model and the data. This measures how error-free the model's predictions are.

Adaptive system A system that can change itself in response to changes in its environment in such a way that its performance improves through a continuing interaction with its surroundings.

Agent-based integrator A system with a number of agents that coordinate business processes.

Agent-based modeling See Intelligent agents.

Agent-based system A system with a number of agents, which include social interactions, such as cooperation, coordination, and negotiation.

Ant colony optimization (ACO) A population-based method for finding approximate solutions to difficult optimization problems. In ACO, a set of software agents called artificial ants searches for good solutions to a given optimization problem.

Antecedent A conditional statement in the IF part of a rule.

Applied AI A system of methods and infrastructure to mimic human intelligence by representing available domain knowledge and inference mechanisms for solving domain-specific problems.

Applied CI Applied computational intelligence is a system of methods and infrastructure that enhance human intelligence by learning and discovering new patterns, relationships, and structures in complex dynamic environments for solving practical problems.

Approximate reasoning A process of inexact solution of a system of relational assignment equations. A characteristic feature of approximate reasoning is the fuzziness and nonuniqueness of consequents of fuzzy premises.

Approximation An inexact result adequate for a given purpose.

Artificial intelligence (AI) The part of computer science concerned with designing computer systems that exhibit characteristics we associate with intelligence in human behavior – understanding language, learning reasoning, solving problems, and so on.

Artificial neural network (ANN) A system of processing elements, called neurons, connected in a network that can learn from examples through adjustments of their weights.

Assumption space The conditions under which valid results can be obtained from a specific modeling technique.

Bio-inspired computing Bio-inspired computing uses methods, mechanisms, and features from biology to develop novel computer systems for solving complex problems.

Black-box model A model that is opaque to its user.

Bottom-up modeling Bottom-up modeling includes piecing together systems to give rise to grander systems, thus making the original systems subsystems of the emergent system. In a bottom-up approach the individual base elements of the system are first specified in great detail. These elements are then linked together to form larger subsystems, which then in turn are linked, sometimes in many levels, until a complete top-level system is formed.

Chromosome A string of genes that represents an individual (entity).

Classification Separation of objects into different groups.

Clustering Clustering algorithms find groups of items that are similar. For example, clustering could be used by an insurance company to group customers according to income, age, types of policies purchased, and prior claims experience.

Co-evolution Evolution of one species caused by its interaction with another.

Combinatorial optimization A branch of optimization where the set of feasible solutions is discrete or can be reduced to a discrete one, and the goal is to find the best possible solution.

Competitive advantage of a research approach Technical superiority that cannot be reproduced by other technologies and can be translated with minimal effort into a position of competitive advantage in the marketplace.

Complex system A system featuring a large number of interactive components whose aggregate activity is nonlinear and typically exhibit self-organization.

Computational intelligence (CI) Computational intelligence is a methodology involving computing that exhibits an ability to learn and/or to deal with new situations, such that the system is perceived to possess one or more attributes of reason, such as generalization, discovery, association, and abstraction.

Confidence limit (interval) The region containing the limits or band of a model parameter with an associated confidence level (for example 95%).

Consequent A conclusion or action in the IF part of a rule.

Convergence In evolutionary algorithms, it is defined as a tendency of individuals in the population to be the same.

Crossover Exchange of genetic material in sexual reproduction.

Curse of dimensionality The problem caused by the exponential increase in a volume associated with adding extra dimensions to a (mathematical) space.

Data Values collected through record keeping or by polling, observing, or measuring, typically organized for analysis or decision making. More simply, data are facts, transactions, and figures.

Data analysis Data analysis is the process of looking at and summarizing data with the intent to extract useful information and develop conclusions.

Data mining An information and knowledge extraction activity whose goal is to discover hidden facts contained in databases. Typical applications include market segmentation, customer profiling, fraud detection, evaluation of retail promotions, and credit risk analysis.

Data record A row in a database.

Decision-making An outcome of mental processes leading to selection among several alternatives.

Degree of membership A numerical value between 0 and 1 representing the degree to which an item belongs to a related set.

Degrees of freedom A number of measurements that are independently available for estimating a model parameter.

Dependent variable The dependent variables (outputs or responses) of a model are the variables predicted by the equation or rules of the model using the independent variables (inputs or predictors). In the typical representation of a model as $y = f(x)$, y is the dependent variable and x is the independent variable.

Deployment cost Includes the following components: the hardware cost for running the application, the run-time license fees, the labor effort to integrate the solution into the existing work processes (or to create a new work procedure), to train the final user, and to deploy the model.

Derivative-free optimization Optimization method that is not based on explicit calculation of a derivative of the objective function.

Design for Six Sigma (DFSS) A methodology for designing new products and redesigning existing products and/or processes with defects at a Six Sigma level.

Design of experiments (DOE) A systematic approach to data collection such that the information obtained from the data is maximized by determining the (cause-and-effect) relationships between factors (controllable inputs) affecting a process and one or more outputs measured from that process.

Designed data Data collected by a design of experiments.

Development cost Includes the following components: the necessary hardware cost (especially if more powerful computational resources than PCs are needed), the development software licenses, and, above all, the labor effort to introduce, improve, maintain, and apply a new technology.

Disruptive technology A technological innovation, product, or service that uses a "disruptive" strategy, rather than an "evolutionary" or "sustaining" strategy, to overturn the existing dominant technologies or status quo products in a market.

Distance to cluster There are various ways to calculate a distance measure between a data point and a cluster. The most frequently used method is called the Squared Euclidean distance and is calculated as the square root of the sum of the squared differences in value for each variable.

Domain expert A person with deep knowledge and strong practical experience in a specific technical domain.

Dynamic model Dynamic models capture the change in the system state and are typically represented with difference equations or differential equations.

Elevator pitch A condensed overview of an idea for a product, service, or project. The name reflects the fact that an elevator pitch can be delivered in the time span of an elevator ride, approximately 60 seconds.

Emerging phenomena Emergent phenomena can appear when a number of simple entities operate in an environment, forming more complex behaviors as a collective.

Emerging technology A general term used to denote significant technological developments that broach new territory in some significant way in their field. Examples of currently emerging technologies include nanotechnology, biotechnology, and computational intelligence.

Empirical model An empirical model is based only on data and is used to predict, not explain, a system. An empirical model consists of a function that captures the trend of the data.

Epoch An epoch is the presentation of the entire training set to the neural network.

Error The difference between the actual and the predicted value from a model.

Euclidean distance See Distance to cluster.

Evolutionary algorithm A population-based optimization algorithm that uses mechanisms, inspired by biological evolutions, such as reproduction, recombination, mutation, and selection.

Evolutionary computation Evolutionary computation automatically generates solutions of a given problem with defined fitness by simulating natural evolution in a computer.

Expert system A computer program capable of performing at the level of a human expert in a specific domain of knowledge.

Expressional complexity A complexity measure for mathematical expressions based on the number of nodes of their tree-structure representation.

Extrapolation Predictive modeling which goes beyond the known and recognized system behavior and the ranges of available data.

False negative Test result that is read as negative but actually is positive.

False positive A positive test that results in a subject that does not possess the attribute for which the test is being conducted.

Feature space An abstract space of mathematically transformed original variables. The function that performs the transformation is called a kernel.

Feedback Process when a portion of the output signal is fed back to the input.

First-principles model A model developed by the laws of Nature.

Fitness function A mathematical function used for calculating the fitness. In complex optimization problems, it measures how good any particular solution is. The lower (or higher) the values of this function, the better the solution.

Fitness landscape Evaluations of a fitness function for all candidate solutions.

Fuzzy logic Unlike crisp or Boolean logic, fuzzy logic is multi-valued and can handle the concept of partial truth (truth values between "completely true" or "completely false").

Fuzzy math Messy or wrong calculations.

Fuzzy set A set with fuzzy boundaries, such as "small", "medium" or "large".

Fuzzy system A qualitative modeling scheme describing system behavior using a natural language and non-crisp fuzzy logic.

Generalization ability The ability of a model to produce correct results from data on which it has not been trained.

Generation One iteration of an evolutionary algorithm.

Genetic algorithm (GA) An evolutionary algorithm that generates a population of possible solutions encoded as chromosomes, evaluates their fitness, and creates a new population of offspring by using genetic operators, such as crossover and mutation.

Genetic programming (GP) An evolutionary algorithm of evolving structures.

Genotype The DNA of an organism.

GIGO 1.0 effect Garbage-In-Garbage-Out.

GIGO 2.0 effect Garbage-In-Gold-Out.

Global minimum The lowest value of a function over the entire range of its parameters.

Gradient A rate of inclination or the slope of a function.

Heuristics Rules of thumb.

Hyperplane A higher-dimensional generalization of the concepts of a line in a two-dimensional plane and a plane in three-dimensional space.

Hypothesis test A statistical algorithm used to choose between the alternative (for or against the hypothesis) which minimizes certain risks.

Ill-defined problem A problem whose structure lacks definition in some respect. The problem has unknowns associated with the ends (set of objectives) and means (set of process actions and decision rules) of the solution, at the outset of the problem-solving process.

Independent variable The independent variables (inputs or predictors) of a model are the variables used in the equation or rules of the model to predict the output (dependent) variable. In the typical representation of a model as $y = f(x)$, y is the dependent variable and x is the independent variable. Statistically, two variables are independent when the probability of both events occurring is equal to the product of the probability of each occurring.

Inductive learning Progressing from particular cases to general relationships represented by models.

Inference mechanism A basic component of an expert system which carries out reasoning that allows the expert system to reach a conclusion (find a solution).

Input See Independent variable.

Intelligent agents Artificial entities that have several intelligent features, such as being autonomous, responding adequately to changes in their environment, persistently pursuing goals, and being flexible, robust, and social by interacting with other agents.

Interpolation Model predictions inside the known ranges of inputs.

Kernel A mathematical function that transforms the original variables into features.

Lack of fit (LOF) Statistical measure that indicates that a model does not fit the data properly.

Laws of Nature A generalization that describes recurring facts or events in Nature, such as the laws of physics.

Laws of numbers The rule or theorem that states that the average of a large number of independent measurements of a random quantity tends toward the theoretical average of that quantity.

Learning Training models (estimating their parameters) based on existing data.

Least squares The most common method of training (estimating) the weights (parameters) of a model by choosing the weights that minimize the sum of the squared deviation of the predicted values of the model from the observed values of the data.

Linear model The term linear model refers to a model that is linear in the parameters, β_k, not the input variables (the x's), i.e. models in which the output is related to the inputs in a nonlinear fashion can still be treated as linear provided that the parameters enter the model in a linear fashion.

Linguistic variable A variable that can have values that are words or phrases.

Local minimum The minimum value of a function over a limited range of its input parameters.

Loss function See Fitness function.

Machine learning A research field for the design and development of algorithms and techniques that allow computers to "learn".

Maintenance cost of a model Involves assessment of the long-term support efforts of model validation, readjustment, and redesign.

Marketing myopia Focusing on products, not on customer benefits.

Membership function A mathematical function that defines a fuzzy set on the universe of discourse.

Model A model can be descriptive or predictive. A descriptive model helps in understanding underlying processes or behavior. A predictive model is an equation or set of rules that makes it possible to predict an unseen or unmeasured value (the dependent variable or output) from other, known values (independent variables or input). In the typical representation of a model as y = f(x), y is the dependent variable and x is the independent variable.

Model complexity Includes different measures related to model type and its structure. Examples are the number of parameters in statistical models and the number of hidden layers and neurons in neural networks.

Model credibility Trustworthy model performance in a wide range of operating conditions. Usually model credibility is based on its principles, performance, and transparency.

Model deployment The simplest meaning is using a developed model. Examples would include using a model for process control or making the model accessible on the Internet for others who will use it periodically. The deployment may be off-line, static use of the model, or be imbedded online as part of a larger system.

Model interpretability Explaining model structure and behavior using process knowledge.

Model maintenance Includes periodic model validation and corrections, such as model re-adjustment or model redesign.

Model overfit Fitting a model that has too many parameters and, as a result, model performance deteriorates sharply on unknown data.

Model performance A measure of a model's ability to reliably predict on known data according to the defined fitness function.

Model readjustment Refitting model parameters on new data.

Model redesign Redevelopment of the model from scratch on new data.

Model selection Choosing a model based on several criteria, mostly based on its performance on test data and its complexity.

Model underfit Fitting a model that has too few parameters to represent properly the functional relationship in the given data.

Multicollinearity Assumes a correlation structure among the model inputs. Multicollinearity among the inputs leads to biased model parameter estimates.

Multilayer perceptron The most popular neural network architecture in which neurons are connected together to form layers. A multilayer perceptron includes a three-layer structure: an input layer, one hidden layer that captures the nonlinear relationship, and an output layer.

Multiobjective optimization The process of simultaneously optimizing two or more conflicting objectives subject to certain constraints. An example in modeling is deriving optimal models on two objectives – model accuracy and model complexity.

Natural selection Process in Nature by which only the fittest organisms tend to survive and transmit their genetic material in increasing number of offspring.

Neural network See Artificial neural network.

Nonlinear model A model in which the dependent variable y depends in a nonlinear fashion on at least one of the parameters in the model.

Objective function See Fitness function.

Objective intelligence Artificial intelligence based on automatically extracted solutions through machine learning, simulated evolution, and emerging phenomena.

Object-oriented programming A high-level programming paradigm that uses objects as a basis for analysis, design, and implementation.

Ontology A description of concepts and relationships among concepts in a domain.

Optimization An iterative process of improving the solution to a problem as effectively or functionally as possible with respect to a specific objective function.

Optimization criterion A positive function of the difference between predictions and data estimates that are chosen so as to optimize the function or criterion.

Optimum The point at which the condition, degree, or amount of something is the most favorable.

Outliers Technically, outliers are data items that did not come from the assumed population of data and fall outside the boundaries that enclose most other data items in the data set.

Output See Dependent variable.

Overfitting A tendency of some modeling techniques to assign importance to random variations in the data by declaring them important patterns.

Pareto-front genetic programming A genetic programming algorithm based on multiobjective selection.

Parsimony The principle that the simplest explanation that explains a phenomenon is the one that should be selected. In empirical modeling, this generally refers to using as few variables as possible for developing a robust model.

Particle swarm optimization (PSO) A population-based stochastic optimization technique inspired by the social behavior of bird flocking and fish schooling.

Pattern Regularities that represent an abstraction. For example, a pattern can be a relationship between two variables.

Pattern recognition The automatic identification of figures, characters, shapes, forms, and patterns without active human participation in the decision process.

Phenotype The set of observable properties of an organism, mostly its body and behavior.

Pheromone A chemical produced and secreted by insects to transmit a message to other members of the same species.

PID control A proportional-integral-derivative controller (PID controller) is a generic control loop feedback mechanism widely used in industrial control systems. A PID controller attempts to correct the error between a measured process variable and a desired setpoint by calculating and then outputting a corrective action that can adjust the process accordingly.

Population A group of individuals that breed together.

Population diversity Includes the structural differences between genotypes and behavioral differences between phenotypes.

Predictive model A mathematical representation of an entity used for analysis and planning.

Premature convergence A population for an optimization problem converges too early.

Principal component analysis (PCA) A technique used to reduce multi-dimensional data sets to lower dimensions for analysis.

Probability A quantitative description of the likelihood of a particular event.

Process monitoring system A centralized system which monitors and controls entire sites, or complex systems spread out over large areas (anything between industrial plant and a country).

Range The range of the data is the difference between the maximum value and the minimum value. Alternatively, range can include the minimum and maximum, as in "The value ranges from 2 to 8".

Record A list of values of features that describe an event. Usually one row of data in a data set, database or spreadsheet. Also called an instance, example, or case.

Recurrent neural network A neural network architecture with a memory and feedback loops that is able to represent dynamic systems.

Regression A technique used to build equation-based predictive models for continuous variables that minimize some measure of the error.

Reinforcement learning A learning mechanism based on trial-and-error of actions and evaluating the rewards.

Residual See Error.

Robust model Robust models are less influenced by a small number of aberrant observations or outliers.

R-squared (R^2) A number between 0 and 1 that measures how well a model fits its training data. One is a perfect fit, zero implies the model has no predictive ability.

Rule A statement expressed by an IF (antecedent) THEN (consequent) form.

Search space The set of all possible solutions to a given problem.

Self-organizing map (SOM) A neural network architecture characterized by the formation of a topographic map of the input patterns in which the coordinates of a neuron are indicative of intrinsic statistical features contained in the input patterns.

Self-organization A process of automatically increased organization of a system without guidance or management from an outside source.

Sigma A measure of the dispersion of random errors about the mean value.

Six Sigma A business management strategy, originally developed by Motorola, that seeks to identify and remove the causes of defects and errors in manufacturing and business processes.

Statistical learning theory A statistical theory of learning and generalization concerns the problem of choosing desired functions on the basis of empirical data.

Statistics A branch of mathematics dealing with the collection, analysis, interpretation, and presentation of masses of numerical data.

Stigmergy A method of indirect communication in which the individuals communicate with each other by modifying their local environment.

Stochastic process The counterpart of a deterministic process.

Subject matter expert (SME) See Domain expert.

Subjective intelligence Artificial intelligence based on expert knowledge and rules of thumb.

Supervised learning A type of machine learning that requires an external teacher, who presents a sequence of training samples to a neural network. As a result, a functional relationship that maps the inputs to the desired outputs is generated.

Support vector machines (SVM) A machine learning method derived from the mathematical analysis of statistical learning theory. The generated models are based on the most informative data points called support vectors.

Swarm intelligence A computational intelligence method that explores the advantages of collective behavior of an artificial flock of computer entities by mimicking the social interactions of animal and human societies.

Symbolic reasoning Reasoning based on symbols that represent different types of knowledge, such as facts, concepts, and rules.

Symbolic regression Method for automatic discovery of both the functional form and the numerical parameters of a mathematical expression.

Test data A data set independent of the training data set, used to fine-tune the estimates of the model parameters (i.e. weights). The test data set is particularly useful in avoiding overfitting (overtraining) in a model.

Total-cost-of-ownership The sum of development, deployment, and maintenance cost of a model.

Training data A data set used to estimate or train a model.

Transductive learning A machine learning method of generating predictions directly from available data without building a predictive model.

Undesigned data Data collected in an unsystematic fashion by Design of Experiments (DOE).

Universal approximation Having the ability to approximate any function to an arbitrary degree of accuracy. Neural nets are universal approximators.

Unsupervised learning A type of machine learning that does not require a teacher and the neural network is self-adjusted by the discovered patterns in the input data.

Index